T0320714

Problems and Solutions in
Mathematical
Olympiad

High School 3

Other Related Titles from World Scientific

Problems and Solutions in Mathematical Olympiad
Secondary 3
by Jun Ge
translated by Huan-Xin Xie
ISBN: 978-981-122-982-4
ISBN: 978-981-123-141-4 (pbk)

Problems and Solutions in Mathematical Olympiad
High School 1
by Bin Xiong and Zhi-Gang Feng
translated by Tian-You Zhou
ISBN: 978-981-122-985-5
ISBN: 978-981-123-142-1 (pbk)

Problems and Solutions in Mathematical Olympiad
High School 2
by Shi-Xiong Liu
translated by Jiu Ding
ISBN: 978-981-122-988-6
ISBN: 978-981-123-143-8 (pbk)

Problems and Solutions in
Mathematical Olympiad

High School 3

Editors-in-Chief

Zun Shan *Nanjing Normal University, China*

Bin Xiong *East China Normal University, China*

Original Author

Hong-Bing Yu *Suzhou University, China*

English Translators

Fang-Fang Lang *Shanghai Qibao Dwight High School, China*

Yi-Chao Ye *Shanghai Qibao High School, China*

Copy Editors

Ming Ni *East China Normal University Press, China*

Ling-Zhi Kong *East China Normal University Press, China*

Lei Rui *East China Normal University Press, China*

 **East China Normal University Press**

 **World Scientific**

Published by

East China Normal University Press
3663 North Zhongshan Road
Shanghai 200062
China

and

World Scientific Publishing Co. Pte. Ltd.
5 Toh Tuck Link, Singapore 596224
USA office: 27 Warren Street, Suite 401-402, Hackensack, NJ 07601
UK office: 57 Shelton Street, Covent Garden, London WC2H 9HE

British Library Cataloguing-in-Publication Data
A catalogue record for this book is available from the British Library.

PROBLEMS AND SOLUTIONS IN MATHEMATICAL OLYMPIAD
High School 3

ISBN 978-981-122-991-6 (hardcover)
ISBN 978-981-123-144-5 (paperback)
ISBN 978-981-122-992-3 (ebook for institutions)
ISBN 978-981-122-993-0 (ebook for individuals)

For any available supplementary material, please visit
https://www.worldscientific.com/worldscibooks/10.1142/12089#t=suppl

Desk Editor: Tan Rok Ting

Typeset by Stallion Press
Email: enquiries@stallionpress.com

Printed in Singapore

Editorial Board

Preface

It is said that in many countries, especially the United States, children are afraid of mathematics and regard mathematics as an "unpopular subject." But in China, the situation is very different. Many children love mathematics, and their math scores are also very good. Indeed, mathematics is a subject that the Chinese are good at. If you see a few Chinese students in elementary and middle schools in the United States, then the top few in the class of mathematics are none other than them.

At the early stage of counting numbers, Chinese children already show their advantages.

Chinese people can express integers from 1 to 10 with one hand, whereas those in other countries would have to use two.

The Chinese have long had the concept of digits, and they use the most convenient decimal system (many countries still have the remnants of base 12 and base 60 systems).

Chinese characters are all single syllables, which are easy to recite. For example, the multiplication table can be quickly mastered by students, and even the "stupid" people know the concept of "three times seven equals twenty one." But for foreigners, as soon as they study multiplication, their heads get bigger. Believe it or not, you could try and memorise the multiplication table in English and then recite it, it is actually much harder to do so in English.

It takes the Chinese one or two minutes to memorize $\pi = 3.14159\cdots$ to the fifth decimal place. However, in order to recite these digits, the Russians wrote a poem. The first sentence contains three words and the second sentence contains one $\cdots$ To recite π, recite poetry first. In our opinion, this just simply asks for trouble, but they treat it as a magical way of memorization.

Application problems for the four arithmetic operations and their arithmetic solutions are also a major feature of Chinese mathematics. Since ancient times, the Chinese have compiled a lot of application questions, which has contact or close relations with reality and daily life. Their solutions are simple and elegant as well as smart and diverse, which helps increase students' interest in learning and enlighten students'. For example:

"There are one hundred monks and one hundred buns. One big monk eats three buns and three little monks eat one bun. How many big monks and how many little monks are there?"

Most foreigners can only solve equations, but Chinese have a variety of arithmetic solutions. As an example, one can turn each big monk into 9 little monks, and 100 buns indicate that there are 300 little monks, which contain 200 added little monks. As each big monk becomes a little monk 8 more little monks are created, so $200/8 = 25$ is the number of big monks, and naturally there are 75 little monks. Another way to solve the problem is to group a big monk and three little monks together, and so each person eats a bun on average, which is exactly equal to the overall average. Thus the big monks and the little monks are not more and less after being organized this way, that is, the number of the big monks is $100/(3 + 1) = 25$.

The Chinese are good at calculating, especially good at mental arithmetic. In ancient times, some people used their fingers to calculate (the so-called "counting by pinching fingers"). At the same time, China has long had computing devices such as counting chips and abaci. The latter can be said to be the prototype of computers.

In the introductory stage of mathematics – the study of arithmetic, our country has obvious advantages, so mathematics is often the subject that our smart children love.

Geometric reasoning was not well-developed in ancient China (but there were many books on the calculation of geometric figures in our country), and it was slightly inferior to the Greeks. However, the Chinese are good at learning from others. At present, the geometric level of middle school students in our country is far ahead of the rest of the world. Once a foreign education delegation came to a junior high school class in our country. They thought that the geometric content taught was too in-depth for students to comprehend, but after attending the class, they had to admit that the content was not only understood by Chinese students, but also well mastered.

The achievements of mathematics education in our country are remarkable. In international mathematics competitions, Chinese contestants have won numerous medals, which is the most powerful proof. Ever since our country officially sent a team to participate in the International Mathematical Olympiad in 1986, the Chinese team has won 14 team championships, which can be described as very impressive. Professor Shiing-Shen Chern, a famous contemporary mathematician, once admired this in particular. He said, "One thing to celebrate this year is that China won the first place in the international math competition $\cdots$ Last year it was also the first place." (Shiing-Shen Chern's speech, *How to Build China into a Mathematical Power*, at Cheng Kung University in Taiwan in October 1990)

Professor Chern also predicted: "China will become a mathematical power in the 21st century."

It is certainly not an easy task to become a mathematical power. It cannot be achieved overnight. It requires unremitting efforts. The purpose of this series of books is: (1) To further popularize the knowledge of mathematics, to make mathematics be loved by more young people, and to help them achieve good results; (2) To enable students who love mathematics to get better development and learn more knowledge and methods through the series of books.

"The important things in the world must be done in detail." We hope and believe that the publication of this series of books will play a role in making our country a mathematical power. This series was first published in 2000. According to the requirements of the curriculum reform, each volume is revised to different degrees.

Well-known mathematician, academician of the Chinese Academy of Sciences, and former chairman of the Chinese Mathematical Olympiad Professor Yuan Wang, served as a consultant to this series of books and wrote inscriptions for young math enthusiasts. We express our heartfelt thanks. We would also like to thank East China Normal University Press, and in particular Mr. Ming Ni and Mr. Lingzhi Kong. Without them, this series of books would not have been possible.

Zun Shan and Bin Xiong
May 2018

Contents

Chapter 1

Permutations and Combinations

Combinatorial mathematics, also known as combinatorial analysis, is an important branch of mathematics with a long history.

Combinatorial mathematics intersects many branches of mathematics, so it is difficult (and unnecessary) to define it formally. The readers can roughly understand the most basic characteristics of combinatorial mathematics involved in this book.

An important problem in combinatorial mathematics is the counting problem, which aims to determine the number of elements that meet certain restrictions. "Permutations" and "combinations" are the simplest and most basic contents in this subject.

Addition Principle and Multiplication Principle

Addition principle and multiplication principle are the basic principles of counting, and also the basis of further research on other combination problems.

1. Addition Principle Suppose the methods to accomplish one objective can be divided into n disjoint classes. In the first class, and there are m_1 different methods, in the second class, there are m_2 different methods, ... and there are m_n different methods in the nth class. Then the total number of the methods of accomplishing the task is

$$m_1 + m_2 + \cdots + m_n.$$

The spirit of the addition principle is that the "whole" is equal to the sum of "parts." Applying the addition principle is to divide the "whole" (the methods of accomplishing one task) into several disjoint classes, so that the number of the elements in each class is easy to calculate. As for how to group the method, of course, it depends on the specific problems.

Remark 1. The addition principle can be expressed in a more general form using the set language:

Let S be a (finite) set and $S_1, S_2, \ldots, S_n$ be a **partition** of S, that is, $S_1, S_2, \ldots, S_n$ are pairwise disjoint and their union is S. Then

$$|S| = |S_1| + |S_2| + \cdots + |S_n|.$$

Here and later, the notation $|X|$ represents the number of elements in the finite set X.

However, if $S_1, S_2, \ldots, S_n$ are not pairwise disjoint, in order to calculate $|S|$, we need a slightly deeper method — the inclusion-exclusion principle, which will be discussed in Chapter 4.

2. Multiplication Principle If there are m_1 ways to do the first thing, m_2 ways to do the second thing after the first thing is done, $\ldots$, and m_n ways to do the nth thing after the $(n-1)$st thing is done, then there are

$$m_1 \times m_2 \times \cdots \times m_n$$

different ways to do the first thing, second thing, $\ldots$, and the nth thing in succession.

The key point of applying the multiplication principle is that the process of accomplishing a task is divided into several steps, and the number of the methods in each step is easy to determine.

The multiplication principle can also be expressed in the set language, but this form is a little abstract, and it is not needed in this book, so we will not discuss it.

Some Basic Counting Problems

In this section, we introduce some typical counting problems in permutations and combinations. Many counting problems can be treated as one of these models.

1. Permutations

(1) **Permutations without repetition** Selecting $k(1 \leq k \leq n)$ elements from n different elements in order and without repetition is called a permutation without replacement of k elements from n different elements, which is called a k **− permutation** for short. The number of all such permutations is denoted as P_n^k.

From the multiplication principle,

$$\mathrm{P}_n^k = n(n-1) \cdots (n-k+1).$$

(There are n methods to select the first element. After selecting the first element, there are $n - 1$ methods to select the second element because the first element cannot be repeated, ..., and finally, there are $n-k+1$ methods to select the kth element.)

In particular, if $k = n$, then we get the **full permutation formula** of n different elements (that is, the number of n-permutations of n different elements):

$$\mathrm{P}_n^n = n \times (n-1) \times \cdots \times 2 \times 1 = n!.$$

For convenience, we define $0! = 1$. Then the above formula can be rewritten as

$$\mathrm{P}_n^k = \frac{n!}{(n-k)!}.$$

(2) **Permutations with repetition** Selecting $k(k \geq 1)$ elements orderly and repeatedly from n different elements is called a k-**permutation with repetition** of n different elements.

It is easy to know from the multiplication principle that the number of k-permutations with repetition of n different elements is n^k. (There are n ways to select the first element. After selecting the first element, there are still n ways to select the second element, ..., and finally, there are n ways to select the kth element.)

(3) **The Full permutations of a finite number of repeated elements** Let n elements be divided into k groups. The elements in the same group are the same as each other, and the elements in different groups are different. Let the number of elements in k groups be $n_1, n_2, \ldots, n_k(n_1 + n_2 + \cdots + n_k = n)$. Then a full permutation of these n elements is called a

full permutation of a finite number of repeated elements, and the number of such permutations is

$$\frac{n!}{n_1! n_2! \ldots n_k!}$$

To prove this, suppose the numbers of $x, y, \ldots, z$ are $n_1, n_2, \ldots, n_k$ respectively. Let's take any one of these permutations. Assume that these n_1 numbers of x are marked with subscripts $1, 2, \ldots, n_1$. Then there are $n_1!$ ways to attach subscripts. If these n_2 numbers of y are also marked with subscripts, then there are $n_2!$ ways to do so. Finally, n_k numbers of z are also marked with subscripts, and the number of ways is $n_k!$.

In this way, according to the multiplication principle, each permutation that meets the requirements produces $n_1! n_2! \ldots n_k!$ permutations marked with subscripts. Also, it is easy to know that no two different permutations generate the same permutation marked with subscripts. In turn, any permutation marked with subscripts can be obtained in this way. Therefore, $n_1! n_2! \ldots n_k!$ times the number of permutations in question is the number of all the permutations marked with subscripts, and the latter is exactly the number of permutations of n different elements, i.e., $n!$, thus the above formula is obtained for the number of the full permutations of a finite number of repeated elements.

Remark 2. The above solution used a very important idea: **correspondence**. However, this correspondence is not a one-to-one correspondence (a satisfactory permutation corresponds to a set consisting of $n_1! n_2! \ldots n_k!$ permutations marked with subscripts).

Correspondence is an important idea to deal with counting and many other combination problems. Please refer to the following content and Chapter 3.

Remark 3. If every n_i is 1 (thus $k = n$), then our formula reduces to the full permutation formula of n different elements.

Remark 4. The full permutation of a finite number of repeated elements is certainly an integer, so we get a "by-product":

(i) Let $n_1, n_2, \ldots, n_k$ all be positive integers. Then

$$\frac{(n_1 + n_2 + \cdots + n_k)!}{n_1! n_2! \ldots n_k!}$$

is an integer.

In particular, take $k = 2$, and let $n_1 = m$ and $n_2 = n$. Then, since

$$\frac{(m+n)!}{m!n!} = \frac{(m+n)\cdots(n+1)}{m!}$$

is an integer, we know that:

(ii) The product of m consecutive positive integers can be divisible by $m!$ (that is, the integer multiple of $m!$; see the definition of exact division in Chapter 6).

Therefore, it is not difficult to deduce a very basic result as follows: The product of m consecutive integers is divisible by $m!$.

In fact, if the m consecutive integers are all positive or negative, then the said conclusion becomes (ii); if there are both positive and negative numbers in the m consecutive integers, then there must be a zero in them, so their product is zero and of course, can be divisible by $m!$.

By the way, it is not difficult to deduce (i) by using (ii) repeatedly.

(4) **Cyclic permutations** Choosing $k(1 \le k \le n)$ elements from n different elements (without repetition) followed by arranging them on a circle, is called a **k-cyclic permutation** of n different elements. If one k-cyclic permutation can be rotated to get another k-cyclic permutation, then the two permutations are considered to be the same.

The number of k-cyclic permutations of n different elements is

$$\frac{\mathrm{P}_n^k}{k} = \frac{n!}{k \cdot (n-k)!}.$$

In particular, the total number of cyclic permutations made of all n different elements is $(n-1)!$.

To prove it, we notice that for every fixed k-cyclic permutation, the circle is cut between any two elements, just producing k different "straight line permutations", that is, k-permutations. The k-permutations generated by different k-cyclic permutations must be different from each other. Any k-permutation can be obtained in this way. Therefore, k times the number of k-cyclic permutations is equal to the number of k-permutations, i.e., P_n^k.

2. Combinations

(5) **Combinations without repetition** Taking $k(1 \le k \le n)$ elements out of n different elements with no order and without repetition is called a combination (without repetition) of k elements out of n different elements, which is called a **k-combination** for short.

The number of combinations of n elements taken k at a time is denoted as $\binom{n}{k}$. Then

$$\binom{n}{k} = \frac{P_n^k}{k!} = \frac{n(n-1)\cdots(n-k+1)}{k!}.$$

In fact, for every fixed k-combination, arranging all its elements will produce $k!$ different k-permutations. Obviously, different k-combinations produce different permutations, and each k-permutation can be obtained in this way. Hence, we get the result $k!\binom{n}{k} = P_n^k$.

Remark 5. Since the number of combinations is certainly an integer, combining this fact with the above results, we once again obtain "the product of k consecutive positive integers can be divided by $k!$," that is, Remark 4 (ii).

(6) **Combinations with repetition** Taking $k(1 \le k \le n)$ elements out of n different elements with no order but repeatedly is called a combination (without repetition) of k elements out of n different elements, which is called a **k-combination with repetition** for short.

The number of k-combinations with repetition of n different elements is $\binom{n+k-1}{k}$. To prove that, suppose the n elements are $1, 2, \ldots, n$. Let the selected k elements be

$$a_1 \le a_2 \le \cdots \le a_k \, (\le n).$$

Then obviously,

$$(1 \le) a_1 + 0 < a_2 + 1 < \cdots < a_k + k - 1 (\le n + k - 1).$$

If $\{a_1, a_2, \ldots, a_k\}$ correspond to $\{a_1 + 0, a_2 + 1, \ldots, a_k + k - 1\}$, then the latter is a k-combination of $1, 2, \ldots, n + k - 1$.

In turn, any k-combination

$$(1 \le) b_1 < b_2 < \cdots < b_k (\le n + k - 1)$$

of $1, 2, \ldots, n + k - 1$ also corresponds to a k-combination with repetition

$$(1 \le) b_1 \le b_2 - 1 \le \cdots \le b_k - (k-1)(\le n)$$

of $1, 2, \ldots, n$.

Therefore, the above correspondence is a one-to-one correspondence, so the number of the k-combinations with repetition is equal to the number of the k-combinations of $1, 2, \ldots, n + k - 1$, i.e., $\binom{n+k-1}{k}$.

Remark 6. See Example 1 and Remark 1 in Chapter 3 for a different solution to this problem.

Illustrative Examples

Example 1. How many possible permutations can be made of nine ones and four zeros, if no two zeros are adjacent?

Solution First arrange nine ones to generate 10 "gaps." Then insert four zeros into the 10 "gaps." Hence there are $\binom{10}{4} = 210$ ways.

Example 2. Find the number of five-digit numbers such that each digit appears at least twice.

Solution There are 9 such five-digit natural numbers if all 5 digits are the same. In other cases, if a five-digit number does not contain 0, then there are exactly two different digits (one appears twice and the other three times) in such a five-digit number, hence there are $\binom{5}{2}\binom{9}{1}\binom{8}{1} = 720$ such numbers. (There are $\binom{5}{2}$ ways to select two of the five positions; there are $\binom{9}{1}$ ways to fill in the same non-zero digits in these two positions; finally, fill in the other same digits in the remaining three positions in $\binom{8}{1}$ ways.)

If a five-digit number contains 0 (note that 0 cannot be placed as the first digit), then the count of the five-digit numbers in which 0 appears twice is $\binom{4}{2}\binom{9}{1} = 54$; the count of the five-digit numbers in which appears three times is $\binom{4}{3}\binom{9}{1} = 36$. So, there are $54 + 36 = 90$ (qualified) five-digit numbers containing digit 0.

Hence, there are $720 + 90 + 9 = 819$ such five-digit numbers in total.

Example 3. The elements in a nonempty set A are all positive integers, and the set satisfies the property: If $a \in A$, then $12 - a \in A$. Determine the number of such sets A.

Solution From the property of A, we know that A must be a union of several of the six sets $\{1,11\}$, $\{2,10\}$, $\{3,9\}$, $\{4,8\}$, $\{5,7\}$ and $\{6\}$, so the number of A satisfying the requirement is

$$\binom{6}{1} + \binom{6}{2} + \binom{6}{3} + \binom{6}{4} + \binom{6}{5} + \binom{6}{6} = 63.$$

Example 4. $A \cup B \cup C = \{1, 2, \ldots, 10\}$ is the union of sets $A, B,$ and C. Find out how many ordered triples (A, B, C) are there.

Solution If $A, B,$ and C meet the requirement, then any element in $\{1, 2, \ldots, 10\}$ only belongs to one of $A, B,$ and C, belongs to two of A,

B, and C, or belongs to all of A, B, and C, so there are $3 + 3 + 1 = 7$ cases in total.

Because $\{1, 2, \ldots, 10\}$ has 10 elements in total, there are 7^{10} ways to select sets A, B, and C (this counting considered the order of the three sets), that is, there are 7^{10} ordered triples that meet the requirement.

Example 5. Take out a_1, a_2, and a_3 from the numbers $1, 2, \ldots, 14$ in ascending order from left to right such that $a_2 - a_1 \geq 3$ and $a_3 - a_2 \geq 3$. Try to find the number of different ways that meet the above requirements.

Solution　The point of this question is to make the correspondence

$$(a_1, a_2, a_3) \to (a_1, a_2 - 2, a_3 - 4).$$

Clearly we know that $a_1 < a_2 - 2 < a_3 - 4$ are three numbers selected from $1, 2, \ldots, 10$, and this correspondence is one-to-one. So, the number of ways equals $\binom{10}{3} = 120$.

Example 6. How many subsets of the set $\{1, 2, \ldots, 100\}$ contain at least one odd number?

Solution　The number of all subsets of the set $\{1, 2, \ldots, 100\}$ is easy to determine, which is 2^{100}, but it is a little troublesome to directly find the number of subsets that meet the requirement. However, the number of undesired subsets, that is subsets of $\{2, 4, \ldots, 98, 100\}$, is clearly known to be 2^{50}. Therefore, the number of the desired subsets is $2^{100} - 2^{50}$.

Remark 7. Let S be a finite set and A be a set of elements with "some properties" in S. When $|S|$ is known (or easy to find), if $|A|$ is not easy to determine, then we can (try to) consider the set $\overline{A}$ (the complement of A with respect to S), that is, the set of elements in S that do not have "some properties". Because $|A| + |\overline{A}| = |S|$, if one finds $|A|$, then $|\overline{A}|$ is obtained (indirectly). This is a quite basic idea in counting problems (Example 6 is only a simple application), and it has been extended to a very important principle in combinatorial mathematics — including-excluding principle (see Chapter 4).

Example 7. Suppose each term of a sequence $a_1, a_2, \ldots, a_n$ is one of $0, 1, \ldots, k - 1$ ($k \geq 2$ is an integer). Find the number of sequences in which 0 appears even times. (Express the answer as a simple function of n and k.)

Solution In this question, it is more suitable to consider the sequences of "0 appears odd times", which can be treated as the "complementary" (or "companion") sequences of "0 appears even times."

Let x_n and y_n be the numbers of the sequences of "0 appears even times" and "0 appears odd times" respectively. For each integer $i \geq 0$, we clearly know that there are $\binom{n}{i}(k-1)^{n-i}$ sequences containing exactly i zeros (in such sequences, there are $\binom{n}{i}$ ways to arrange i zeros in n positions; then fill in the remaining positions with $1, 2, \ldots, k-1$, and each position has $k-1$ ways). Therefore,

$$x_n = \binom{n}{0}(k-1)^n + \binom{n}{2}(k-1)^{n-2} + \binom{n}{4}(k-1)^{n-4} + \cdots \quad (1)$$

and

$$y_n = \binom{n}{1}(k-1)^{n-1} + \binom{n}{3}(k-1)^{n-3} + \binom{n}{5}(k-1)^{n-5} + \cdots . \quad (2)$$

Add (1) and (2), subtract (2) from (1), and apply the binomial theorem (see Chapter 2) to get

$$\begin{cases} x_n + y_n = k^n, & (3) \\ x_n - y_n = (k-2)^n. & (4) \end{cases}$$

Therefore, $x_n = \frac{1}{2}[k^n + (k-2)^n]$.

Remark 8. The correctness of the equality "(3)" can be seen from the combinatorial significance (it is not necessary to use (1) and (2)). After finding out $x_n + y_n$, in order to determine x_n, we can try to deduce another equivalent relation between x_n and y_n (the simplest, of course, is to consider $x_n - y_n$). However, without deriving (1) and (2), such an equality seems to be difficult to establish.

This problem can also be solved by the recursive method (it is not necessary to use (1) and (2) or the binomial theorem), which is left for the reader to complete (Exercise 10 in Chapter 3).

Remark 9. "Complementary" (or "companion") quantities generally have good behaviors as a whole. (Refer to Remark 7 and Equation (3) above.) The introduction of quantities that are complementary (in some sense) to the quantities under consideration is often a basis for solving some problems. There are also such problems later in this book.

Exercises

Group A

1. How many 3-digit numbers can be constructed from the digits 1, 2, 3, 4, 5, 6 if each digit may be used (i) only once; (ii) as often as desired?

2. How many 7-digit numbers can be constructed from the digits 1, 2, 3, 4, 5, 6, 7, 8, 9 if each digit can be used only once, and 8 and 9 must not be adjacent?

3. How many ways are there to arrange a program list with 6 singing programs and 4 dancing programs, so that any two dance programs are not adjacent?

4. There are 9 rooms, of which 2 will be painted white, 3 will be painted green, and 4 will be painted yellow. How many plans are there?

5. Use the letters $a, b,$ and c to form five-letter words. In each word, a appears at most twice, b appears at most once, and c appears at most three times. Find the number of such words.

6. Suppose n boys and n girls sit in a circle such that no two boys are next to each other, and so do any two girls. How many arrangements are there?

7. Put $n+1$ different balls into n different boxes. How many ways are there if each box is not empty?

8. How many different results can be produced by rolling k same dice at the same time?

Group B

9. Take out any three different numbers from 1 to 300 such that the sum of the three numbers is divisible by 3. How many options are there in total?

10. Given five points in a plane, it is known that the straight lines connecting these points are not parallel, perpendicular to each other, or coincident with each other. Construct perpendicular lines through each point to the lines between any two of the other four points. Find the maximual number of the intersection points of these perpendicular lines (excluding the known five points).

11. Find the number of ordered non-empty set pairs A and B satisfying the following conditions:

(1) $A \cup B = \{1, 2, \ldots, 12\}$;

(2) $A \cap B = \emptyset$;

(3) the number of elements of A is not an element of A, and the number of elements of B is also not an element of B.

12. Given the set $S = \{1, 2, \ldots, 10\}$, find the number of unordered pairs of its non-empty subsets A and B satisfying $A \cap B = \emptyset$.

Chapter 2

Binomial Coefficients

Combination numbers $\binom{n}{k}$, also known as binomial coefficients, is an important role in combinatorial mathematics. This chapter briefly discusses play basic properties of binomial coefficients and some identities expressed by them.

Basic Properties of Binomial Coefficients

Combination numbers $\binom{n}{k}$ have the following three aspects:

(1) Combinatorial significance: the number of k-combinations of n different elements.
(2) Explicit representation: $\binom{n}{k} = \frac{n(n-1)\cdots(n-k+1)}{k!} = \frac{n!}{k!(n-k)!}$.
(3) Coefficients of the binomial expansion

$$(x+y)^n = \sum_{k=0}^{n} \binom{n}{k} x^k y^{n-k} \quad \text{or} \quad (1+x)^n = \sum_{k=0}^{n} \binom{n}{k} x^k.$$

The above identity is called the binomial theorem, so we also call $\binom{n}{k}$ the **binomial coefficients**.

Remark 1. Generally speaking, n and k in $\binom{n}{k}$ are positive integers. For convenience, we agree on $\binom{n}{0} = 1$ when $n \geq 0$. (Note that the "0-element set" is the empty set, and there is only one way to take 0 element out of n elements); when k is a negative integer or $0 \leq n < k$, we have $\binom{n}{k} = 0$ (Because k elements cannot be taken out of an n-element set). Clearly we see that these conventions are consistent with the above (2) and (3).

Based on "counting," it is not difficult to give a proof of the binomial theorem. From

$$(x+y)^n = \underbrace{(x+y)\cdots(x+y)}_{n \text{ times}},$$

it can be seen that the expansion of the right side is obtained by taking any term in the first parentheses on the right side, taking any term in the second parentheses, ..., and taking any term in the nth parentheses, and then multiplying these terms, and finally adding all the results. Hence, the expansion of $(x+y)^n$ has the form

$$(x+y)^n = \sum_{k=0}^{n} a_{n,k}x^k y^{n-k},$$

where $a_{n,k}$ is the number of times that each term $x^k y^{n-k}$ appears in the above expansion, which is obviously the number of ways to (non-repeatedly) take k numbers of x from n parentheses, which is by definition $\binom{n}{k}$ (note that if x is selected from k parentheses, it means that y must be automatically selected from the remaining $n-k$ parentheses). This proves the binomial theorem in (3).

There are three basic properties of the binomial coefficients:

(4) Symmetry: For integers k and $n \geq 0$, we have $\binom{n}{k} = \binom{n}{n-k}$.

(5) Recurrence relation: For integers k and $n \geq 1$, it holds that $\binom{n}{k} = \binom{n-1}{k} + \binom{n-1}{k-1}$.

(6) Unimodality: If n is even, then

$$\binom{n}{0} < \binom{n}{1} < \cdots < \binom{n}{\frac{n}{2}} > \cdots > \binom{n}{n-1} > \binom{n}{n};$$

if n is odd, then

$$\binom{n}{0} < \binom{n}{1} < \cdots < \binom{n}{\frac{n-1}{2}} = \binom{n}{\frac{n+1}{2}} > \cdots > \binom{n}{n-1} > \binom{n}{n}.$$

In particular, the maximum of the nth degree binomial coefficients $\binom{n}{0}$, $\binom{n}{1}, \ldots, \binom{n}{n}$ is $\binom{n}{[\frac{n}{2}]}$. (Here and later $[x]$ represents the maximal integer not exceeding the real number x).

To prove (4), (5), and (6), let n and k be positive integers. (In other cases, the conclusion can be known from the agreement in Remark 1).

Both (4) and (5) can be easily derived from the explicit representation (2) of $\binom{n}{k}$, and the details are omitted here. The following is a demonstration based on counting:

Let X be an n-element set. Obviously, the complementarity correspondence of X is a one-to-one correspondence from the set of k-element subsets of X to the set of $(n-k)$-element subsets. Because there are $\binom{n}{k}$ and $\binom{n}{n-k}$ subsets of k elements and $(n-k)$ elements in X respectively, we obtain (4).

To prove (5), we note that on one hand, there are $\binom{n}{k}$ subsets of k elements in X. On the other hand, these k-element subsets can be divided into two (disjoint) $k+1$ classes: one class does not contain a fixed element a in X, while the other class contains a. The former has $\binom{n-1}{k}$ subsets of k elements, and the latter has $\binom{n-1}{k-1}$ subsets of k elements. According to the addition principle, the total number of k-elements subsets of X is $\binom{n-1}{k} + \binom{n-1}{k-1}$. Combining the two results, we obtain (5).

Remark 2. The argument derived by considering the combinatorial significance as above is often called a combinatorial argument. Its basic idea is to establish a one-to-one correspondence between two (finite) sets, so that the numbers of elements of the two sets are equal; or count the number of elements of the same set in two different ways to generate the required equality. There are also some examples in this chapter.

The results in (6) can be proved by the explicit representation (2) (the details are omitted here), or by mathematical induction on n using (5): When $n = 1$ and 2, (6) is obviously true. Assume that (6) is true for n. Now consider the case of $n + 1(n \geq 2)$. We just need to prove:

$$\binom{n+1}{k} < \binom{n+1}{k+1} \text{ is true for } 2 \leq k+1 \leq \left[\frac{n+1}{2}\right].$$

From (5), we have $\binom{n+1}{k} = \binom{n}{k} + \binom{n}{k-1}$ and $\binom{n+1}{k+1} = \binom{n}{k+1} + \binom{n}{k}$. Hence, the problem is equivalent to proving that the inequality $\binom{n}{k-1} < \binom{n}{k+1}$ is true for $2 \leq k+1 \leq \left\lceil\frac{n+1}{2}\right\rceil$.

When n is even, $k + 1 \leq \left[\frac{n+1}{2}\right] = \frac{n}{2}$; when n is odd,

$$k + 1 \leq \left[\frac{n+1}{2}\right] = \frac{n+1}{2}.$$

Therefore, by the principle of mathematical induction, the above inequality is true.

Now let's talk about some basic equalities involving the sum of the binomial coefficients:

(7) $\binom{n}{0} + \binom{n+1}{1} + \cdots + \binom{n+k}{k} = \binom{n+k+1}{k}$.

(8) $\binom{n}{0} + \binom{n}{1} + \cdots + \binom{n}{n} = 2^n$.

(9) $\binom{n}{0} - \binom{n}{1} + \binom{n}{2} - \cdots + (-1)^n \binom{n}{n} = 0$.

(10) $\binom{m}{0}\binom{n}{k} + \binom{m}{1}\binom{n}{k-1} + \binom{m}{2}\binom{n}{k-2} + \cdots + \binom{m}{k}\binom{n}{0} = \binom{m+n}{k}$.

We prove the above basic equalities in the following.

Proof of (7): The first proof is based on recursion. Using the recurrence formula (5) repeatedly, we have

$$\binom{n}{0} + \binom{n+1}{1} = \binom{n+1}{0} + \binom{n+1}{1} = \binom{n+2}{1},$$

$$\binom{n+2}{1} + \binom{n+2}{2} = \binom{n+3}{2},$$

$$\binom{n+3}{2} + \binom{n+3}{3} = \binom{n+4}{3}, \cdots,$$

then we can obtain (7).

The more formal expression is to let the left side of (7) be a_k. Then (for $k \geq 0$)

$$a_{k+1} - a_k = \binom{n+k+1}{k+1},$$

it can be easily proved by induction and the formula (5).

The second method is based on counting: Let X be a set of $n + k + 1$ elements, and divide its $\binom{n+k+1}{k}$ subsets of k elements into $k+1$ (disjoint)

$k+1$ classes (let $a_1, \ldots, a_k$ be the specified k elements of X): the first class excluding a_1; the second class including a_1 but excluding a_2; the third class including a_1 and a_2 but excluding a_3; the $(k+1)$st class including $a_1, a_2, \ldots, a_k$. Clearly, the first class, the second class, $\ldots$, the kth class, and the $(k+1)$st class have $\binom{n+k}{k}$, $\binom{n+k-1}{k-1}, \ldots, \binom{n+1}{1}$, and $\binom{n}{0} = 1$ k-element subsets respectively. By the addition principle, we obtain (7).

This argument is a generalization of the proof of (5).

Proof of (8): The first method is to use the binomial theorem: Take $x = 1$ and $y = 1$ in the identity (3).

The second method is based on counting the number of subsets of an n-element set X. On one hand, all subsets of X can be classified into n classes: 0-element subsets (i.e., the empty set), 1-element subsets, 2-element subsets, $\ldots$, and n-element subsets. By the addition principle, the set X has $\binom{n}{0} + \binom{n}{1} + \cdots + \binom{n}{n}$ subsets.

On the other hand, each subset of X is composed of several elements selected from X. Therefore each subset corresponds to a selection of these n elements. Since each element can be selected or not, there are $2 \times 2 \times \cdots \times 2 = 2^n$ ways to select elements from X to form a subset (based on the multiplication principle). Comparing the two ways of counting, we obtain (8).

Proof of (9): The equality (9) is a direct corollary of the binomial theorem (taking $x = 1$ and $y = -1$), which can also be proved by the idea of combination. For this purpose, we rewrite (9) as

$$\binom{n}{0} + \binom{n}{2} + \binom{n}{4} + \cdots = \binom{n}{1} + \binom{n}{3} + \binom{n}{5} + \cdots.$$

This is equivalent to: The number of subsets of an n-element set X with an even number of elements is equal to the number of subsets of X with an odd number of elements.

Now, let's take an element a from X, and for any subset A of X that has an even number of elements, if $a \in A$, then match A to $A \backslash \{a\}$; if $a \notin A$, then match A to $A \cup \{a\}$. Clearly we see that this is a correspondence from the subsets of X with an odd number of elements to the subsets of X with an even number of elements, which is a one-to-one correspondence. Thus, the numbers of elements in these two sets are equal, which leads to the equality.

Proof of (10): There are $\binom{m+n}{k}$ ways to select a group of k people from m men and n women. In addition, such groups of k people can be divided into

$k+1$ classes: the ith class consists of i men and $k-i$ women ($i = 0, 1, \ldots, k$). Obviously, there are $\binom{m}{i}\binom{n}{k-i}$ groups in the ith class, so there are $\sum_{i=0}^{k}\binom{m}{i}\binom{n}{k-i}$ groups of k people. We can deduce (10) by using the addition principle.

Remark 3. The equality (7) can also be proved by (2) (the essence is the same as the first proof of (7)), but it is not easy to prove (8), (9), and (10) by (2).

Remark 4. The equality (10) is usually called **Vandermonde's identity**. See Example 1 in Chapter 16 for another proof.

Vandermonde's identity can be seen as the expansion of $\binom{m+n}{k}$ about $\binom{m}{i}$ and $\binom{n}{j}$ ($i, j \leq k$), which is similar to the binomial theorem.

Taking some special values of m, n, and k in the equality (10), we can get many interesting results. For example, let $m = n = k$ and use (4). Then

$$\binom{n}{0}^2 + \binom{n}{1}^2 + \cdots + \binom{n}{n}^2 = \binom{2n}{n}.$$

Combinatorial Identities

Combinatorial identities form an important part of combinatorial mathematics, among which the most common ones involve the sum of the binomial coefficients. In the previous section, we have introduced some of these basic identities and several proof methods (induction and recursion, application of the binomial theorem, and combinatorial demonstration). Here are a few more examples demonstrated by the identical deformations.

The most basic deformation is to deform a single term of the sum, so as to transform the sum into a form that is easy to deal with.

The following two identities are useful in this kind of identical deformation.

(11) $\binom{n}{k} = \frac{n}{k}\binom{n-1}{k-1}$.

(12) $\binom{n}{k}\binom{k}{m} = \binom{n}{m}\binom{n-m}{k-m}$.

In order to prove (11) and (12), the explicit representation of the binomial coefficients is very useful. We can easily get (11) by (2). And since

$$\binom{n}{k}\binom{k}{m} = \frac{n!}{k!(n-k)!}\frac{k!}{m!(k-m)!}$$

$$= \frac{n!}{m!(n-m)!}\frac{(n-m)!}{(n-k)!(k-m)!}$$

$$= \binom{n}{m}\binom{n-m}{k-m},$$

we have (12) (note that taking $m = 1$ in (12) leads to (11)).

We now give another proof of (12) by the combinatorial method: Consider the number of ways to select k people from n people to form a committee and then select m people from these k people to form a group.

Based on the multiplication principle, on one hand, the number of selection methods is $\binom{n}{k}\binom{k}{m}$; on the other hand, if the group of m people is selected first, and then $k - m$ people are supplemented from the remaining $n - m$ people to form the committee of k, then the number of methods is $\binom{n}{m}\binom{n-m}{k-m}$. Combining the two results, we obtain the equality (12).

Illustrative Examples

Example 1. Prove:

(i) $\displaystyle\sum_{k=0}^{n} k\binom{n}{k} = n \cdot 2^{n-1}$;

(ii) $\displaystyle\sum_{k=0}^{n} k^2\binom{n}{k} = n(n+1)2^{n-2}$.

Proof. (i) From (11), $k\binom{n}{k} = n\binom{n-1}{k-1}$. Therefore,

$$\sum_{k=0}^{n} k\binom{n}{k} = n\sum_{k=1}^{n}\binom{n-1}{k-1} = n\sum_{k=0}^{n-1}\binom{n-1}{k} = n \cdot 2^{n-1}.$$

(The basic equality (8) was used in the proof.) We can see that the effect of (11) is to "absorb" the summing target k into the binomial coefficients, so as to transform the sum into a known sum.

Please also note that k in the last sum above is actually $k - 1$ in the previous one, so the range of summation changes from $1 \leq k \leq n$ to $0 \leq k \leq n - 1$. In the proof of identies, this method of "substitution" is often used, which does not set a new variable, but only changes the range of the summation.

(ii) Using $k(k-1)\binom{n}{k} = n(n-1)\binom{n-2}{k-2}$ and $k^2 = k(k-1) + k$, we can transform the sum into

$$\sum_{k=1}^{n} k(k-1)\binom{n}{k} + \sum_{k=1}^{n} k\binom{n}{k}$$

$$= \sum_{k=2}^{n} n(n-1)\binom{n-2}{k-2} + n \cdot 2^{n-1} \text{ (using(i))}$$

$$= n(n-1)\sum_{k=0}^{n-2}\binom{n-2}{k} + n \cdot 2^{n-1}$$

$$= n(n-1)2^{n-2} + n \cdot 2^{n-1} = n(n+1)2^{n-2}.$$

Example 1 can also be proved by counting.

First we prove (i). Consider the number of ways to select several people from n people to form a committee, and elect a chairman from the committee. For $k = 0, 1, \ldots, n$, the number of ways to elect a committee of k members and then choose a chairman from them is $\binom{n}{k}k$. According to the addition principle, the number of methods is equal to the sum of the left side of (i). On the other hand, if we first choose the chairman and then calculate the number of all committees including the chairman, then the number of methods is equal to the right side of (i). Combining the two results, (i) is proved. (Refer to the combinatorial proof of (8)).

A combinatorial proof of (ii) may be derived if a secretary is co-opted in each of the above committees (the chairman and the secretary may be held by the same person). In fact, for $k = 0, 1, \ldots, n$, the number of ways to elect a committee of k members first and then choose the chairman and the secretary from them is $\binom{n}{k} \cdot k^2$, so, the total number of methods is equal to the left side of (ii).

Now choose the chairman and the secretary first. There are n ways for the chairman and the secretary to be the same person, and 2^{n-1} ways to choose a committee including the chairman; there are $n(n-1)$ ways for the chairman and the secretary to be two persons, and 2^{n-2} ways to choose a committee including these two persons. Therefore, the total number of the selections is $n \cdot 2^{n-1} + n(n-1)2^{n-2} = n(n+1)2^{n-2}$. Combining the two results, we obtain (ii).   □

Example 2. Prove $\sum\limits_{k=m}^{n} \binom{n}{k}\binom{k}{m} = 2^{n-m}\binom{n}{m}$.

Proof. This is easy to prove from (12). We have

$$\sum_{k=m}^{n} \binom{n}{k}\binom{k}{m} = \sum_{k=m}^{n} \binom{n}{m}\binom{n-m}{k-m}$$

$$= \binom{n}{m}\sum_{k=0}^{n-m}\binom{n-m}{k} = 2^{n-m}\binom{n}{m}. \qquad \square$$

Proof 2. Consider the number of methods to select m formal representatives and a (unlimited) number of informal representatives from n people. On one hand, for $m \le k \le n$, first select k people, and then select m formal representatives from them, so there are $\binom{n}{k}\binom{k}{m}$ methods. Hence, the total number of methods is $\sum_{k=m}^{n}\binom{n}{k}\binom{k}{m}$.

On the other hand, there are $\binom{n}{m}$ ways to select m formal representatives from n people first; and then there are 2^{n-m} ways to select informal representatives from the rest of the $n-m$ people. Hence there are $\binom{n}{m}2^{n-m}$ ways in total. The conclusion is proved by combining the two results. $\qquad \square$

Example 3. Prove the identity: $\sum_{i=0}^{n}\binom{n}{i}\binom{n+i}{i} = \sum_{i=0}^{n} 2^{i}\binom{n}{i}^{2}$.

Proof. From (4) and Vandermonde's identity, the left side of the equality becomes

$$\sum_{i=0}^{n}\binom{n}{i}\binom{n+i}{n} = \sum_{i=0}^{n}\left[\binom{n}{i}\sum_{k=0}^{n}\binom{n}{k}\binom{i}{n-k}\right]$$

$$= \sum_{k=0}^{n}\left[\binom{n}{k}\sum_{i=0}^{n}\binom{n}{i}\binom{i}{n-k}\right]$$

$$= \sum_{k=0}^{n}\left[\binom{n}{k}\sum_{i=n-k}^{n}\binom{n}{i}\binom{i}{n-k}\right]$$

$$\left(\text{since } \binom{i}{n-k} = 0 \quad \text{if } i < n-k\right)$$

$$= \sum_{k=0}^{n} \binom{n}{k} 2^k \binom{n}{n-k} \text{ (using Example 2)}$$

$$= \sum_{k=0}^{n} 2^k \binom{n}{k}^2 = \sum_{i=0}^{n} 2^i \binom{n}{i}^2 \text{ (let } i = k\text{).} \qquad \square$$

(Note that in the second step above, we changed the order of summations of i and k to make it easy to find the new inner sum. See Remark 4 in Chapter 15 for these.)

Proof 2. This is a combinatorial proof. Consider the following counting problem:

There are n men and n women, from which n people are to be elected to form a committee, and then from the n people, elect several women to form a leading group (where the number of women in the leading group is unspecified, possibly zero). How many selections are there in total?

The first counting method: For $0 \le i \le n$, there are $\binom{n}{n-i}$ ways to elect $n-i$ women from all the women to form a leading group, and $\binom{n+i}{i}$ ways to elect i people from the remaining $n+i$ men and women to make up the committee of n members. Therefore, the n-member committee with the leading group composed of $n-i$ women has $\binom{n}{n-i}\binom{n+i}{i} = \binom{n}{i}\binom{n+i}{i}$ selections methods.

Summing over $i = 0, 1, \ldots, n$, and by the addition principle, we see that there are $\sum_{i=0}^{n} \binom{n}{i}\binom{n+i}{i}$ selection methods.

The second counting method: For $0 \le i \le n$, there are $\binom{n}{i}$ and $\binom{n}{n-i}$ ways to elect i from women and $n-i$ from men respectively, so there are $\binom{n}{i}\binom{n}{n-i} = \binom{n}{i}^2$ n-person committee composed of i women. In each of these committees, there are obviously 2^i ways to elect a leading group from i women. Therefore, there are $2^i \binom{n}{i}^2$ ways of choosing an n-member committee with i women and the leading group. Sum over i. Then there are $\sum_{i=0}^{n} 2^i \binom{n}{i}^2$ selection methods.

Combining the results of the above two kinds of counting, we have proved the identity. $\qquad \square$

Remark 5. In elementary mathematics, in addition to the induction and recursion method, the identical deformation method, and the combinatorial counting method mentioned above, there is a more powerful method that we will introduce in Chapter 16: the generating function method.

Induction, recursion, and identical deformation are universal methods, which appear almost everywhere in algebraic and trigonometric summations; the combinatorial counting method and the generating function method are the methods that can reflect the characteristics of combinatorial mathematics.

Remark 6. In combinatorial mathematics, the combinatorial proof of a combinatorial identity (refer to Remark 2) is often the most admired. However, it is usually not easy to make such a proof. In fact, many of the equalities that have been proved by other methods have not been proved by an combinatorial argument. This chapter introduced a lot of examples of combinatorial arguments. It is hoped that the readers will experience the most basic ideas and techniques in combinatorial mathematics, which may be helpful for further research on other topics.

Exercises

Group A

1. Prove: $\sum\limits_{i=0}^{k}(-1)^i \binom{n}{i} = (-1)^k \binom{n-1}{k}$.

2. Prove: $\sum\limits_{k=0}^{n} \frac{1}{2^k} \binom{n+k}{k} = 2^n$.

3. Prove: $\sum\limits_{k=0}^{\lceil\frac{n}{2}\rceil} \frac{(-1)^k}{4^k} \binom{n\ k}{k} = \frac{n+1}{2^n}$.

4. Prove:

   (i) $\sum\limits_{k=0}^{n} \frac{1}{k+1} \binom{n}{k} = \frac{1}{n+1}(2^{n+1} - 1)$.

   (ii) $\sum\limits_{k=0}^{n} \frac{(-1)^k}{k+1} \binom{n}{k} = \frac{1}{n+1}$.

5. Prove: $\sum\limits_{k=1}^{n} \frac{(-1)^{k+1}}{k} \binom{n}{k} = 1 + \frac{1}{2} + \cdots + \frac{1}{n}$.

6. Prove: $\sum\limits_{k=m}^{n} (-1)^{k+m} \binom{n}{k} \binom{k}{m} = \begin{cases} 1, & \text{if } m = n, \\ 0, & \text{if } m \neq n. \end{cases}$

7. Prove: $\sum\limits_{k=0}^{n} \binom{2n}{k} = 2^{2n-1} + \frac{1}{2} \binom{2n}{n}$.

Group B

8. Prove: $\displaystyle\sum_{k=0}^{\left[\frac{n}{2}\right]}\left\{\binom{n}{k}-\binom{n}{k-1}\right\}^2 = \frac{1}{n+1}\binom{2n}{n}$.

9. Suppose n and k are positive integers and $k+3 \le n$. Prove that $\binom{n}{k}$, $\binom{n}{k+1}$, $\binom{n}{k+2}$, and $\binom{n}{k+3}$ can't form an arithmetic sequence.

10. Give a combinatorial proof of $\binom{2n}{2} = 2\binom{n}{2} + n^2$.

11. By using properties of combination numbers and the combinatorial method, give two proofs of the equality

$$\sum_{k=1}^{n} k\binom{n}{k}^2 = n\binom{2n-1}{n-1}.$$

Chapter 3

Counting: Correspondence and Recursion

Correspondence and recursion are very basic ideas and methods widely used in mathematics. In this chapter, we show some applications of correspondence and recursion in counting problems through several examples.

Correspondence

Let X and Y be two sets. Suppose for each $x \in X$, there is a unique element y in Y corresponding to it. Then we get a **mapping** from the set X to the set Y, denoted as

$$f : X \to Y.$$

Generally, we call y the **image** of x (under the mapping f) and call x the **preimage** of y (under the mapping f), denoted as $y = f(x)$.

If for any two different elements x and x' in the set X, there always holds that

$$f(x) \neq f(x'),$$

namely the images of different elements are different, then f is an **injection** from X to Y.

If there is (at least) one $x \in X$ for each element y in the set Y such that

$$f(x) = y,$$

that is, every element in the set Y is an image of some elements in the set X, then f is called a **surjection** from X onto Y.

If the mapping f is both injective and surjective, then it is called a **one-to-one mapping** (or a one-to-one correspondence).

Now let X be a finite set and $|X|$ be the number of elements in the set X. The basic idea of finding $|X|$ using the correspondence method is to try to make a one-to-one mapping from X to a set Y, and $|Y|$ is easy to determine. Because there is a one-to-one correspondence between (finite) sets X and Y, we have $|X| = |Y|$. From this we can (indirectly) find $|X|$. Sometimes, this process needs to be carried out several times, that is, to establish a series of one-to-one mappings $X \to Y \to Y' \to \cdots \to Y''$ such that $|Y''|$ is easy to find, and $|X| = |Y''|$. (By the way, if the numbers of elements two finite sets X and Y are equal, then there must be a one-to-one correspondence between X and Y. We only need to number the elements of X and Y respectively (arbitrarily) first, and then let the first element of X correspond to the first element of Y, the second element of X correspond to the second element of Y, etc. This correspondence is obviously one-to-one.)

The correspondence method is not suitable for every counting problem, and there is no simple rule to follow for what kind of problems can be dealt with by correspondence. The examples in this chapter and the book show some basic applications of the correspondence methods in elementary problems.

Example 1. Let $n \geq 1$ and $k \geq 0$. Find the number of (ordered) non-negative integer solutions of the indeterminate equation

$$x_1 + \cdots + x_n = k. \tag{1}$$

Solution   First, make the substitution $x_i = y_i - 1\,(i = 1, \ldots, n)$. This obviously gives a one-to-one correspondence between the non-negative integer solution set of (1) and the positive integer solution set of the equation

$$y_1 + \cdots + y_n = n + k. \tag{2}$$

Each group of positive integer solutions to equation (2) can be expressed intuitively as follows: If $n + k$ identical balls are arranged in a row, then there is a gap between two adjacent balls, and there are $n + k - 1$ gaps in total. Take any $n - 1$ gaps and insert a "board" into each gap. The $n - 1$ "boards" divide $n + k$ balls into n sections. If the numbers of balls in each section from left to right are $y_1, \ldots, y_n$, then y_i is a positive integer $(1 \leq i \leq n)$, and $y_1 + \cdots + y_n = n + k$, that is, $(y_1, \ldots, y_n)$ is a group of positive integer solutions of (2). In turn, for any group of positive integer solutions $(y_1, \ldots, y_n)$ of (2), it obviously corresponds to a way of inserting "boards". (That is, we insert a "board" between the (y_1)th and $(y_1 + 1)$st balls (starting from the left side),..., and between the $(y_1 + \cdots + y_{n-1})$th and $(y_1 + \cdots + y_{n-1} + 1)$st balls respectively.) Therefore, a positive integer solution set of (2) corresponds to a way of inserting $n - 1$ "boards" into the above $n + k - 1$ gaps one by one. The latter obviously has $\binom{n+k-1}{n-1} = \binom{n+k-1}{k}$ methods. Therefore, the number of non-negative integer solutions to equation (1) is $\binom{n+k-1}{k}$.

It is not hard to see from the above argument that the number of (ordered) positive integer solutions $(x_1, \ldots, x_n)$ to equation (1) is $\binom{k-1}{n-1}$.

Remark 1. It is no accident that the result of this problem is the same as the number of k-combinations with repetition of n elements (in Chapter 1). In fact, it is easy to point out that there is a one-to-one correspondence between the two problems, so the essences of the two are the same. Let $a_1, \ldots, a_n$ be n distinct elements. Suppose in any k-combination with repetition, a_i repeats x_i times $(i = 1, \ldots, n)$. Then $(x_1, \ldots, x_n)$ is a non-negative integer solutions to (1). In turn, any group of (ordered) non-negative integer solution to (1) also corresponds to a k-combination with repetition of $a_1, \ldots, a_n$.

Therefore, this problem can be treated as an intuitive expression of the k-combination with repetition problem of n elements, which is sometimes more convenient for applications.

Note that if it can be guessed in advance that the number of ordered non-negative integer solutions to equation (1) is $\binom{n+k-1}{k}$, then it is not difficult to give a proof by induction on n:

When $n = 1$, the conclusion is obviously true (there is a total of one solution to (1)). Assume that conclusion is true when the number of unknowns is less than n. Next, consider the number of solutions to equation (1) with n unknowns, and rewrite (1) as

$$x_1 + \cdots + x_{n-1} = k - x_n.$$

Since $0 \le x_n \le k$, the values of the right side of the above equation are all non-negative integers not exceeding k. Let $r = k - x_n$. It can be seen from the addition principle and the inductive hypothesis that the number of (ordered) non-negative integer solutions to equation (1) is

$$\sum_{r=0}^{k} \binom{n-1+r-1}{r} = \sum_{r=0}^{k} \binom{n-2+r}{r} = \binom{n+k-1}{k}.$$

(The last step applied the identity (7) in Chapter 2.)

Example 2. Let S be a finite set, $|S| = n$, and k be a positive integer. Find the number of ordered k-subset families $(S_1, \ldots, S_k)$ of S satisfying $S_1 \cap \cdots \cap S_k = \varnothing$.

Solution The solution here is to match a family of subsets $S_1, \ldots, S_k$ of $S = \{x_1, \ldots, x_n\}$ to a $k \times n$ numerical table composed of 0 and 1:

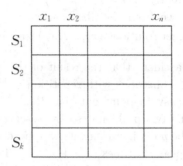

Figure 3.1

As shown in Figure 3.1, if the element x_j belongs to the set S_i, the number 1 will be marked in the cross grid of row i and column j in the table. Otherwise, the cross grid will be marked with 0. Obviously, $\bigcap\limits_{i=1}^{k} S_i = \varnothing$ is equivalent to that any column in this table is not all 1. Let T denote the set of the above $k \times n$ numerical tables that each column in the table is not all 1. Clearly we that the set of ordered subsets $(S_1, \ldots, S_k)$ that meet the requirement corresponds to the set T one by one.

Because there are two ways to fill in every grid in the numerical table, the 1st, $2^{nd}, \ldots, n$th columns of any numerical table in T each has $2^k - 1$ selections, and the selections of each column are independent, so $|T| = (2^k - 1)^n$. Therefore, the number is $(2^k - 1)^n$.

Remark 2. The numerical table described in the solution directly describes the relation of subordination between the sets and the elements, which is very useful.

There are two purposes of the numerical table filling method. One is to distinguish whether the elements belong to a subset (this is all we need in this question, so the numbers 0 and 1 can be replaced by any other two different numbers). The other is to make sure that the sum of the numbers filled in row i of the numerical table is equal to the number of elements of

S_i and the sum of the numbers filled in column j is equal to the number of subsets containing x_j in $S_1, \ldots, S_k$. Both aspects are of clear combinatorial significance.

Sometimes, for specific problems, other number filling methods can also be used, as long as the "horizontal addition" and "vertical addition" (of numbers in the numerical table) are properly explained.

Remark 3. For other two solutions to this problem, see Example 4 below and Example 5 in Chapter 4.

Example 3. Let n be a positive integer. In a permutation $(x_1, x_2, \ldots, x_{2n})$ of the set $\{1, 2, \ldots, 2n\}$, if $|x_i - x_{i+1}| = n$ for a certain $i\,(1 \leq i \leq 2n - 1)$, then this permutation is said to have property **P**. Prove that there are more permutations with property **P** than those without property **P**.

Proof. Let S denote the set of permutations with property **P**; S' denotes the set of permutations such that there is only one i, satisfying $|x_i - x_{i+1}| = n$. Then $S' \subset S$. And let T denote the set of permutations without property **P**.

First of all, let's note that if $\pi = (x_1, \ldots, x_{2n})$ is any permutation of the set $\{1, 2, \ldots, 2n\}$, then for any x_i, there must be a unique x_k in $x_1, \ldots, x_{2n}$, such that $|x_k - x_i| = n$ (this is because x_k is one of $x_i + n$ and $x_i - n$, and exactly one of these two numbers belongs to $\{1, 2, \ldots, 2n\}$).

For any $\pi \in T$, the (unique) number x_k satisfying $|x_k - x_1| = n$ cannot be x_2. Therefore,

$$f : (x_1, x_2, \ldots, x_{2n}) \to (x_2, \ldots, x_{k-1}, x_1, x_k, \ldots, x_{2n})$$

is a correspondence from T to S'.

If the images of two elements $\pi = (x_1, \ldots, x_{2n})$ and $\pi' = (x'_1, \ldots, x'_{2n})$ in T are the same under the mapping f, that is

$$(x_2, \ldots, x_{k-1}, x_1, x_k, \ldots, x_{2n}) = (x'_2, \ldots, x'_{l-1}, x'_1, x'_l, \ldots, x'_{2n})$$

(here $|x_k - x_1| = n$ and $|x'_l - x'_1| = n$), then from the definition of S, we know that $k = l$, so $x_i = x'_i(i = 1, \ldots, 2n)$, i.e., $\pi = \pi'$, hence f is an injection. Therefore, $|T| \leq |S'| < |S|$.　　　　　□

Remark 4. Injection is also very important in combinatorial mathematics.

Let X and Y be finite sets. If we can establish an injection from X to Y, then $|X| \leq |Y|$. This is a basic method to generate inequalities in combinatorial mathematics. (See the above solution of Example 3.)

Remark 5. By the way, if finite sets X and Y satisfy $|X| > |Y|$, then we can see from Remark 4 that any mapping f from X to Y must not be injective, so (by definition) there exist x, $x' \in X$ with $x \neq x'$ such that $f(x) = f(x')$.

This fact can be treated as an extension of the famous "pigeonhole principle," which is widely used in combinatorics.

Remark 6. Example 3 can also be solved by the inclusion-exclusion principle. Please refer to Example 7 in Chapter 4.

Recursion

Many counting problems are related to recursion. Generally speaking, a recurrence relation is a special relation of quantities with integer parameters. Depending on this relation and the initial values of the given quantity, the quantity can be calculated step by step.

The most common case is that the quantity that we count only depends on an integer parameter n — let the quantity be a_n, and a_n satisfies a recurrence relation of order k: There is a (fixed) positive integer k and a function f of $k + 1$ variables such that for all $n \geq k$,

$$a_n = f(a_{n-1}, \ldots, a_{n-k}, n) \tag{3}$$

(the initial values $a_0, a_1, \ldots, a_{k-1}$ are known), so $a_n (n \geq k)$ is uniquely determined by the recurrence formula (3) and the initial values.

In elementary problems, the above positive integer k is usually very small while f has a very simple form. In these cases, it is possible for us to obtain a_n (in an explicit expression of n) from (3).

Example 4. Use the recurrence method to solve Example 2 in this chapter.

Solution We regard k as a fixed integer, and let a_n denote the number of ordered subset groups $(S_1, \ldots, S_k)$ satisfying $S_1 \cap \cdots \cap S_k = \emptyset$ of the n-element set S. Then $a_1 = 2^k - 1$.

Let a be a fixed element of S. For any subset group $(S_1, \ldots, S_k)$ of S that meets the requirements, let $T_i = S_i \backslash \{a\} (i = 1, \ldots, k)$. Then $(T_1, \ldots, T_k)$ is a subset group of the $(n-1)$-element set $S \backslash \{a\}$. In turn, for any subset group $(T_1, \ldots, T_k)$ of $S \backslash \{a\}$ that meets the requirements, whether or not adding a (but excluding the case of adding a to all) to each T_i produces exactly $2^k - 1$ subset groups of S that meet the requirements.

From the above (many to one) correspondence,

$$a_n = (2^k - 1)a_{n-1}. \tag{4}$$

This is the simplest (first order) linear homogeneous recurrence formula with constant coefficients. Combining it with $a_1 = 2^k - 1$, we obtain $a_n = (2^k - 1)^n$.

Example 5. Let a circle be separated into n sectors $S_1, \ldots, S_n$ (as shown in Figure 3.2). Now k different colors are used to color these n sectors ($n \geq 3$ and $k \geq 3$). Each sector is colored with one color, and adjacent (i.e., the ones with a common edge) sectors are colored with different colors. Find the total number of coloring methods.

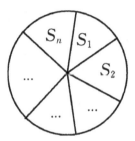

Figure 3.2

Solution   Let a_n denote the number of satisfactory coloring methods for n sectors. It follows that $a_3 = k(k-1)(k-2)$.

Let $n \geq 4$. There are k different coloring methods for sector S_1; (after coloring S_1) there are $k-1$ coloring methods for S_2, so that S_2 and S_1 have different colors; there are $k-1$ coloring methods for S_3, so that S_3 and S_2 have different colors; $\cdots$; there are $k-1$ coloring methods for S_n, so that S_n and S_{n-1} have different colors. Hence there are $k(k-1)^{n-1}$ different coloring methods in total.

However, the above coloring methods include the case of S_1 and S_n being in the same color, and the number of such coloring methods is exactly a_{n-1}. (S_1 and S_n can be combined as one sector, and then the said coloring method is to color $n-1$ sectors with k colors, so the adjacent sectors have different colors.)

Thus we obtain

$$a_n = k(k-1)^{n-1} - a_{n-1}(n \geq 4).$$

This is a first order linear non-homogeneous recurrence formula with constant coefficients. We can obtain a_n by the following method. From the above equation,

$$a_n - (k-1)^n = -[a_{n-1} - (k-1)^{n-1}] \, (n \geq 4).$$

Therefore, $a_n - (k-1)^n$ is a geometric sequence with the first term $a_3 - (k-1)^3$ and common ratio -1. Hence, it's easy to find that $a_n = (k-1)^n + (-1)^n(k-1) \, (n \geq 3)$.

Example 6. How many n-digit numbers can be constructed by 1, 2, and 3, so that there are no two adjacent 1?

Solution Let a_n be the number of n-digit numbers that meet the requirement. Easily we know that $a_1 = 3$ and $a_2 = 8$, and when $n \geq 3$, we divide the n-digit numbers that meet the requirement into two categories according to the rightest digit: If the rightest digit is 2 or 3, then there are a_{n-1} ways to construct the n-digit numbers; if the rightest digit is 1, then the second rightest digit can only be 2 or 3, so there are a_{n-2} ways to construct the n-digit numbers. Thus, (from the addition principle) we get

$$a_n = 2a_{n-1} + 2a_{n-2} \, (n \geq 3). \tag{5}$$

This is a (second order) linear homogeneous recurrence formula with constant coefficients. It can be obtained by the well-known method of characteristic equation that

$$a_n = \frac{2+\sqrt{3}}{2\sqrt{3}}(1+\sqrt{3})^n + \frac{-2+\sqrt{3}}{2\sqrt{3}}(1-\sqrt{3})^n \, (n \geq 1). \tag{6}$$

The following Example 7 is a classic problem, which we will discuss again in Chapter 4.

Example 7. In a permutation $\{a_1, \ldots, a_n\}$ of the set $\{1, 2, \ldots, n\}$, if $a_i \neq i \, (i = 1, \ldots, n)$, it is called a **dislocation arrangement** (also called a derangement). (In other words, in a dislocation arrangement, no element is in its natural position.) Find the number D_n of dislocation arrangements.

Solution Obviously, $D_1 = 0$ and $D_2 = 1$. For any dislocation arrangement

$$(a_1, a_2, \ldots, a_n) \tag{7}$$

of $\{1, 2, \ldots, n\}$ with $n \geq 3$, the first position a_1 can be any other $n - 1$ numbers except 1. If $a_1 = k \, (k \neq 1)$ is taken, then the arrangement (7) can be classified in two categories according to whether a_k is 1: If $a_k = 1$, then number of these dislocation arrangements is the number of dislocation arrangements of $n - 2$ elements, so there are D_{n-2} in total; if $a_k \neq 1$, then this dislocation arrangement is the arrangement of the elements $1, 2, \ldots, k - 1, k + 1, \ldots, n$ in the 2nd to the nth of these $n - 1$ positions, where 1 is not in the kth position and all other elements are not in their own marked positions. This arrangement is equivalent to a dislocation arrangement of the $n - 1$ elements marked as $2, 3, \ldots, n$, so there are D_{n-1} in total. Hence (from the addition principle)

$$D_n = (n - 1)(D_{n-1} + D_{n-2}), \, (n \geq 3). \tag{8}$$

This is a (linear homogeneous) recurrence relation with non-constant coefficients (the coefficients are related to n), which can't be solved as (5).

We suppose $E_n = \frac{D_n}{n!}$. Then (8) can be transformed into

$$E_n = \left(1 - \frac{1}{n}\right) E_{n-1} + \frac{1}{n} E_{n-2},$$

that is

$$E_n - E_{n-1} = \left(-\frac{1}{n}\right)(E_{n-1} - E_{n-2}).$$

Using this recurrence relation repeatedly, we can get (notice that $E_1 = 0$ and $E_2 = \frac{1}{2}$): For $n \geq 2$,

$$E_n - E_{n-1} = \left(\frac{-1}{n}\right)\left(\frac{-1}{n-1}\right)(E_{n-2} - E_{n-3})$$

$$= \cdots = \left(\frac{-1}{n}\right)\left(\frac{-1}{n-1}\right)\cdots\left(\frac{-1}{3}\right)(E_2 - E_1)$$

$$= \frac{(-1)^n}{n!}.$$

Thus $E_n = \frac{(-1)^n}{n!} + \frac{(-1)^{n-1}}{(n-1)!} + \cdots + \frac{(-1)^2}{2!}$, so

$$D_n = n!\left(1 - \frac{1}{1!} + \frac{1}{2!} - \cdots + (-1)^n \frac{1}{n!}\right). \tag{9}$$

In the above example, a so-called unary recurrence relation (i.e., a one-subscript sequence) is established. In combinatorial mathematics, binary (and multivariate) recurrence relations are often involved. The most typical example is the binomial coefficients, which we are familiar with. If we let $C(n, k) = \binom{n}{k}$, then there is a binary recurrence relationship

$$C(n, k) = C(n - 1, k) + C(n - 1, k - 1). \qquad (10)$$

Generally speaking, binary recurrence relations are not easy to deal with. In more elementary problems, they are often related to binomial coefficients. Let's give an example.

Example 8. Find the number of k-element subsets of the set $\{1, 2, \ldots, n\}$ that don't contain consecutive integers.

Solution   Let $f(n, k)$ denote the number. Then $f(n, 1) = n$ and $f(n, n) = 0$ (for $n > 1$).

When $1 < k < n$, the subsets that meet the requirement are classified in two categories: The first kind of subsets contain the element 1, so they must not contain the element 2, from which there are $f(n - 2, k - 1)$ such subsets. The other kind of subsets don't contain the element 1, so there are $f(n - 1, k)$ such subsets. Thus, we get the recurrence formula

$$f(n, k) = f(n - 1, k) + f(n - 2, k - 1). \qquad (11)$$

It is not very simple to find the explicit representation of $f(n, k)$ directly from (11). If it is noted that (11) is similar to the recurrence formula (10) (thus the result should be related to the binomial coefficients) and the initial values of (previously mentioned) $f(n, k)$ are taken into account, then it is easy to find that

$$f(n, k) = \binom{n - k + 1}{k}. \qquad (12)$$

(A complete demonstration should be based on the double mathematical induction method, i.e., assume that $f(n - 1, k) = \binom{n - k}{k}$ and $f(n - 2, k - 1) = \binom{n - k}{k - 1}$, and then prove $f(n, k) = \binom{n - k + 1}{k}$ by (11).)

Remark 7. A proper classification of the set under consideration and an application of correspondence are the basic methods for establishing recurrence relations, as in the previous examples.

Remark 8. The first step of the recurrence method is to establish a recurrence relation. Due to different focuses, different recurrence relations may be derived for the same problem, but not every such relation is suitable for solving the problem. Let's take Example 2 of this chapter as an example:

Let $g(n, k)$ denote the number of ordered subsets $(S_1, \ldots, S_k)$ of the n-element set S satisfying $S_1 \cap \cdots \cap S_k = \emptyset$. Suppose

$$S_1 \cap \cdots \cap S_{k-1} = T, \quad |T| = i.$$

Then for any given T, the number of ordered subset groups $(S_1, \ldots, S_{k-1})$ established by the above equation is equal to the number of ordered subset groups $(T_1, \ldots, T_{k-1})$ of the $(n-i)$-element set $S \backslash T$ satisfying $\bigcap\limits_{i=1}^{k-1} T_i = \emptyset$, i.e., $g(n - i, k - 1)$. Now S_k can be any subset of $S \backslash T$, so there are 2^{n-i} in total, and because T has $\binom{n}{i}$ selections, it is concluded that

$$g(n, k) = \sum_{i=0}^{n} \binom{n}{i} 2^{n-i} g(n - i, k - 1), \tag{13}$$

where $g(0, k) = 1$.

Although $g(n, k) = (2^k - 1)^n$ can be obtained from the above recurrence relation by using the generating function method, it is not easy to derive this solution by elementary methods. Therefore, we may be helpless in the face of (13) and can only seek other solutions. (On the other hand, if we can guess the solution, then (13) will come into use. It is not difficult to prove our results by induction.)

Remark 9. The second step of the recurrence method is to solve the established recurrence relation. Generally speaking, this is rather difficult. We will not discuss it for the purpose of this book.

On the other hand, we remind the readers that a recursive relation may be more suitable for studying some problems because it provides the basis of an inductive argument. For example, it is easy to prove by the recurrence relation (5) and mathematical induction that a_n (for $n \geq 3$) in Example 6 are all even numbers, but this is not clear from the general term formula (6). Later in this book, there are some examples that use a recurrence formula derived from a general term formula for the purpose of a proof.

Exercises

Group A

1. Find the number of monotone nondecreasing mappings that map the set $\{1, 2, \ldots, n\}$ to itself.
2. There are k different kinds of postcards, i.e., there are a_i postcards of the ith kind ($i = 1, \ldots, k$). How many different ways are there to send

them to n people? (Allow the same postcard to be sent to the same person.)

3. The positive integer n is written as the sum of positive integers, which is called a partition of n. If the orders of addends are different, they are regarded as different partitions. Find the number of ordered partitions of n. (e.g., $3 = 3 = 2 + 1 = 1 + 2 = 1 + 1 + 1$, so the integer 3 has 4 different ordered partitions.)

4. There are $n(\geq 3)$ points on a circle. Every two points are connected by a chord. If there are no three chords intersecting at a point (except the end points), find the number of intersection points of these chords in the circle.

5. Determine how many parts that a plane is separated into by n lines of general position (i.e., no two lines are parallel or three lines share one point).

6. Find the number of subsets of the set $\{1, 2, \ldots, n\}$ that do not contain consecutive integers.

7. Let $a_i = 0$ or 1 $(i = 1, 2, \ldots, n)$, and try to find the number of sequences $(a_1, a_2, \ldots, a_n)$ that satisfy $a_1 \leq a_2 \geq a_3 \leq a_4 \geq a_5 \leq \cdots$.

Group B

8. Solve Example 8 of this chapter by using the proof method of (6) in Chapter 1.

9. Let k be a fixed positive integer with $1 \leq k \leq n$. Determine how many integer sequences $1 \leq a_1 < a_2 < \cdots < a_k \leq n$ satisfy that a_i and i have the same parity $(i = 1, 2, \ldots, n)$.

10. Use the recurrence method to solve Example 7 in Chapter 1.

11. Let S be a finite set, $|S| = n$, and k be a positive integer. Find the number of subset groups $(S_1, S_2, \ldots, S_k)$ that satisfy $S_1 \subseteq S_2 \subseteq \cdots \subseteq S_k$.

12. Determine the number of sequences consisting of 0 and 1 that meet the following conditions:

   (1) there are n zeros and n ones in the sequence;
   (2) when counted from left to right, the number of 1 is never less than that of 0.

Chapter 4

Counting: Inclusion-Exclusion Principle

Many counting problems can be expressed in the following form:

Let S be a known finite set, and suppose $\mathbf{P}_1, \ldots, \mathbf{P}_n$ are n properties related to the elements in S (for each element x in S and any property $\mathbf{P}_i$, it can be determined whether x has property $\mathbf{P}_i$). Find the number N of elements in S that do not have any property $\mathbf{P}_i (i = 1, \ldots, n)$.

Suppose $S_i = \{x | x \in S$ and x has property $\mathbf{P}_i\}$ $(i = 1, \ldots, n)$. Then the above problem can be converted into a more mathematical form:

Let $S_1, \ldots, S_n$ be n subsets of S. Find $|\overline{S}_1 \cap \cdots \cap \overline{S}_n|$, where $\overline{S}_i$ represents the complementary set of S_i with regard to S.

If $|\overline{S}_1 \cap \cdots \cap \overline{S}_n|$ (i.e., N) is not easy to find directly, but for any positive integer k and any non-empty subset $\{i_1, \ldots, i_k\}$ of $\{1, \ldots, n\}$, the number of elements of the set $S_{i_1} \cap \cdots \cap S_{i_k}$ is easy to determine (note that these elements may also have properties other than $\mathbf{P}_{i_1}, \ldots, \mathbf{P}_{i_k}$), then we can find $|\overline{S}_1 \cap \cdots \cap \overline{S}_n|$ from the information. This is the basic idea of the inclusion-exclusion principle. To be exact, we have

Theorem 1 (Inclusion-exclusion principle). *Suppose $S_1, \ldots, S_n$ are subsets of S. Then*

$$\left| \overline{S}_1 \bigcap \cdots \bigcap \overline{S}_n \right| = |S| - \sum_{i=1}^{n} |S_i| + \sum_{1 \le i_1 < i_2 \le n} \left| S_{i_1} \bigcap S_{i_2} \right| - \cdots$$

$$+ (-1)^k \sum_{1 \le i_1 < \cdots < i_k \le n} \left| S_{i_1} \bigcap \cdots \bigcap S_{i_k} \right| + \cdots$$

$$+ (-1)^n \left| S_1 \bigcap \cdots \bigcap S_n \right|. \tag{1}$$

Remark 1. In the above formula, the number N of elements that do not have any property $\mathbf{P}_i$ can be obtained by counting according to the following process: In S, the elements with property $\mathbf{P}_i$ are excluded first, and the number of remaining elements $|S| - \sum_{i=1}^{n} |S_i|$ is obviously less than N, because those elements with multiple properties are repeatedly excluded many times; Therefore, every two properties $\mathbf{P}_{i_1}$ and $\mathbf{P}_{i_2}$ $(1 \le i_1 < i_2 \le n)$ are supplemented by the elements with the both properties $\mathbf{P}_{i_1}$ and $\mathbf{P}_{i_2}$, so the number of elements $|S| - \sum_{i=1}^{n} |S_i| + \sum_{1 \le i_1 < i_2 \le n} |S_{i_1} \cap S_{i_2}|$ is greater than N, because the elements with $k > 2$ properties are counted $1 + \binom{k}{2}$ times repeatedly, and excluded only $k = \binom{k}{1}$ times. Therefore, it must exclude the elements with all properties $\mathbf{P}_{i_1}$, $\mathbf{P}_{i_2}$, and $\mathbf{P}_{i_3}$ for every three properties $\mathbf{P}_{i_1}$, $\mathbf{P}_{i_2}$, and $\mathbf{P}_{i_3}$ $(1 \le i_1 < i_2 < i_3 \le n)$, hence the obtained number

$$|S| - \sum_{i=1}^{n} |S_i| + \sum_{1 \le i_1 < i_2 \le n} |S_{i_1} \bigcap S_{i_2}| - \sum_{1 \le i_1 < i_2 < i_3 \le n} |S_{i_1} \bigcap S_{i_2} \bigcap S_{i_3}|$$

is less than N. So supplement and exclude repeatedly (i.e., "include" and "exclude"), and finally get the required N. Therefore, the formula in Theorem 1 is called the inclusion-exclusion principle, also known as the successive sweep principle.

Proof of Theorem 1 The method of proof has been suggested in Remark 1. For any $x \in S$, if x does not belong to any subset $S_i (i = 1, \ldots, n)$, then x is counted once in $|S|$ on the right side of (1), and 0 time in the rest terms, so such x is counted once on the right side of (1).

If x belongs to k sets in $S_1, \ldots, S_n$, where $k \ge 1$, then x is counted 1 time, $\binom{k}{1}$ times, $\binom{k}{2}$ times, $\ldots$, $\binom{k}{k}$ times, 0 time, $\ldots$, 0 time in the first term, the first summation, the second summation, $\ldots$, the kth summation, $\ldots$, and the nth summation on the right side of (1), respectively. Therefore, this x is counted

$$1 - \binom{k}{1} + \binom{k}{2} - \cdots + (-1)^k \binom{k}{k} = 0 \text{ times}$$

on the right side of (1). Based on the above discussion, the right side of (1) exactly counts the number of elements in S that do not belong to any S_i, that is, the left side of (1). $\qquad \square$

For any subsets $S_1, \ldots, S_n$ of S, since $\overline{S_1 \cup \cdots \cup S_n} = \overline{S}_1 \cap \cdots \cap \overline{S}_n$ (the duality principle of intersections and unions), we have $|S_1 \cup \cdots \cup S_n| + |\overline{S}_1 \cap$

$\cdots \cap \overline{S}_n| = |S|$. Therefore, the following Theorem 2 can be derived from Theorem 1, which is sometimes easier to apply.

Theorem 2 (Dual form of the inclusion-exclusion principle). *We have*

$$\left|S_1 \bigcup \cdots \bigcup S_n\right| = \sum_{1 \leq i \leq n} |S_i| - \sum_{1 \leq i_1 < i_2 \leq n} \left|S_{i_1} \bigcap S_{i_2}\right| + \cdots$$

$$+ (-1)^{k-1} \sum_{1 \leq i_1 < \cdots < i_k \leq n} \left|S_{i_1} \bigcap \cdots \bigcap S_{i_k}\right| + \cdots$$

$$+ (-1)^{n-1} \left|S_1 \bigcap \cdots \bigcap S_n\right|.$$

In particular, if $S_1, \ldots, S_n$ form a partition of S, then Theorem 2 becomes the equality in Remark 1 of Chapter 1.

Now we turn to applications of the inclusion-exclusion principle. Let's give two basically straightforward problems to help the readers get familiar with the inclusion-exclusion principle.

Example 1. In positive integers less than 1000, determine how many integers that are neither divisible by 5 nor by 7.

Solution Let S be the set of positive integers less than 1000 and S_k be the set of integers divisible by positive integers k in S. Then $\overline{S}_5 \cap \overline{S}_7$ is the set of integers in S that are neither divisible by 5 nor by 7. From Theorem 1,

$$\left|\overline{S}_5 \bigcap \overline{S}_7\right| = |S| - |S_5| - |S_7| + \left|S_5 \bigcap S_7\right|.$$

Clearly we know that $|S| = 999$, $|S_5| = \left[\frac{999}{5}\right] = 199$, $|S_7| = \left[\frac{999}{7}\right] = 142$, and $|S_5 \cap S_7| = |S_{35}| = \left[\frac{999}{35}\right] = 28$. Therefore, $|\overline{S}_5 \cap \overline{S}_7| = 686$.

Example 2. Use the digits 1, 2, and 3 to construct n-digit numbers, so that all the digits 1, 2, and 3 appear at least once in an n-digit number. Find the number of such n-digit numbers.

Solution Let S be the set of all the n-digit numbers consisting of $1, 2,$ and 3, and denote

$$S_k = \{x | x \in S, \text{ there is no digit } k \text{ in } x\}, \ k = 1, 2, 3.$$

Then $|S| = 3^n$ and $|S_k| = 2^n \ (k = 1, 2, 3)$. Since there is only one n-digit number in S without digits k and $l \ (1 \leq k < l \leq 3)$, i.e., $|S_k \cap S_l| = 1$, obviously

$$|S_1 \cap S_2 \cap S_3| = 0.$$

Then from Theorem 1, the desired number is

$$\left|\overline{S}_1 \cap \overline{S}_2 \cap \overline{S}_3\right| = |S| - (|S_1| + |S_2| + |S_3|) + \left|S_1 \cap S_2\right|$$
$$+ \left|S_2 \cap S_3\right| + \left|S_3 \cap S_1\right| - \left|S_1 \cap S_2 \cap S_3\right|$$
$$= 3^n - 3 \cdot 2^n + 3.$$

Example 3. There are 100 math questions solved by three people A, B, and C, and each of them solves exactly 60 of them. The questions that are solved by only one person are called difficult questions, and the questions can be solved by all the three people are called easy questions. Determine whether there are more difficult questions than easy questions? How many more?

Solution We use A, B, and C to represent the set of questions solved by A, B, and C respectively. Then $|A| = |B| = |C| = 60$. By the inclusion-exclusion principle,

$$100 = \left|A \cup B \cup C\right| = |A| + |B| + |C| - \left|A \cap B\right| - \left|B \cap C\right| - \left|C \cap A\right|$$
$$+ \left|A \cap B \cap C\right|.$$

Thus,

$$\left|A \cap B\right| + \left|B \cap C\right| + \left|C \cap A\right| - \left|A \cap B \cap C\right| = 80.$$

Let α be the number of questions solved only by A. It is still obtained by the inclusion-exclusion principle that

$$\alpha = |A| - \left|A \cap B\right| - \left|A \cap C\right| + \left|A \cap B \cap C\right|.$$

In the same way, let β denote the number of questions solved only by B and γ denote the number of questions solved only by C, which have similar equalities with regard to α. Thus, it's easy for us to know that

$$\alpha + \beta + \gamma = |A| + |B| + |C| - 2\left(\left|A \cap B\right| + \left|B \cap C\right|\right.$$
$$+ \left|C \cap A\right| - \left|A \cap B \cap C\right|\right) + \left|A \cap B \cap C\right|$$
$$= 180 - 2 \times 80 + \left|A \cap B \cap C\right|,$$

that is, $\alpha + \beta + \gamma - |A \cap B \cap C| = 20$. Therefore, there are 20 more difficult questions than easy ones.

The following Examples 4 and 5 are Examples 7 and 2 in Chapter 3, respectively. We now use the inclusion-exclusion principle to solve them.

Example 4. Find the number D_n of the dislocation arrangements of the set $\{1, 2, \ldots, n\}$.

Solution Let S be the set of all permutations of $\{1, 2, \ldots, n\}$, and S_k be the set of all permutations $(a_1, \ldots, a_n)$ satisfying $a_k = k$ in S (i.e., k is in its natural position) for $k = 1, \ldots, n$.

Obviously, $|S| = n!$. For $1 \le k \le n$ and any k-element subset $1 \le i_1 < \cdots < i_k \le n$ of $\{1, 2, \ldots, n\}$ (there are $\binom{n}{k}$ in total), $|S_{i_1} \cap \cdots \cap S_{i_k}| = (n - k)!$. Then we can derive from Theorem 1 that

$$D_n = \left| \overline{S}_1 \cap \cdots \cap \overline{S}_n \right|$$

$$= n! - \binom{n}{1}(n-1)! + \binom{n}{2}(n-2)! - \cdots + (-1)^n \binom{n}{n}(n-n)!$$

$$= n! \left(1 - \frac{1}{1!} + \frac{1}{2!} - \cdots + (-1)^n \frac{1}{n!} \right). \tag{2}$$

Example 5. Let S be a finite set with $|S| = n$ and k be a positive integer. Find the number of k ordered subset groups $(S_1, \ldots, S_k)$ of S satisfying $S_1 \cap \cdots \cap S_k = \varnothing$.

Solution Let $S = \{1, 2, \ldots, n\}$ and let A be the set of all groups of k ordered subsets $(S_1, \ldots, S_k)$ of S. Then $|A| = (2^n)^k = 2^{nk}$. Suppose

$$A_i = \left\{ (S_1, \ldots, S_k) \in A \,\middle|\, i \in \bigcap_{l=1}^{k} S_l \right\}, \quad i = 1, 2, \ldots, n.$$

For any $1 \le r \le n$ and any r-element subset $B = \{i_1, \ldots, i_r\}$ of S, the ordered group $(S_1, \ldots, S_k) \in A_{i_1} \cap \cdots \cap A_{i_r}$ is equivalent to $B \subseteq \bigcap_{l=1}^{k} S_l$, which is obviously equivalent to $B \subseteq S_l (l = 1, \ldots, k)$. Therefore, each S_l has 2^{n-r} selections. Hence $|A_{i_1} \cap \cdots \cap A_{i_r}| = (2^{n-r})^k$. By the inclusion-exclusion principle, the desired number is

$$\left|\overline{A}_1 \cap \cdots \cap \overline{A}_n\right| = \sum_{r=0}^{n} (-1)^r \binom{n}{r} (2^{n-r})^k \tag{3}$$

$$= \sum_{r=0}^{n} (-1)^r \binom{n}{r} (2^k)^{n-r} = (2^k - 1)^n. \tag{4}$$

(In the last step, the binomial theorem was applied.)

Remark 2. When we use the inclusion-exclusion principle to solve problems, we often get complicated results. In this way, we make some comments on the solution of a counting problem.

The solution of a counting problem is usually an explicit expression of an integer parameters (e.g., both (2) and (3) are regarded as solutions of the problems). But for the same problem, due to different methods of solving, solutions may be derived in very different forms, and which one is better should not be generalized. However, for answers, usually the simpler expression is better. Therefore, we should simplify the results of complex forms as much as possible (e.g., reduce the above formula (3) to formula (4)), but it's not always easy. (If you like, you can try to "simplify" (2).)

On the other hand, $\binom{k}{2}$ identities are generated by k different forms of solutions of a counting problem. Some combinatorial identities can be proved based on this point (Refer to Remarks 2 and 6 in Chapter 2). Therefore, in this sense, the more complex the form of a solution, the more likely it is to deduce interesting "by-products." See Example 6 below.

Example 6. Let X and Y denote two sets of n and m elements respectively. Find the number of surjections from X to Y. Here m and n are positive integers.

Solution 1 Suppose $Y = \{y_1, \ldots, y_m\}$. For $i = 1, 2, \ldots, m$, let S_i denote the set of mappings from X to Y that do not take value y_i. The set of surjections from X to Y is the same as the set of mappings from X to Y that do not belong to any S_i, so the total number is $|\overline{S}_1 \cap \cdots \cap \overline{S}_m|$.

There are m^n mappings from X to Y (because the image of any element in X can be any element in Y).

S_i is the number of mappings from X to $Y \setminus \{y_i\}$, so $|S_i| = (m-1)^n$. On the other hands $S_i \cap S_j$ is the number of mappings from X to $Y \setminus \{y_i, y_j\}$, so $|S_i \cap S_j| = (m-2)^n$. In general, for $1 \leq i_1 < \cdots < i_k \leq n$ with $1 \leq k \leq m-1$,

$$\left| S_{i_1} \cap \cdots \cap S_{i_k} \right| = (m-k)^n.$$

Note that a mapping from X to Y takes at least one element of Y, so $S_1 \cap \ldots \cap S_m = \emptyset$.

Therefore, according to the inclusion-exclusion principle (Theorem 1), the number of surjections is

$$\left| \overline{S}_1 \cap \cdots \cap \overline{S}_m \right| = m^n - \binom{m}{1}(m-1)^n + \binom{m}{2}(m-2)^n - \cdots$$

$$+ (-1)^{m-1} \binom{m}{m-1}$$

$$= \sum_{k=0}^{m} (-1)^k \binom{m}{k} (m-k)^n. \tag{5}$$

Solution 2 When $n \leq m$, we can get a simple result from another angle.

In fact, if $n < m$, that is, $|X| < |Y|$, then the number of surjections from X to Y is certainly 0.

If $n = m$, that is, $|X| = |Y|$, then the number of surjections from X to Y is the number of one-to-one mappings from X to Y, that is, the number of full permutations of elements in Y, which is $n!$.

Combining the results of the above two countings, we get a "by-product" (note that $\binom{m}{k} = \binom{m}{m-k}$, and replace $m-k$ with i in (5)):

$$\sum_{i=0}^{m} (-1)^i \binom{m}{i} i^n = \begin{cases} 0, & \text{if } n < m; \\ (-1)^n n!, & \text{if } n = m. \end{cases} \tag{6}$$

(6) is called Euler's identity, which combines binomial coefficients and integer powers, and it is a very basic result. (In Chapter 14, we will give a different proof of (6).)

The inclusion-exclusion principle can also deduce inequalities. Let's give an example of this.

Example 7. Let n be a positive integer. In a permutation $(x_1, x_2, \ldots, x_{2n})$ of the set $\{1, 2, \ldots, 2n\}$, if $|x_i - x_{i+1}| = n$ for a certain $i\,(1 \leq i \leq 2n-1)$, then this permutation is said to have property **P**. Prove that there are more permutations with property **P** than those without property **P**.

Proof. See Example 3 in Chapter 3 for a solution to this problem. The solution here (using the inclusion-exclusion principle) is more natural and general.

We prove that the number of permutations with property **P** is more than half of all permutations, that is, $\frac{1}{2}(2n)!$.

For $k \in \{1, 2, \ldots, n\}$, let S_k denote the set of permutations such that k and $k + n$ are adjacent. Then $S = \cup_{k=1}^{n} S_k$ is the set of all permutations with property **P**. It is easy to get from Theorem 2 (refer to Remark 1 and Theorem 4 below) that

$$|S| \geq \sum_{k=1}^{n} |S_k| - \sum_{1 \leq k < l \leq n} \left| S_k \bigcap S_l \right|. \tag{7}$$

Also $|S_k| = 2 \cdot (2n - 1)!$ (if we combine k and $k + n$ as one element, then there are $(2n - 1)!$ permutations, and the positions of k and $k + n$ can be exchanged) and $|S_k \cap S_l| = 2^2 \cdot (2n - 2)!$. The reason is the same as above, that is, combining k and $k + n$ and combining l and $l + n$, we see that there are $(2n - 2)!$ permutations, and if the positions of k and $k + n$, and l and $l + n$ are exchanged, then there are two possibilities respectively. Hence, we can get from (7) that

$$|S| \geq n \cdot 2 \cdot (2n - 1)! - \binom{n}{2} \cdot 2^2 \cdot (2n - 2)!$$

$$= 2n \cdot (2n - 2)! \cdot n > \frac{1}{2} \cdot (2n)!. \qquad \square$$

The number of elements of some sets can be estimated by the inclusion-exclusion principle (see Example 7 above). We give a brief introduction to its basic principle.

Let $S_1, \ldots, S_n$ be subsets of a finite set S. For convenience, suppose

$$\sigma_0 = |S|, \sigma_k = \sum_{1 \leq i_1 < \cdots < i_k \leq n} \left| S_{i_1} \bigcap \cdots \bigcap S_{i_k} \right| \ (1 \leq k \leq n).$$

We want to give a more accurate upper (lower) bound estimate of $|\overline{S}_1 \cap \cdots \cap \overline{S}_n|$ or $|S_1 \cup \cdots \cup S_n|$ using (known) $\sigma_i \ (i = 0, 1, \ldots, n)$.

For this purpose, directly using the formula

$$\left| \overline{S}_1 \bigcap \cdots \bigcap \overline{S}_n \right| = \sigma_0 - \sigma_1 + \sigma_2 - \cdots + (-1)^n \sigma_n \tag{8}$$

in Theorem 1 is generally not suitable, because there are too many terms in (8) and it is not easy to simplify (refer to Remark 2). We will prove the following Theorem 3.

Theorem 3. *Let k be a positive integer and $k \leq n$. If k is an even number, then*

$$\left| \overline{S}_1 \cap \cdots \cap \overline{S}_n \right| \leq \sigma_0 - \sigma_1 + \sigma_2 - \cdots + (-1)^k \sigma_k; \qquad (9)$$

if k is odd, then

$$\left| \overline{S}_1 \cap \cdots \cap \overline{S}_n \right| \geq \sigma_0 - \sigma_1 + \sigma_2 - \cdots + (-1)^k \sigma_k. \qquad (10)$$

In other words, the partial sums $\sigma_0 - \sigma_1 + \sigma_2 - \cdots + (-1)^k \sigma_k$ ($k = 0, 1, \ldots, n$) are alternately not less than and not more than the actual value of $|\overline{S}_1 \cap \cdots \cap \overline{S}_n|$ (this can be seen as a mathematical representation of the process described in Remark 1).

Note that Theorem 3 is equivalent to the following theorem (refer to Theorem 2).

Theorem 4. *Let k be a positive integer and $k \leq n$. If k is even, then*

$$\left| S_1 \cup \cdots \cup S_n \right| \geq \sigma_1 - \sigma_2 + \cdots + (-1)^k \sigma_k; \qquad (11)$$

if k is odd, then

$$\left| S_1 \cup \cdots \cup S_n \right| \leq \sigma_1 - \sigma_2 + \cdots + (-1)^k \sigma_k. \qquad (12)$$

In order to use Theorem 3 (or 4) to estimate the upper and lower bounds of $|\overline{S}_1 \cap \cdots \cap \overline{S}_n|$ (or $|S_1 \cup \cdots \cup S_n|$), the parameter k should be selected properly. On one hand, it makes the sums in (9)–(12) easy to deal with, while on the other hand, it makes the upper and lower bounds as accurate as possible. However, in primary problems, it is usually enough to take a small value of k (formula (7) of Example 7 is equivalent to taking $k = 2$ in (11)).

Proof of Theorem 3. This can be similar to the proof of Theorem 1. For any $x \in S$, if x does not belong to any set $S_i (i = 1, \cdots, n)$, then x is counted once on the left and right sides of (9) and (10) respectively. If x belongs to r sets ($r \geq 1$) in $S_1, \ldots, S_n$, then x is counted 0 time on the left side of (9) and (10), and it is counted

$$1 - \binom{r}{1} + \binom{r}{2} - \ldots + (-1)^r \binom{r}{k}$$

times on the right side. Easily we know that the sum in the above is equal to $(-1)^k \binom{r-1}{k}$ (see Exercise 1 in Chapter 2), which is not less than 0 when k is an even number, and not more than 0 when k is an odd number. Therefore, (9) and (10) are proved.

Note that if $\sigma_1, \ldots, \sigma_n$ satisfy

$$\sigma_1 \geq \sigma_2 \geq \cdots \geq \sigma_n, \tag{13}$$

then when k is even, by the obvious inequality

$$(-1)^{k+1} \sigma_{k+1} + (-1)^{k+2} \sigma_{k+2} + \cdots + (-1)^n \sigma_n$$
$$= -\sigma_{k+1} + \sigma_{k+2} - \cdots + (-1)^n \sigma^n$$
$$= -(\sigma_{k+1} - \sigma_{k+2}) - (\sigma_{k+3} - \sigma_{k+4}) - \cdots \leq 0$$

and the equation (8), we can obtain (9). When k is odd, (10) can be similarly proved. (Refer to the following Remark 3.)

By the way, although the formula (13) is not always true, it is satisfied in many cases (Please verify that the inequality (13) is true for Example 7). In these cases, Theorem 3 (and Theorem 4) can be established more simply and directly.

Remark 3. To estimate the alternating sums, we can try the simple principle below:

Suppose $a_1, \ldots, a_n$ are non-negative real numbers. If $a_1 \geq \cdots \geq a_n$, then

$$0 \leq a_1 - a_2 + \cdots + (-1)^{n-1} a^n \leq a_1.$$

In fact, let S_n denote the sum mentioned in the above formula. Then, when n is even,

$$S_n = (a_1 - a_2) + \cdots + (a_{n-1} - a_n)$$
$$= a_1 - (a_2 - a_3) - \cdots - (a_{n-2} - a_{n-1}) - a_n.$$

From $a_k \geq a_{k+1}$ and $a_n \geq 0$, we know that $0 \leq S_n \leq a_1$.

When n is odd,

$$S_n = (a_1 - a_2) + \cdots + (a_{n-2} - a_{n-1}) + a_n$$
$$= a_1 - (a_2 - a_3) - \cdots - (a_{n-1} - a_n).$$

Therefore, as above, $0 \leq S_n \leq a_1$.

Exercises

Group A

1. Find the number of positive integers not exceeding 1000 that are divisible by at least one of 2, 3, and 5.
2. In the full permutations of a, b, c, d, e, and f, find the number of permutations in which abc and ef don't appear.
3. Find the number of permutations with property $\mathbf{P}$ in Example 7.
4. Prove that the number of permutations with property $\mathbf{P}$ in Example 7 is less than $\frac{2}{3} \cdot (2n)!$, but greater than $\frac{5}{8} \cdot (2n)!$.

Group B

5. Find the number of (ordered) integer solutions to the equation $x_1 + x_2 + x_3 + x_4 = 20$ that satisfy the conditions that $1 \leq x_1 \leq 6, 0 \leq x_2 \leq 7$, $4 \leq x_3 \leq 8$, and $2 \leq x_4 \leq 6$.
6. Let n be a given positive integer. Find the number of solutions $(x_1, \ldots, x_n)$ to the equation

$$x_1 + \cdots + x_n = n,$$

where $x_i = 0, 1$, or 2 $(i = 1, \ldots, n)$.

Chapter 5

Combinatorial Problems

The problems of combinatorial mathematics are various and flexible. In this section, some basic methods of combinatorial mathematics will be shown through several combinatorial problems in mathematical competitions.

Example 1. Seven vertices of a cube are marked with 0, and one vertex is marked with 1. Now do the following operation: Add 1 to the number of two endpoints of an edge. Can you repeat the above operation so that (i) the eight numbers are equal; (ii) the eight numbers are all multiples of three?

Solution The answer is "no." To prove that the eight numbers can't be equal, we consider the sum S of the numbers of the eight vertices. This is an even number because each operation increases S by 2. Therefore, the operations do not change the parity of S. At the beginning, $S = 1$ is an odd number, so S is always an odd number. But if the eight numbers are equal, then S is an even number, which is impossible.

However, we note that considering S is not sufficient to show that (ii) is impossible. (Because during the operations, the value of S can be a multiple of 3.)

To solve (ii), we select four vertices so that any two of them are not the two endpoints of an edge. Let a be the sum of the numbers of the four vertices, and b be the sum of the numbers of the four remaining vertices. Consider $T = a - b$. It can be seen from the rule of the operation that each operation does not change the value of T. At the beginning, $T = \pm 1$, so T is always ± 1. If the numbers of all eight vertices are divisible by 3, then T

should be divisible by 3, but it's impossible. (Notice that considering the value of T can also solve (i).)

Remark 1. Match the object that we consider to a certain quantity, and solve the problem by studying the latter, which is a universal method in mathematics.

The quantities (S and T) involved in Example 1, which remain unchanged in the changing process, are often called invariants. Invariant is a very important concept in mathematics; In some problems, we often use an invariant to prove impossibility, or make a further argument based on it.

Example 2. Let $n \geq 2$. Mark points $A_1, A_2, \ldots, A_n$ on a circle in the counterclockwise order and place a plate at each point A_i. Take a plate at any two different points A_i and A_j, and move them to the adjacent points in the opposite directions (the plates can overlap). For what value of n, after several such moves, can we make n plates stack at one point?

Solution When n is an odd number, the answer is "yes" (please point out an operation procedure by yourself).

The following proves that when n is even, the answer is "no".

Let a_i be the number of plates at point A_i (each a_i is equal to 1 at the beginning). The main point of the argument is to consider the following sum

$$S = a_1 + 2a_2 + \cdots + na_n$$

after each move of the plate.

For any two points A_i and A_j, where $a_i > 0$ and $a_j > 0$, let one plate at A_i and at A_j move to the adjacent points the counterclockwise and clockwise directions, respectively. There are four situations:

(1) $i \neq n$ and $j \neq 1$. The plates at A_i and A_j are moved to A_{i+1} and A_{j-1} respectively. The change of the sum S after moving is

$$(i+1)(a_{i+1}+1) + i(a_i - 1) + (j-1)(a_{j-1}+1) + j(a_j - 1)$$
$$- (i+1)a_{i+1} - ia_i - (j-1)a_{j-1} - ja_j = 0,$$

that is, S is unchanged.

(2) $i = n$ and $j \neq 1$. The plates at A_i and A_j are moved to A_1 and A_{j-1} respectively. The change of the sum S after moving is

$$1 \times (a_1 + 1) + n(a_n - 1) + (j - 1)(a_{j-1} + 1) + j(a_j - 1)$$

$$-a_1 - na_n - (j - 1)a_{j-1} - ja_j = -n,$$

thus S decreases by n.

(3) $i \neq n$ and $j = 1$. Similarly, we know that S increases by n.

(4) $i = n$ and $j = 1$. Obviously, S is unchanged.

To conclude, the corresponding S will either not change or increase or decrease by n.

At the beginning, $S = 1 + 2 + \cdots + n = \frac{n(n+1)}{2}$. If n plates can finally be stacked together, then $S = ni$, where i is an integer ($1 \leq i \leq n$), that is, the final S is a multiple of n. Therefore, from the change law of S above, we can see that at the beginning S should also be a multiple of n; but when n is even, that's clearly not true. This proves that when n is even, it is impossible to stack all plates at one point.

Remark 2. The value of S in this problem may change during the move of the plate, but the remainder on the division by n is a constant. In the language of congruence (see Chapter 8), S is invariant modulo n. (The S in question (i) in Example 1 is similar to this, wheres is an invariant modulo 2.)

Example 3. n students who take a math exam. The exam consists of m questions. Consider the following statistics:

Let α be a real number ($0 < \alpha < 1$). At least αm problems are difficult problems (a difficult problem is that at least αn students have not solved it); at least αn students pass (a student passes means that he (or she) solves at least αm problems). Try to determine whether the above situation is possible with $\alpha = \frac{2}{3}, \frac{3}{4}$, and $\frac{7}{10}$.

Solution (1) It is possible when $\alpha = \frac{2}{3}$. Suppose there are three students S_1, S_2, and S_3 and three questions P_1, P_2, and P_3. Assume that S_1 worked out P_1 and P_2, S_2 worked out P_1, and P_3, S_3 couldn't work out P_2 and P_3. Then the students S_1 and S_2 passed the exam, and P_2 and P_3 are difficult problems.

(2) It is impossible when $\alpha = \frac{3}{4}$. To prove it, let r be the number of difficult problems and t be the number of students who have passed the exam. Let $P_1, \ldots, P_r, \ldots, P_m$ be all the questions (the first r problems are difficult

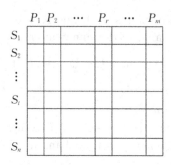

Figure 5.1

problems), and $S_1, \ldots, S_t, \ldots, S_n$ be all the students (the first t students are those who passed the exam). Construct a numerical table shown in Figure 5.1. If S_i worked out P_j, fill in 1 in the grid where line i intersects column j; otherwise fill in 0. Then the table is filled with mn numbers.

On the other hand, the number of integers in the table can be estimated in the following way: Because the students who passed the exam have solved at least αm problems, there are at least αm numbers of 1 in each of the first t rows, so the number of 1 in the table is $\geq t \cdot \alpha m > \alpha^2 mn$. In addition, there are at least αn students who have not solved each problem, so there are at least αn numbers of 0 in each of the first r columns, and the number of 0 in the table is $\geq r \cdot \alpha n \geq \alpha^2 mn$. Therefore, we come to the conclusion that

$$mn = (\text{number of } 1) + (\text{number of } 0) \geq 2\alpha^2 mn,$$

which is not true when $\alpha = \frac{3}{4}$.

(3) It is also impossible when $\alpha = \frac{7}{10}$. Since $2 \times \left(\frac{7}{10}\right)^2 < 1$, the above argument no longer works.

Let's consider the $t \times r$ table in the upper left corner of Figure 5.1 (that is, the table corresponding to whether the students who passed worked out the problem), which has a total of rt numbers.

For any student S_i who passed, suppose he (or she) worked out x difficult problems and y easy problems. Then $x + y \geq \alpha m$. It is also obvious that $y \leq m - r$, so combining $\alpha - 1 < 0$ and $m \leq \frac{r}{\alpha}$, we easily get $x \geq \left(2 - \frac{1}{\alpha}\right) r$. Therefore (looking by row), the number of 1 in the above $t \times r$ table is $\geq \left(2 - \frac{1}{\alpha}\right) rt$.

Similarly, for any difficult problem, if x' students who have passed didn't work it out and y' students who have failed didn't work it out, then the inequality $x' \geq \left(2 - \frac{1}{\alpha}\right) t$ is obtained from $x' + y' \geq \alpha n$, $y' \leq n - t$, and $n \leq \frac{t}{\alpha}$, so that the number of 0 in the $t \times r$ table is greater than or equal to $\left(2 - \frac{1}{\alpha}\right) rt$.

Combining the above results, we obtain that

$$rt \geq \left(2 - \frac{1}{\alpha}\right) rt + \left(2 - \frac{1}{\alpha}\right) rt = 2 \left(2 - \frac{1}{\alpha}\right) rt,$$

which is not true when $\alpha = \frac{7}{10}$. (Notice that the method of (3) can also solve the problem in (2).)

Remark 3. The main contribution of the numerical table in the solution is to make the statement of the argument intuitive and clear, and it's not essential. The filling method of the numerical table in this problem is only to distinguish whether an element belongs to a certain set (that is, whether an element has a certain property), so the numbers 0 and 1 can be replaced by other two different numbers.

The main point of this argument is to calculate (estimate) the same quantity in two different ways, and then produce applicable results. This is a fairly basic way to deal with many combinatorial problems. The above two examples are done in this way, and there are many others (see Remark 2 in Chapter 2).

Example 4. There are n $(n \geq 3)$ people in a party. Prove that there are two people in the party such that at least $\left[\frac{n}{2}\right] - 1$ of the remaining $n - 2$ people know either both of them or neither of them (it is assumed that "knowing" is a symmetric relationship).

Proof. Use $X_1, \ldots, X_n$ to denote the n people. Let each X_i correspond to the $\binom{n-1}{2}$ pairs of two persons formed by the $n - 1$ people other than X: divide these pairs into class A and class B: Class A is where neither person in the pair knows X_i or both two persons know X_i, and class B is where exactly one of them knows X_i. The conclusion of this problem is to prove that there exists a pair in class A such that at least $\left[\frac{n}{2}\right] - 1$ people correspond to it.

We first consider the number of pairs in class B that X_i corresponds to. Suppose X_i knows k people and doesn't know the rest $n - 1 - k$ people. Then X_i corresponds to $k(n-k-1)$ pairs in class B. By the basic inequality, it can be seen that this number is less than or equal to $\frac{(n-1)^2}{4}$, that is, the number of correspondences that from any person to the pairs in class B is less than or equal to $\frac{(n-1)^2}{4}$.

Construct the following table (where $Y_1, \ldots, Y_{\frac{n(n-1)}{2}}$ represent $\binom{n}{2} = \frac{n(n-1)}{2}$ pairs consisting of n people): If Y_j is a two-person pair in class B that X_i corresponds to, fill in 1 in the grid where row i crosses column j; otherwise fill in 0. (Please draw a diagram by yourself).

Therefore, the sum of numbers in row i of the numerical table is the number of pairs in class B that X_i corresponds to; the sum of numbers in column j is the number of persons in (one of) two-person pairs in class B corresponding to Y_j among n persons.

Consider the sum N of all numbers in the table. On one hand,
$$N = \text{the sum of } n \text{ rows} \le n \cdot \frac{(n-1)^2}{4}.$$
On the other hand, N is also equal to the sum of $\frac{n(n-1)}{2}$, columns, so there must be a column sum no more than
$$\frac{N}{\frac{n(n-1)}{2}} \le \frac{\frac{n(n-1)^2}{4}}{\frac{n(n-1)}{2}} \le \frac{n-1}{2},$$
that is, there is one column sum less than or equal to $\left[\frac{n-1}{2}\right]$. Hence there is a two-person pair Y_j in class B such that no more than $\left[\frac{n-1}{2}\right]$ people correspond to it. Therefore, among the remaining $n-2$ people (excluding two people in Y_j), at least
$$n - 2 - \left[\frac{n-1}{2}\right] = \left[\frac{n}{2}\right] - 1$$
people correspond to Y_j, which is a pair in class A.   □

Remark 4. It is convenient to focus on "two-person pairs of class B" firstly, because the number of "two-person pairs of class B" that each person corresponds to is easy to express and estimate. (See Remark 7 of Chapter 1.)

The basic idea of the above argument is: First of all, use the information of the sum of each row to produce the result of the whole quantity N; then apply this result to derive the information of the sum of a column, and then solve the problem. This method, from a single to the whole, and then from the whole to a single, is widely used in combination problems. There are some examples later in this book.

Example 5. At a party of 33 people, each participant was asked, "How many of the rest of you have the same surname as you, and how many have the same age as you?" As a result, the answer includes all integers from 0 to 10. Prove that there must be two people with the same surname and the same age.

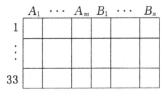

Figure 5.2

Proof. Mark 33 people as $1, 2, \ldots, 33$. Let the different ages of these 33 people be $A_1, \ldots, A_m$, and the different surnames be $B_1, \ldots, B_n$. Construct a numerical table as shown in Figure 5.2. For the kth person, if his/her age is A_i, fill 1 in the cross grid of the kth row and the column where A_i is located; otherwise fill in 0. If his/her surname is B_j, fill 1 in the cross grid of the kth row and the column where B_j is located; otherwise fill in 0.

Obviously, there are two 1 in each row. Suppose there are a_i numbers of 1 in the column where a_i is located, and there are b_j numbers of 1 in the column where B_j is located ($1 \leq i \leq m$ and $1 \leq j \leq n$). The number of 1 in the numerical table shall be calculated by row and by column respectively, and the results shall be the same, hence

$$a_1 + \cdots + a_m + b_1 + \cdots + b_n = 2 \times 33 = 66. \tag{1}$$

One the other hand, a_i is the number of people aged A_i, and b_j is the number of people with the same surname B_j. From the known conditions, we see that $a_1, \ldots, a_m, b_1, \ldots, b_n$ contain the numbers $1, 2, \ldots, 11$. Hence

$$a_1 + \cdots + a_m + b_1 + \cdots + b_n \geq 1 + 2 + \cdots + 11 = 66.$$

Combining it with formula (1), we see that $a_1, \ldots, a_m, b_1, \ldots, b_n$ must be an arrangement of $1, 2, \ldots, 11$, so $m + n = 11$. Hence $m, n \leq 10$.

Suppose $a_1 = 11$, that is, there are 11 people of age A_1. Because the number of surnames $n \leq 10$, there must be two of these 11 people with the same surname.

The starting point of this argument is to consider the same quantity $a_1 + \cdots + a_m + b_1 + \cdots + b_n$ in two ways, so as to infer that the situation in problem is actually an extreme situation (i.e., $a_1, \ldots, a_m, b_1, \ldots, b_n$ form exactly an arrangement of $1, 2, \ldots, 11$). $\qquad \square$

In many problems in mathematics competitions, the test builder intentionally conceals some extreme situation and asks the solvers to reveal it. The following Example 6 is also such a problem.

Example 6. There are $2n$ players in a chess match, and there are two rounds. In every round, each player has to compete with the others one by one. The scoring rule is: 1 point for the winner, $\frac{1}{2}$ point for each draw, and 0 point for the loser. Suppose for each player, the difference between his scores in the first round and in the second round is at least n points. Prove that the difference is exactly n points.

Proof. Suppose the ith player got a_i points in the first round and b_i points in the second round $(1 \leq i \leq 2n)$. Let's suppose the 1st, 2nd, ... , kth players scored less in the first round than in the second, while the $(k+1)$st, ..., $(2n)$th players scored more in the first round than in the second. Since $|a_i - b_i| \geq n \, (i = 1, \ldots, 2n)$,

$$b_i - a_i \geq n \, (1 \leq i \leq k), \, a_i - b_i \geq n \, (k+1 \leq i \leq 2n). \tag{2}$$

Suppose set A consists of the $1^{\text{st}}, 2^{\text{nd}}, \ldots, k$th players, set B consists of the $(k+1)$st, ..., $(2n)$th players, and let

$$a = \sum_{i=1}^{k} a_i, a' = \sum_{i=k+1}^{2n} a_i, \, b = \sum_{i=1}^{k} b_i, \, b' = \sum_{i=k+1}^{2n} b_i. \tag{3}$$

Let's consider the total score of players in set A, and compute the difference between the second round and the first round. It can be seen from (3) that the difference is $b-a$. On the other hand, we should note that the sum of the scores of the two players in each game is 1, so the contribution of the scores of the players in A obtained in the competitions between each other to $b-a$ is, that is to say, $b - a$ is actually obtained in the competitions between players in A and players in B. Obviously, in the first round, the total score obtained in the competitions between players in A and players in B is ≥ 0; In the second round, the total score obtained in the competitions between players in A and players in B is $\leq k(2n - k)$ (as there are $k(2n - k)$ games, each game can contribute up to 1 point). Therefore, $b - a \leq k(2n - k)$. For the same reason, $a' - b' \leq k(2n - k)$.

Furthermore, $b_i - a_i \geq n \, (1 \leq i \leq k)$ in (2) are added to get $b - a \geq kn$. Similarly, $a' - b' \geq (2n - k)n$. Hence

$$\begin{cases} kn \leq b - a \leq k(2n - k), \\ (2n - k)n \leq a' - b' \leq k(2n - k). \end{cases} \tag{4}$$

In particular,

$$kn \leq k(2n - k) \text{ and } (2n - k)n \leq k(2n - k).$$

Thus $n \leq 2n - k$, and $n \leq k$, so $k = n$. Combining it with (4), we know that $b - a = n^2$ and $a' - b' = n^2$. Hence, from (3),

$$\sum_{i=1}^{n}(b_i - a_i) = n^2, \quad \sum_{i=n+1}^{2n}(a_i - b_i) = n^2.$$

Therefore, the inequalities in (2) must all hold as equalities, that is $b_i - a_i = n \, (i = 1, \ldots, n)$ and $a_i - b_i = n \, (i = n + 1, \ldots, 2n)$, hence $|a_i - b_i| = n \, (i = 1, 2, \ldots, 2n)$. □

Example 7. At a party, each participant has at most three acquaintances. Prove that the participants can be classified in two groups, so that each person has at least one acquaintance in his/her group.

Proof. First, the participants were randomly classified in two groups (group I and group II). Let S denote the sum of the numbers of acquaintances that each participant has in his/her own group. Suppose there is a person A in group I, who has at least two acquaintances in the group. Then we transfer A to group II, so that A has at most one acquaintance in group II. Now consider the change of S due to this action. In group I, at least two people lose one acquaintance in their group. In group II, at most one person gains an acquaintance in his group. Also A himself has fewer acquaintances in his group. Therefore, the value of S after adjusting is strictly decreasing. Since the sequence of strictly decreasing natural numbers can only have a finite number of terms, the above decrease will stop after a finite number of times. Thus, the grouping method that meets the requirement is given (since no more adjustment can be done). □

Remark 5. The following proof can also be used for Example 7 (the same idea as above): Since there are only a limited number of grouping methods, S has only a limited number of possible values, among which there must be a minimum value. It is easy to prove (by contradiction) that the grouping corresponding to the minimum S meets the requirement. (Note that in the previous solution, when the adjustment process stops, S does not necessarily reach the minimum value.)

In mathematics, quantities with extreme properties (maximum, minimum, most, least, etc.) are often very important. In some problems, quantities with extreme properties give the desired solutions; in other cases, we can make further arguments based on them.

Example 8. Let S be a set of $2n$ points in a plane, where any three points are not collinear. Now, color n points red and n points blue. Prove that n

straight line segments that are pairwise disjoint can be found, in which two ends of each segment are heterochromatic points in S.

Proof. Because there are only n pairs of points, there are finitely many ways to match red with blue points. For each pairing method π, consider the sum $S(\pi)$ of the lengths of the n line segments. There must be a certain π to minimize $S(\pi)$. Suppose there is a line segment RB intersecting $R'B'$ in π (here R and R' represent the red points, and B and B' represent the blue points). Then replace the 2 line segments with RB' and $R'B$. Since the sum of any two sides of a triangle is larger than the third side, the sum $S(\pi')$ of the line segment lengths of its corresponding pairing method π' must be smaller than $S(\pi)$, which contradicts our selection of π. Hence the n line segments in π are pairwise non-intersecting. $\qquad\square$

In this solution, each pairing method corresponds to a (numerical) quantity, and the quantity with extreme properties corresponds to the pairing that meets the requirement of the problem.

Example 9. Several people attend a party, and some of them know each other. It is known that (i) if two people have an equal number of acquaintances, then they do not have common acquaintances; (ii) there is one person with at least 100 acquaintances. Prove that there must be someone who has exactly 100 acquaintances in the party.

Proof. Consider any person A (as an undetermined parameter). Let the acquaintances of A be $B_1, \ldots, B_n$ $(n \geq 1)$. Because A is the common acquaintance of any two of $B_1, \ldots, B_n$, according to the condition (i), the numbers of people known by $B_1, \ldots, B_n$ are different from each other. However, from this information alone, it can't be asserted that there are exactly 100 of these numbers of people.

Now we suppose A is (one of) the person who has the most acquaintances. Then the number n of people he/she knows controls the numbers of people that $B_1, \ldots, B_n$ know — they are all not greater than n. And $B_1, \ldots, B_n$ all know at least one person (i.e., A). Therefore, combined with the previous results, the numbers of people that n people $B_1, \ldots, B_n$ know are exactly an arrangement of $1, 2, \ldots, n$. And condition (ii) shows that $n \geq 100$, so 100 appears in the above sequence, that is, there is someone in $B_1, \ldots, B_n$ who knows exactly 100 people. $\qquad\square$

Example 10. Can you fill in a real number in every small grid of a 17×17 grid table, so that these real numbers are not all zero, and the number in each grid equals the sum of the numbers in all adjacent grids (grids with common sides are called adjacent grids)?

Solution The answer is "yes." We construct a grid table that meets the requirement:

Fill in

$$1, 1, 0, -1, -1, 0, 1, 1, 0, -1, -1, 0, 1, 1, 0, -1, -1$$

in the first row (from left to right) of the grid table in turn. Thus, each number in the first row is equal to the sum of the numbers in its adjacent grids in that row. Fill in 0 in each grid of the second row. In this way, each number in the first row meets the requirement.

In order to make each number in the second row meet the requirement of the problem, the opposite number of the corresponding number in the first row should be filled in each grid of the third row (from left to right), that is, fill in $-1, -1, 0, 1, 1, \ldots, 0, 1, 1$. Then, fill in all in the fourth row, so that the numbers in the third row meet the requirement. In this way, the 5th row to the 8th row are the same as the first four rows, $\cdots$, etc., and the last row is the same as the first row.

Remark 6. To prove the existence of something with a certain property, we often use the construction method, that is, actually construct objects that meet the requirement. (Of course, we can also use the proof by contradiction or other methods.)

There is often more than one thing that meets the requirement of the problem. We can choose something with special properties to try, that is, make it meet appropriate sufficient conditions to ensure that it meets the requirement of the question. This kind of idea of retreating in order to advance is often used in construction.

Example 11. There are n $(n \geq 2)$ lines that are placed in a plane, where no two lines are parallel and no three lines share one point. These lines divide the plane into several regions. Show that it is possible to fill in a non-zero integer with absolute value less than or equal to n in each region, such that the sum of the numbers in all regions on the same side of each line is 0.

Solution When $n = 2$, the problem is easy to solve: Fill $+1$ and -1 in the regions as shown in Figure 5.3, which obviously meets the requirement.

In general, consider all the intersection points of n lines. Each intersection point can be regarded as the case when $n = 2$. For any intersection

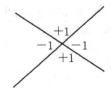

Figure 5.3

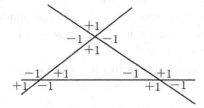

Figure 5.4

point, write $+1$ and -1 in each region near the vertex (just as in the previous case). Then the sum of the numbers on the same side of each line is apparently 0, since it is true at each vertex (Figure 5.4 shows a filling method when $n = 3$).

Now for each region, we write the sum of the numbers at all its vertices. Clearly we know from the condition of this question that any region has at most n "vertices," so the absolute value of this number is less than or equal to n, and the sum of the written numbers in the region on the same side of any straight line is obviously 0. But unfortunately, this filling method may make the numbers filled in some regions equal to 0. Therefore, we haven't fully obtained the filling method that meets the requirement.

Now we modify the above idea to get the filling method that meets all the requirements. Let's fill in the numbers so that the numbers in the same region all have the same sign. If this can be done, then the sum of the numbers in each region is nonzero, whose absolute value is less than or equal to n, and all the required conditions are satisfied.

We need to prove that this method of filling ± 1 is compatible, that is, the number filled in each region can be uniquely determined. It is not hard to see that this is equivalent to the following question:

For the regions determined by n straight lines in the problem, we can color them in black and white, so that any two regions with common edges have different colors.

This can be proved by mathematical induction. The conclusion is obviously true for $n = 2$. Assume that the proposition is true for $n = k$. When one more straight line is added, it divides some regions into two parts. We will change the color of each region on one side of the straight line (that is, the original black region will be colored white, while the original white region will be colored black), while the region on the other side will retain the original color. Combined with the inductive hypothesis, it is known that such a coloring makes the colors of any two regions with common edges different, i.e., the regions connected "diagonally" are colored with the same color. Therefore, the conclusion is also true for $n = k + 1$.

Exercises

Group A

1. There are 100 numbers on the blackboard:

$$1, \frac{1}{2}, \frac{1}{3}, \ldots, \frac{1}{100}.$$

Each operation allows to erase two numbers a and b arbitrarily selected on the blackboard, and then write another number $a + b + ab$. After 99 such operations, suppose only one number is left on the blackboard. Find the number.

2. Given a quadratic trinomial $f(x)$, we can replace $f(x)$ with $x^2 f\left(\frac{1}{x} + 1\right)$ or $(x-1)^2 f\left(\frac{1}{x-1}\right)$. Can we get $x^2 + 10x + 9$ from $x^2 + 4x + 3$ after such operations?

3. Fill in the numbers in the 3×3 grid table shown in Figure 5.5. Now, the table is operated as follows: At each step, we may add a same number to two adjacent cells (adjacent numbers refer to the numbers in two small grids with common sides). After several operations, is it possible that (i) all the numbers in the grid table are 0? (ii) the numbers at four corners of the grid table are 1, and the rest numbers are 0?

0	3	2
6	7	0
4	9	5

Figure 5.5

4. Suppose there are n points in a plane, any three of which are the vertices of a right triangle. Prove: $n \leq 4$.

5. Let S be a finite set of real numbers with $|S| \geq 2$. Let $A = \{(a, b)|a, b \in S, a+b \in S\}$ and $B = \{(a,b)|a, b \in S, a-b \in S\}$. Prove: $|A| = |B|$.

6. Prove that it is possible to arrange all n-digit numbers consisting of the only digits 1 and 2 into a circle, so that there is only one digit difference between any two adjacent numbers.

Group B

7. There are six questions in a math test, and it is known that exactly 100 students have answered each question correctly. For any two students, there is at least one question that neither of them can work out. Determine at least how many students take the test.

8. In an $n \times n$ real numerical table, the absolute value of each number is less than or equal to 1, and the sum of all numbers in the numerical table is 0. Prove that the numbers in each column can be exchanged appropriately (i.e., exchange positions), so that the absolute value of the sum of the numbers in each row of the new numerical table is not greater than 2.

9. Prove that any $2n$ (ordered) pairs (a_i, b_i) of positive numbers can be classified into two groups with n pairs in each group, so that the difference between the sums of a_i in the two groups does not exceed the maximum number in a_i, and the difference between the sums of b_i in the two groups does not exceed the maximum number in b_i.

Chapter 6

Exact Division

Number theory, known as the queen of mathematics, is a fascinating and important branch of mathematics.

Elementary number theory problems often appear in mathematics competitions, which are flexible and diverse.

Solving problems of number theory certainly requires relevant knowledge in this field, but the more important and difficult thing is to understand and put into applications. This chapter and the following two chapters will briefly introduce the most basic content of number theory, and show and explain some basic ideas and techniques in elementary number theory through some simple examples.

Exact Division

The basic objects of concern in elementary number theory are the natural number set and the integer set. We know that addition, subtraction, and multiplication can be performed in the set of integers, and these operations satisfy some laws (i.e., the associative law and commutative law of addition and multiplication, and the distributive law of addition and multiplication), but in general, division cannot be performed, that is to say, even though a and b are integers with $b \neq 0$, the number $\frac{a}{b}$ is not necessarily an integer, thus leading to the first basic concept in elementary number theory: exact division.

Let a and b be integers with $b \neq 0$. If there exists an integer c such that $a = bc$, then b **exactly divides** a, denoted as $b|a$. We say that b is a **divisor** (or **factor**) of a and a is a **multiple** of b. If there is no such an integer c, then b **doesn't exactly divide** a, denoted as $b \nmid a$.

From the definition of exact division, it is easy to deduce several simple properties of exact division (the details are left for the readers to complete):

(1) Let a, b, and c be integers. If $b|c$ and $c|a$, then $b|a$. That is, exact division is transitive.

(2) Let a, b, and c be integers. If $b|a$ and $b|c$, then $b|(a \pm c)$. That is, the integers that are multiples of b are closed under addition and subtraction. Therefore, if integers $a_1, \ldots, a_n$ are all multiples of b, then $b|(a_1 + \cdots + a_n)$.

(3) Let a and b be integers. If $b|a$, then either $a = 0$ or $|a| \geq |b|$. Therefore, if $b|a$ and $a|b$, then $|a| = |b|$.

Remark 1. In problems of exact division, sometimes it is not easy to directly prove $b|a$. We can (try to) choose an appropriate "intermediate quantity" c to prove $b|c$ and $c|a$, and then we can draw a conclusion according to the property (1).

This method is similar to the scaling method for inequalities: In order to prove $B \leq A$, first "amplify" B to a proper C, and then (expect to) prove $C \leq A$.

Remark 2. Property (2) provides a basic idea to prove $b| (a_1 + \cdots + a_n)$: We can try to prove a stronger (and often easier to prove) proposition:

each a_i is a multiple of b $(i = 1, \ldots n)$.

Of course, this proposition is not necessarily true. Even when it is not true, the above idea still has a (often effective) modification: Properly group the sum $a_1 + \cdots + a_n$ into $c_1 + \cdots + c_k$, and show that $b|c_1, \ldots, b|c_k$, which of course can lead to $b|(a_1 + \cdots + a_n)$.

Remark 3. Property (3) provides a basic way to a transition from exact division to inequality. (3) can also be replaced with the following two statements:

(i) If an integer $a \neq 0$ and $|a| < |b|$, then b doesn't exactly divide a.

(ii) If $b|a$ and $|a| < |b|$, then $a = 0$.

For any two integers a and b $(b > 0)$, it is not necessarily that a is divisible by b, but we have the following conclusion, called the division with remainder, which is the most basic and important result in elementary number theory.

(4) (**Division with remainder**) If a and b are integers and $b > 0$, then there exist integers q and r such that

$$a = bq + r, \quad \text{where } 0 \le r < b; \qquad (1)$$

the integers q and r are uniquely determined by the above condition.

In fact, suppose bq is the maximum multiple of b not exceeding a, that is, the integer q satisfies $bq \le a < b(q+1)$. Let $r = a - bq$. Then naturally $0 \le r < b$. Therefore, we get a representation similar to (1).

If there are integers q_1 and r_1 to make $a = bq_1 + r_1$ and $0 \le r_1 < b$, then $b(q - q_1) = r_1 - r$ from (1), thus $b|(r - r_1)$. But $0 \le |r - r_1| < b$, so $r - r_1 = 0$, that is, $r = r_1$ (refer to Remark 3(ii)), hence $q = q_1$. This proves the uniqueness.

Remark 4. The integer q in the equality (1) is called the (incomplete if $r \ne 0$) **quotient** of a on the division by b, and the number r is called the **remainder** of a on the division by b. If $r = 0$, then a is divisible by b.

The core of the division with remainder is the inequality with regard to the remainder r, which plays an important role in an argument.

Remark 5. The above argument shows that the quotient q is actually $\left[\frac{a}{b}\right]$ (the largest integer not exceeding $\frac{a}{b}$), which is sometimes very useful.

Greatest Common Divisor

Let $a_1, a_2, \ldots, a_n$ be integers that are not all zeros. An integer (e.g., ± 1) that can exactly divide all $a_1, a_2, \ldots, a_n$ is called their **common divisor** (or **common factor**). From the property (3), a nonzero integer has only finitely many divisors, so there are only a finite number of common divisors of $a_1, a_2, \ldots, a_n$, among which there must be a unique maximum. We call it the **greatest common divisor** (GCD) of $a_1, a_2, \ldots, a_n$, denoted as $(a_1, a_2, \ldots, a_n)$. Note that if d is a common divisor of $a_1, a_2, \ldots, a_n$, then $-d$ is also their common divisor, so the GCD must be a positive integer.

If $(a_1, a_2, \ldots, a_n) = 1$, then $a_1, a_2, \ldots, a_n$ are relatively prime (coprime); in other words, $a_1, a_2, \ldots, a_n$ are relatively prime means that their common divisors are only ± 1. If any two numbers in $a_1, a_2, \ldots, a_n$ are relatively prime, then $a_1, a_2, \ldots, a_n$ are pairwise coprime. It is obvious that if the numbers are pairwise coprime, then they must also be relatively prime, but the converse is not necessarily correct.

We mainly deal with the GCD (a, b) of two integers, where $b \neq 0$. Because $(a, b) = (a, -b)$, we can suppose $b > 0$ in the following discussion.

(5) Let a and b be integers, $b > 0$, and $b|a$. Then $(a, b) = b$.
(6) There is the equality $(a, b) = (a + b, b)$, that is (a, b) is a function with regard to a that has a period of b. In general, $(a, b) = (a + bq, b)$ for any integer q.

Both (5) and (6) can be derived from the definition, of which (5) is trivial. To prove (6), let $d = (a, b)$ and $d' = (a + b, b)$. Then $d|a$ and $d|b$, so $d|a + b$ (property (2)), that is, d is a common divisor of $a + b$ and b, hence $d \leq d'$. Similarly, $d' \leq d$, thus $d = d'$. The general situation can be derived by using this method repeatedly, which is not hard to prove directly as above.

In order to find the greatest common divisor effectively and obtain its nontrivial properties, we apply the following basic facts, which are called **Euclidean (division) algorithm.**

(7) (**Euclidean algorithm**) Let a and b be integers with $b > 0$. Repeat the division with remainder in the following way. After a finite number of steps, it must stop (i.e., the remainder is zero):

$$\left.\begin{array}{l} a \text{ divided by } b : a = bq_0 + r_0, 0 < r_0 < b; \\ b \text{ divided by } r_0 : b = r_0 q_1 + r_1 \ 0 < r_1 < r_0; \\ r_0 \text{ divided by } r_1 : r_0 = r_1 q_2 + r_2, \ 0 < r_2 < r_1; \\ \qquad \cdots \\ r_{n-2} \text{ divided by } r_{n-1} : r_{n-2} = r_{n-1} q_n + r_n, \ 0 < r_n < r_{n-1}; \\ r_{n-1} \text{ divided by } r_n : r_{n-1} = r_n q_{n+1}. \end{array}\right\} \quad (2)$$

Then $(a, b) = r_n$.

In fact, since the remainders $r_0, r_1, \ldots, r_n$ are integers, and

$$r_0 > r_1 > \cdots > r_{n-1} > \cdots > 0,$$

after a finite number of steps of the above divisions with remainders, the last remainder must be zero. On the other hand, from the properties (6) and (5),

$$(a, b) = (bq_0 + r_0, b) = (b, r_0) = (r_0, r_1) = \cdots$$

$$= (r_{n-1}, r_n) = (r_n q_{n+1}, r_n) = r_n.$$

Given a and b, Euclidean algorithm can not only solve (a, b) in a finite number of steps, but also prove that the equation

$$ax + by = (a, b) \tag{3}$$

has integer solutions x and y, and provides an algorithm to find a group of integer solutions. To be specific, we do things in the reverse order. From the penultimate line in (2), we obtain

$$(a, b) = r_n = r_{n-2} - r_{n-1} q_n,$$

which represents (a, b) in the form of "$r_{n-1} \times$ integer $+ r_{n-2} \times$ integer"; substitute $r_{n-3} = r_{n-2} q_{n-1} + r_{n-1}$, the third line from the bottom in (2), into the above formula, and eliminating r_{n-1} to get

$$(a, b) = r_{n-2}(1 + q_{n-1} q_n) - r_{n-3} q_n,$$

which represents (a, b) in the form of "$r_{n-2} \times$ integer $+ r_{n-3} \times$ integer". In this way, a group of integer solutions x and y of (3) is finally obtained.

Now we can easily deduce the following:

(8) Two integers a and b is coprime if and only if there exist integers x and y such that

$$ax + by = 1. \tag{4}$$

In fact, if $(a, b) = 1$, then the above result shows that (4) has integer solutions. Conversely, if x and y make (4) true, then obviously (a, b) exactly divides the left side of (4), so $(a, b) \mid 1$, that is, $(a, b) = 1$.

Equalities (3) and (4) are called **Bézout's identities**, which are widely used.

Now we prove some important properties of the GCD:

(9) Any common divisor of a and b is a divisor of their GCD.

(10) Let m be a positive integer. Then $m(a, b) = (ma, mb)$.

(11) Let $(a, b) = d$. Then $\left(\frac{a}{d}, \frac{b}{d}\right) = 1$.

(12) Let a and b both be coprime with m. Then ab is also coprime with m, that is, integers coprime with m are closed under multiplication (try to compare it with property (2)). In general, if $a_1, \ldots, a_n$ are all coprime with m, then $a_1 \cdots a_n$ is coprime with m.

(13) Suppose $b \mid ac$. If $(b, c) = 1$, then $b \mid a$.

Property (9) is the most essential property of the GCD, which is a direct consequence of the equality (3).

To prove (10), let $d = (a, b)$ and $d' = (ma, mb)$. Then md exactly divides ma and mb, so $md \leq d'$. On the other hand, if both sides of (3) are multiplied by m, we obtain $d' | md$, so $d' \leq md$. Thus $d' = md$.

(11) is a corollary of (10): From $d\left(\frac{a}{d}, \frac{b}{d}\right) = \left(\frac{a}{d} \cdot d, \frac{b}{d} \cdot d\right) = (a, b) = d$, we have

$$\left(\frac{a}{d}, \frac{b}{d}\right) = 1.$$

It is shown in (11) that coprime integers can be naturally generated from integers that are not relatively prime.

Proof of (12): From $(a, m) = 1$ and (8), there exist integers x and y such that $ax + my = 1$; similarly, there exist x' and y' such that $bx' + my' = 1$. Multiply the two formulas to get

$$ab(xx') + m(axy' + bx'y + myy') = 1,$$

thus $(ab, m) = 1$ (still using (8)). The general situation is not difficult to prove by mathematical induction.

Proof of (13): From $(b, c) = 1$, there exist x and y, such that $bx + cy = 1$, so $abx + acy = a$, the left side of which is a multiple of b (since $b|ac$). Therefore b exactly divides a that is on the right side of the above equality.

Remark 6. The above properties are proved by Bézout's identity, which eventually relies on the division with remainder (refer to (7) and Remark 4).

Remark 7. From (9), we can easily prove that

$$(a_1, a_2, \ldots, a_n) = ((a_1, a_2), \ldots, a_n).$$

Therefore, it is not difficult to deduce that (9), (10), and (11) are also true for the greatest common divisor of more than two integers. For the purpose of this book, we omit the details.

Least Common Multiple

Let $a_1, a_2, \ldots, a_n$ be non-zero integers. A number that is a multiple of all $a_1, a_2, \ldots, a_n$ is called their common multiple. The least positive common multiple is called the **least common multiple** (LCM) of $a_1, a_2, \ldots, a_n$, written as $[a_1, a_2, \ldots, a_n]$.

We mainly deal with the LCM $[a, b]$ of two integers.

(14) Let m be an integer, $a|m$ and $b|m$. Then $[a,b]|m$. That is, any common multiple of a and b is a multiple of their LCM.

(15) Let m be an integer. Then $m[a,b] = [ma, mb]$.

(16) This is the equality $(a,b)[a,b] = |ab|$. In particular, if $(a,b) = 1$, then $[a,b] = |ab|$.

(17) Let $b_1, b_2, \ldots, b_n$ be integers that are pairwise coprime, and a is any arbitrary integer. If $b_i|a (i = 1, 2, \ldots, n)$, then $b_1 \cdots b_n|a$. Therefore, in particular, the LCM of positive integers that are pairwise coprime is equal to their product.

Proof of (14): Let $D = [a,b]$. It is known from the division with remainder that there exist integers q and r such that $m = Dq + r$ with $0 \le r < D$. Obviously, r is also a common multiple of a and b, so from $0 \le r < D$ and the minimality of D, we obtain $r = 0$, that is, $D|m$.

Proof of (15): Let $D = [ma, mb]$ and $D' = [a,b]$. We can deduce that $a \mid \frac{D}{m}$ and $b \mid \frac{D}{m}$ from $ma|D$ and $mb|D$. Therefore, $D' \le \frac{D}{m}$ by definition. On the other hand, mD' is a multiple of ma and mb. By definition, $D \le mD'$. Combining the two results, we see that $D = mD'$.

Proof of (16): Let a and b be positive integers. First, consider the case of $(a,b) = 1$. From $a|[a,b]$ and $b|[a,b]$, there exist integers x and y such that $[a,b] = ax$ and $[a,b] = by$, thus $b|ax$. But $(a,b) = 1$, so from (13) we know that $b|x$ and $ab|ax$, i.e., $ab|[a,b]$. But ab is a common multiple of a and b, so $[a,b] \le ab$. Hence $[a,b] = ab$.

The general case can be reduced to the above special case. Let $d = (a,b)$. Then from (11), we know $\left(\frac{a}{d}, \frac{b}{d}\right) = 1$, so from the proven results, $\left[\frac{a}{d}, \frac{b}{d}\right] = \frac{a}{d} \cdot \frac{b}{d} = \frac{ab}{d^2}$. Then, (using (10) and (15)), we get that

$$(a,b)[a,b] = d \cdot \left(\frac{a}{d}, \frac{b}{d}\right) \cdot d \left[\frac{a}{d}, \frac{b}{d}\right] = d^2 \cdot \frac{ab}{d^2} = ab.$$

Proof of (17): Since a is a multiple of b_1 and b_2, from (14) and (16), we can deduce $[b_1, b_2]|a$, and then $b_1 b_2|a$. Because b_3 and $b_1 b_2$ are coprime (using (12)), and a is a multiple of $b_1 b_2$ and b_3, there is $b_1 b_2 \cdot b_3|a$ similarly. If we repeat this argument, we can get $b_1 \cdots b_n \mid a$. In addition, if we take $a = [b_1, \ldots, b_n]$, then $[b_1, \ldots, b_n] = b_1 \cdots b_n$.

Remark 8. Property (17) provides another way to decompose an exact division problem: In order to prove $b|a$, decompose b into the product of several integers $b_1, \ldots, b_n$ that are pairwise coprime, and prove $b_i|a (i = 1, \ldots, n)$ (refer to Remark 2). For a standard application of this technique, please refer to Remark 5 of Chapter 7.

Remark 9. Clearly, we see that properties (14) and (15) are still true for the least common multiple of more than two integers, but (16) is not so, that is

$$(a_1, a_2, \ldots, a_n) [a_1, a_2, \ldots, a_n] = |a_1 a_2 \ldots a_n|$$

is not true in general (it is not difficult for the readers to give a counterexample).

Illustrative Examples

Let's give some examples involving exact division.

The basic technique of proving $b|a$ is (by definition) to decompose a into the product of b and an integer. The decomposition often arises from taking special values in some algebraic factorization. The following two decompositions are very useful.

If n is a positive integer, then

$$x^n - y^n = (x - y)(x^{n-1} + x^{n-2}y + \cdots + xy^{n-2} + y^{n-1}). \tag{5}$$

If n is a positive odd integer, then (replace y with $-y$ in (5))

$$x^n + y^n = (x + y)(x^{n-1} - x^{n-2}y + \cdots - xy^{n-2} + y^{n-1}). \tag{6}$$

Example 1. Let $F_k = 2^{2^k} + 1$ with $k \geq 0$. Prove: If $m > n$, then $F_n \,|\, (F_m - 2)$.

Proof. Take $x = 2^{2^{n+1}}$ and $y = 1$ in (5), and replace n with 2^{m-n-1}. Then we obtain $(2^{2^{n+1}} - 1) \,|\, (2^{2^m} - 1)$. And since $2^{2^{n+1}} - 1 = (2^{2^n} - 1)$ $(2^{2^n} + 1)$, we have $(2^{2^n} + 1) \,|\, (2^{2^{n+1}} - 1)$. Therefore, $(2^{2^n} + 1)|(2^{2^m} - 1)$. (Refer to Remark 1.)

The problem can also be demonstrated as follows: Using the formula for the difference of squares repeatedly, we get

$$2^{2^m} - 1 = (2^{2^0} - 1)(2^{2^0} + 1)(2^{2^1} + 1) \cdots (2^{2^{m-1}} + 1),$$

that is, F_k satisfy the following recurrence relation:

$$F_m - 2 = F_0 \cdot F_1 \cdot \cdots \cdot F_{m-2} \cdot F_{m-1}.$$

Because $m > n$, we see that F_n appears on the right side of the above formula, so we can draw the conclusion. $\square$

Example 2. Suppose $n > 1$ is an odd number. Prove:

$$n \left| \left(1 + \frac{1}{2} + \cdots + \frac{1}{n-1} \right) (n-1)! \right.$$

Proof. Note that $n \mid \frac{1}{k}(n-1)!$ is not always true for $1 \le k \le n-1$.

The skill of the argument is to deform $\left(1 + \frac{1}{2} + \cdots + \frac{1}{n-1}\right)(n-1)!$ by pairing. For $1 \le k \le \frac{n-1}{2}$, since n is odd, $n - k \ne k$, thus $\frac{(n-1)!}{k(n-k)}$ is an integer. Therefore,

$$\left(\frac{1}{k} + \frac{1}{n-k}\right)(n-1)! = \frac{n}{k(n-k)}(n-1)!$$

is a multiple of n, thus their sum is also a multiple of n. (Refer to Remark 2.) □

Example 3. Suppose m and n are positive integers with $m > 2$. Prove: $(2^m - 1) \nmid (2^n + 1)$.

Proof. First, it is easy to prove that the conclusion is true for $n \le m$. In fact, when $m = n$, the conclusion is trivial; when $n < m$, the result can be deduced from $2^n + 1 \le 2^{m-1} + 1 < 2^m - 1$ (note that $m > 2$). (Refer to Remark 3.)

Finally, the case of $n > m$ can be reduced to the above special case by the division with remainder: Suppose $n = mq + r$ with $0 \le r < m$ and $q \ge 0$. Because

$$2^n + 1 = (2^{mq} - 1)\, 2^r + 2^r + 1,$$

while $(2^m - 1) \mid (2^{mq} - 1)$ (using (5)), we have $(2^m - 1) \nmid (2^r + 1)$ (since $0 \le r < m$). Therefore, $(2^m - 1) \nmid (2^n + 1)$ when $n > m$. □

Example 4. Let n be a positive integer.

   (i) Prove: $(21n + 4, 14n + 3) = 1$.
  (ii) Prove: $(n! + 1, (n+1)! + 1) = 1$.
 (iii) Let $F_k = 2^{2^k} + 1$ with $k \ge 0$. Prove: For $m \ne n$, we have $(F_m, F_n) = 1$.

Proof. To prove (i), only one equality is needed: $3(14n + 3) - 2(21n + 4) = 1$.

To prove (ii), we note that

$$(n+1)(n! + 1) - ((n+1)! + 1) = n. \tag{7}$$

Suppose d is the GCD of $n! + 1$ and $(n+1)! + 1$. Then $d \mid n$ from the above formula. Furthermore, $d \mid n!$ since $d \mid n$. Combining it with $d \mid (n! + 1)$, we get $d \mid 1$, so $d = 1$.

To prove (iii), we may assume that $m > n$. From Example 1, there exists an integer x such that

$$F_m + xF_n = 2. \tag{8}$$

Suppose $d = (F_m, F_n)$. Then the above formula implies that $d|2$, so $d = 1$ or 2. But F_n is obviously odd, thus necessarily $d = 1$.   □

Remark 10. Being coprime is very important in elementary number theory (refer to properties (12), (13), (16), and (17)). The basic starting point of proving that two integers a and b are coprime is to establish Bézout's identity (4) (refer to property (8)), as the solution of problem (i) in Example 4.

In general, Bézout's identity (4) satisfied by a and b is not easy to obtain, so we often adopt the following variation method: Make an equation similar to (4) (refer to (7) and (8)):

$$ax + by = r, \tag{9}$$

where r is an appropriate integer. Now let $(a, b) = d$. Then $d|r$ from the above equation. Here r being appropriate means that $d = 1$ can be derived from $d|r$ through a further demonstration (as (ii) of Example 4), or the divisors of r are few, and the result can be proved by the exclusion method.

In addition, (9) is equivalent to $a|(by - r)$ or $b|(ax - r)$. Sometimes, these are more easily derived by exact division (refer to the solution of (iii) in Example 4).

Example 5. Suppose m, a, and b are all positive integers with $m > 1$. Prove:

$$(m^a - 1, m^b - 1) = m^{(a, b)} - 1.$$

Proof. We use Euclidean division on the exponents. From (2) (notice that $q_0, q_1, \ldots$ are all non-negative integers now),

$$(m^a - 1, m^b - 1) = (m^{bq_0 + r_0} - 1, m^b - 1)$$
$$= ((m^{bq_0} - 1)m^{r_0} + m^{r_0} - 1, m^b - 1)$$
$$= (m^{r_0} - 1, m^b - 1),$$

where $(m^b - 1)|(m^{bq_0} - 1)$ and property (6) were applied. Therefore, repeat the above argument to get

$$(m^a - 1, m^b - 1) = (m^b - 1, m^{r_0} - 1)$$
$$= (m^{r_0} - 1, m^{r_1} - 1)$$

$$= \cdots$$
$$= (m^{r_{n-1}} - 1, m^{r_n} - 1)$$
$$= (m^{r_n q_{n+1}} - 1, m^{r_n} - 1)$$
$$= m^{r_n} - 1 = m^{(a,\,b)} - 1.$$
□

Example 6. Let k be a positive odd number. Prove: $\frac{n(n+1)}{2} \Big| (1^k + 2^k + \cdots + n^k)$.

Proof. The problem is equivalent to proving $n(n+1) \mid 2(1^k + 2^k + \cdots + n^k)$. Since n and $n+1$ are coprime, we decompose the problem into proving

$$n \mid 2(1^k + 2^k + \cdots + n^k) \quad \text{and} \quad (n+1) \mid 2(1^k + 2^k + \cdots + n^k).$$

(Refer to Remark 8.) Now it is easy to know from formula (6) that (notice k is a positive odd number)

$$2(1^k + 2^k + \cdots + n^k) = [1^k + (n-1)^k] + \cdots + [(n-1)^k + 1^k] + 2n^k$$

is a multiple of n. Similarly, $2(1^k + \cdots + n^k)$ is a multiple of $n+1$. This proves the conclusion. (Refer to Remark 2 and Example 2.) □

Example 7. Prove that for any positive integer k less than $n!$, it is possible to select at most n numbers (not necessarily distinct) from the positive divisors of $n!$, such that their sum is equal to k.

Proof. We prove it by induction on n. The conclusion is obviously true for $n = 2$. Assume that the statement is true for $n - 1$. Now consider the case of n. Suppose $1 < k < n!$. We want to prove that k can be expressed as the sum of not more than n positive divisors of $n!$. When $k \leq n$, because $k \mid n!$, the conclusion is established trivially. When $k > n$, do the division with remainder. Suppose $k = k'n + r$ with $0 \leq r < n$. Then since $n < k < n!$,

$$1 \leq k' \leq \frac{k}{n} < \frac{n!}{n} = (n-1)!,$$

so from the inductive hypothesis, there exist positive integers $d_1' < \cdots < d_m'$ with $m \leq n-1$ and $d_i' \mid (n-1)! \ (1 \leq i \leq m)$, such that $k' = d_1' + \cdots + d_m'$. Therefore,

$$k = k'n + r = nd_1' + \cdots + nd_m' + r. \tag{10}$$

Obviously, $nd_i' \ (1 \leq i \leq m)$ are distinct positive divisors of $n!$. If $r = 0$, then (10) is a satisfactory representation of k (note that $m \leq n-1 < n$); if

$r > 0$, then from $r < n$, we have $r | n!$, and since $r < n \leq n d_1' < \cdots < n d_m'$, the $m + 1 \leq n$ numbers on the right side of (10) are the distinct positive divisors of $n!$. Therefore, (10) is a satisfactory representation of k.   □

Exercises

Group A

1. Let n and k be positive numbers.
   Prove: In $1, 2, \ldots, n$, exactly $\left[\frac{n}{k}\right]$ numbers can be divisible by k.

2. Let a, b, c, and d be integers $a - c | ab + cd$. Prove $a - c | ad + bc$.

3. Let a and b be positive integers. Suppose $a^2 + ab + 1$ can be divisible by $b^2 + ab + 1$. Prove that $a = b$.

4. Let a and b be positive integers, and $\frac{b+1}{a} + \frac{a+1}{b}$ is an integer. Prove: $(a, b) \leq \sqrt{a + b}$.

5. Let m and n be positive integers and m be an odd number. Prove: $(2^m - 1, 2^n + 1) = 1$.

6. Let a, m and n be positive integers such that $a > 1$ and $a^m + 1 | a^n + 1$. Prove: $m | n$.

7. Let a and b be positive integers, and c be an integer. A necessary and sufficient condition for the equation

$$ax + by = c \qquad (1)$$

   to have an integer solution is $(a, b) | c$. When this condition is satisfied, let $x = x_0$ and $y = y_0$ be a group of integer solutions of (1). Prove: all integer solutions of (1) are

$$x = x_0 + \frac{b}{(a, b)} t, \ y = y_0 - \frac{a}{(a, b)} t,$$

   where t is an arbitrary integer.

8. Let a and b be positive integers greater than 1 with $(a, b) = 1$. Prove: There exists a unique pair of integers r and s satisfying $ar - bs = 1$ with $0 < r < b$ and $0 < s < a$.

9. Let a and b be positive integers with $(a, b) = 1$. Prove: For an integer $n > ab - a - b$, the equation

$$ax + by = n \qquad (2)$$

   has a non-negative integer solution; but when $n = ab - a - b$, (2) doesn't have a non-negative integer solution.

Group B

10. (i) Let a and b be coprime integers, and m and n be any positive integers. Prove that $(a^m, b^n) = 1$.
    (ii) Prove: For any positive integer n, we have $(a^n, b^n) = (a, b)^n$.

11. Let the product ab of positive integers a and b be the kth power of an integer $(k \geq 2)$. If a and b are coprime, prove that a and b both are kth powers of integers.
    To be exact, if $ab = x^k$, then $a = (a, x)^k$ and $b = (b, x)^k$.

12. Let k be a positive integer $(k \geq 2)$. If the kth power of a rational number is an integer, prove that the rational number is an integer. Consequently, deduce that for integers a and b, if $b^k | a^k$, then $b | a$.

13. Let m and n be positive integers with $(m, n) = 1$. Prove: Any positive divisor of mn can be uniquely expressed in the form of $d_1 d_2$, where d_1 and d_2 are positive divisors of m and n respectively.

14. Suppose an integer $q > 1$. Prove:
    (i) Each positive integer n can be uniquely expressed as

    $$n = a_k q^k + \cdots + a_1 q + a_0,$$

    where a_i are a integers satisfying $0 \leq a_i \leq q - 1$ $(0 \leq i \leq k)$ and $a_k \neq 0$. This expression is called the **base-q representation** of n, and the numbers a_i are called the digital codes in the q-ary representation.
    (ii) There are the equalities

    $$a_i = \left[\frac{n}{q^i} \right] - q \left[\frac{n}{q^{i+1}} \right] \quad \text{for } i = 0, 1, \ldots, k.$$

15. Suppose $n > 1$ and $m = 2^{n-1}(2^n - 1)$. Prove: Any integer k $(1 \leq k \leq m)$ can be expressed as the sum of some distinct divisors of m.

16. Let $a > 1$ be a given integer, $A_n = 1 + a + \cdots + a^n$ $(n \geq 1)$. Find all positive integers that can exactly divide a certain term in the sequence $\{A_n\}$.

Chapter 7

Prime Numbers

Prime numbers constitute a subset of the natural number set, which is the most powerful one. This chapter mainly introduces basic properties of prime numbers and the unique factorization theorem of positive integers.

Prime Numbers and the Unique Factorization Theorem

An integer n $(n > 1)$ has at least two different positive divisors: 1 and n. If n has no proper factor (that is, it doesn't have a divisor greater than 1 but less than n), then n is a **prime number**, and we usually use the letter p for a prime; if n has a proper factor, that is, n can be expressed in the form of $a \cdot b$ (a and b are integers greater than 1), then n is called a **composite number**.

As a result, positive integers are classified in the three categories: 1, prime numbers and composite numbers.

The following simple facts can be easily drawn from the definition:

(1) An integer greater than 1 must have a prime factor.
(2) Let p be a prime number and n be any integer. Then either p divides n exactly, or p is coprime with n.

In fact, an integer greater than 1 has positive divisors greater than 1. Among them, the minimum must not have a proper factor, so it is a prime number. This proves (1).

As for (2), we note that (p, n) is a divisor of p, so it is known from the definition of prime numbers, that either $(p, n) = 1$ or $(p, n) = p$, that is, either p and n is coprime or p exactly divides n.

Remark 1. In number theory, it is often needed to prove that a positive integer is 1. (For example, prove that the greatest common divisor of integers is 1, referring to Remark 10 of Chapter 6.) Property (1) provides a number theoretic characterization of whether an integer is 1. Therefore, we often (prove by contradiction) assume that the number under consideration has a prime factor, so that we can use the sharp properties of prime numbers (see (2) and (3)) for further demonstrations to derive contradictions.

The most important properties of prime numbers are as follows:

(3) Let p be a prime number and a and b be integers. If $p \mid ab$, then at least one of a and b is divisible by p.

In fact, if a and b are not divisible by p, then from (2), p is coprime with both a and b, so p and ab are coprime (the property (12) of Chapter 6), a contradiction!

One of the most classical results of prime numbers was proved by Euclid in the third century B.C.:

(4) There are infinitely many prime numbers.

Prove by contradiction: Suppose there is only a finite number of prime numbers. Let $p_1 = 2, p_2 = 3, \ldots, p_k$ be all the prime numbers. Consider $N = p_1 \cdots p_k + 1$. Obviously $N > 1$, so N has a prime factor p (property (1)). Because $p_1, \ldots, p_k$ are all the prime numbers, so p must be a certain p_i ($1 \leq i \leq k$), so p exactly divides $N - p_1 \cdots p_k = 1$, which is impossible. Hence there are infinitely many prime numbers.

(Please note that $p_1 \cdots p_k + 1$ is not necessarily prime.)

Now we prove the most basic result of integers, called the unique factorization theorem of positive integers, or the fundamental theorem of arithmetic.

(5) (**Unique factorization theorem**) (i) Every integer greater than 1 can be decomposed into the product of a finite number of prime numbers.
(ii) If the order of prime factors in the product is ignored, then the factorization in (i) is unique.

Part (i) is a direct corollary of the definition of prime numbers. Let $n > 1$. If n is a prime number, then there is no need to prove it. If n is a composite number, then $n = ab$, where a and b are greater than 1 (and therefore less than n). If a or b still has a proper factor, repeat the above decomposition. Because there is only a finite number of divisors of a positive integer, we can decompose n into the product of prime numbers after a finite number of steps.

The uniqueness of factorization is far from trivial. Let $n > 1$ have two factorizations

$$n = p_1 \cdots p_k = q_1 \cdots q_l. \tag{1}$$

Then we need to prove that $k = l$ and the prime numbers $p_1, \ldots, p_k$ form a permutation of $q_1, \ldots, q_l$. It can be seen from (1) that $q_1(q_2 \cdots q_l)$ is divisible by p_1; by property (3), p_1 exactly divides q_1 or $q_2 \cdots q_l$. If $p_1 \mid q_1$, then because p_1 and q_1 are prime numbers, necessarily $p_1 = q_1$; if $p_1 \mid q_2 \cdots q_l$, repeat this argument to know that p_1 must be one of $q_2, \ldots, q_l$, so we can cancel p_1 on both sides of (1). Then consider p_2 and repeat the above process. Finally, we see that the two kinds of factorizations of n are exactly the same. $\qquad\square$

By collecting the same prime factors in the factorization of n, we know that each integer n greater than 1 can be uniquely written as

$$n = p_1^{\alpha_1} \cdots p_k^{\alpha_k}, \tag{2}$$

where $p_1, \ldots, p_k$ are prime numbers that are different from each other, and $\alpha_1, \ldots, \alpha_k$ are positive integers. The formula (2) is called the **standard factorization** of n.

From the standard factorization of positive integers, several basic results about their divisors can be easily obtained:

(6) Let the standard factorization of n be in the form of (2). Then a necessary and sufficient condition for a positive integer d to be a divisor of n is that its standard factorization is

$$d = p_1^{\beta_1} \cdots p_k^{\beta_k}, 0 \le \beta_i \le \alpha_i (i = 1, \ldots, k). \tag{3}$$

(7) Let $\tau(n)$ denote the number of positive divisors of n, and $\sigma(n)$ denote the sum of positive divisors of n. Let the standard factorization of n be in the form of (2). Then

$$\tau(n) = (\alpha_1 + 1) \cdots (\alpha_k + 1), \tag{4}$$

$$\sigma(n) = \frac{p_1^{\alpha_1+1} - 1}{p_1 - 1} \cdots \frac{p_k^{\alpha_k+1} - 1}{p_k - 1}. \tag{5}$$

(8) An positive integer n is a square number if and only if $\tau(n)$ is an odd number.

Let's prove (6) first. The sufficiency is obvious. The following proves the necessity. If $d = 1$, then this is the form of (3) (all $\beta_i = 0$); if $d > 1$, then from $d \mid n$, we get $n = dd'$ (d' is a positive integer). By the unique factorization theorem, the product of the prime factor decompositions of d and d' is the same as that of factorization (2), so the standard factorization of d has the form of (3).

Proof of (7): Obviously, the exponent β_i in (3) has $\alpha_i + 1$ selections and is independent of each other. And from the unique factorization theorem, all possible selections of exponents $\beta_1, \ldots, \beta_n$ produce distinct divisors of n. Therefore, from the multiplication principle,

$$\tau(n) = (\alpha_1 + 1) \cdots (\alpha_k + 1).$$

In addition, $\sigma(n)$ can be expressed as

$$\sigma(n) = (1 + p_1 + \cdots + p_1^{\alpha_1}) \cdots (1 + p_k + \cdots + p_k^{\alpha_k}), \tag{6}$$

because after multiplying the right side of the above formula term by term, each number in the form of (3) just appears once, that is, the process gives all positive divisors of n. The formula (5) can be obtained from (6) by the sum of the first n terms of a geometric sequence.

Proof of (8): Let $n > 1$ and let the standard factorization of n be in the form of (2). Then we can derive from the formula (4) that $\tau(n)$ is an odd number, which is equivalent to that all α_i in (2) are even, that is, n is a square number.

Another proof: Obviously, when d passes through all positive divisors of n, the quotient $\frac{n}{d}$ also passes through all positive divisors of n. We make the divisor d of n ($< \sqrt{n}$) correspond to the divisor $\frac{n}{d}$ of $n (> \sqrt{n})$. Clearly this is a one-to-one correspondence.

Therefore, when $\sqrt{n}$ is an integer, that is, n is a square number, $\tau(n)$ is an odd number (now $\sqrt{n} \mid n$); otherwise $\tau(n)$ is even.

If the standard factorizations of positive integers a and b are given, then it is easy to write their GCD and LCM.

(9) Let $a = p_1^{\alpha_1} \cdots p_k^{\alpha_k}$ and $b = p_1^{\beta_1} \cdots p_k^{\beta_k}$ be the standard factorizations of the positive integers a and b, where $p_1, \ldots, p_k$ are distinct prime numbers. In order to make the two factorizations contain the same prime numbers formally, we allow the exponents α_i and β_i to be 0, that is, α_i and β_i are non-negative integers ($i = 1, \ldots, k$). Then

$$(a, b) = p_1^{a_1} \cdots p_k^{a_k}, \ a_i = \min(\alpha_i, \beta_i), \ i = 1, \ldots, k;$$

$$[a, b] = p_1^{b_1} \cdots p_k^{b_k}, \ b_i = \max(\alpha_i, \beta_i), \ i = 1, \ldots, k.$$

The proof is similar to that of (6), and the details are left to the readers.

Remark 2. For more than two integers, the GCD and LCM have similar results. The significance of these formulas is mainly theoretical, which makes it more convenient to prove some properties of the GCD and LCM (refer to Example 3 below). It is not a simple tool for a practical calculation, because it is usually very difficult to find the standard factorization of a positive integer.

Distinguishing Method of Prime Numbers and the Standard Factorization of $n!$

Although prime numbers have very good properties, their distribution in natural numbers is very irregular (see Exercise 6 in this chapter). Given a positive integer n, it is very difficult to judge whether n is prime or not. The following (10) is an old distinguishing method, which is more practical when n is not very large.

(10) If $n > 1$ and n is not divisible by prime numbers that are not greater than $\sqrt{n}$, then n is a prime number.

Therefore, to apply this method, it is necessary to know all prime numbers not greater than $\sqrt{n}$.

The proof is very simple: If n is not a prime number, then $n = ab$, where a and b are both greater than 1. We may assume that $a \geq b$. Then $b \leq \sqrt{n}$. Because $b > 1$, the number b has a prime factor p, and since $b \mid n$, so $p \mid n$. But $p \leq b \leq \sqrt{n}$, a contradiction!

It can be imagined that it is more difficult to determine the standard factorization of a positive integer effectively. But interestingly, for any positive integer n, we can find the standard factorization of $n!$.

(11) For any positive integer m and prime number p, the notation $p^\alpha \| m$ represents $p^\alpha \mid m$ but $p^{\alpha+1} \nmid m$, that is, p^α is the power of p appearing in the standard factorization of m.

For any $n > 1$ and prime p, let $p^{\alpha_p} \| n!$. Then

$$\alpha_p = \sum_{l=1}^{\infty} \left[\frac{n}{p^l} \right] \left(= \left[\frac{n}{p} \right] + \left[\frac{n}{p^2} \right] + \left[\frac{n}{p^3} \right] + \cdots \right). \tag{7}$$

(Since $\left[\frac{n}{p^l} \right] = 0$, when $p^l > n$, there is only a finite number of non-zero terms in the sum of (7).)

To prove it, we construct a numerical table as shown in Table 7.1:

Table 7.1

	1	2	k	$\cdots$	n
p					
p^2					
$\vdots$					
p^l					

If $p^l \mid k$, fill in 1 in the cross grid of row l and column k in Table 7.1; otherwise fill in 0. Therefore, the sum of the numbers in the kth column is the power of p contained in the standard factorization of k, and the sum of the numbers in all n columns in the table is α_p.

On the other hand, for $l \geq 1$, the number of 1 in row l is equal to the number of terms that are divisible by p^l in $1, 2, \ldots, n$, i.e., $\left[\frac{n}{p^l}\right]$ (Exercise 1 in Chapter 6). Since the sum of the column-sums is equal to the sum of the row-sums, the equality (7) is deduced.

Remark 3. Formula (7) has a simple relationship with the base-p representation of n, which is more applicable to some problems:

Let the p-ary representation of n be (refer to Exercise 14 in Chapter 6)

$$n = (a_k \cdots a_0)_p = a_0 + a_1 p + \cdots + a_k p^k,$$

where $0 \leq a_i < p$ for $i = 1, \ldots, k$ with $a_k \neq 0$. Let $S_p(n) = a_0 + \cdots + a_k$. Then

$$\alpha_p = \frac{n - S_p(n)}{p - 1}.$$

To prove the above, notice that $0 \leq a_i \leq p - 1$. Then for $l = 1, 2, \ldots, k$,

$$0 \leq a_{l-1} p^{l-1} + \cdots + a_1 p + a_0 < p^l.$$

Therefore $\left[\frac{n}{p^l}\right] = a_k p^{k-l} + \cdots + a_{l+1} p + a_l$. When $l > k$, we have $\left[\frac{n}{p^l}\right] = 0$, thus from (7), we get that

$$\alpha_p = (a_k p^{k-1} + \cdots + a_2 p + a_1)$$
$$+ (a_k p^{k-2} + \cdots + a_2)$$

$$\cdots$$

$$+ (a_k p + a_{k-1})$$

$$+ a_k$$

$$= a_k(1 + p + \cdots + p^{k-1}) + \cdots + a_2(1 + p) + a_1$$

$$= \sum_{l=1}^{k} a_l \frac{p^l - 1}{p - 1} = \frac{1}{p - 1} \sum_{l=0}^{k} (a_l p^l - a_l)$$

$$= \frac{1}{p - 1} (n - S_p(n)).$$

This is the desired result.

Illustrative Examples

Example 1.

(i) Let m be a positive integer and $2^m + 1$ be prime. Prove: m is a power of 2.

(ii) Let a and m be integers greater than 1 such that $a^m - 1$ is prime. Prove: $a = 2$ and m is prime.

Proof. To prove (i), we just need to show that if m is not a power of 2, then $2^m + 1$ is not prime.

A basic technique of proving that a number greater than 1 is not prime is to point out a proper factor of it.

Since m is not a power of 2, it has odd factors greater than 1. Suppose $m = kr$ with $k \geq 1$, $2 \nmid r$ and $r > 1$. It can be seen from formula (6) in Chapter 6 that $2^m + 1$ has a proper factor $2^k + 1$, so $2^m + 1$ is not prime.

The idea of proving (ii) is similar to (i). If $a > 2$, then $a^m - 1$ has a proper factor $a - 1$; if m is not prime, let $m = rs$ (r and s are greater than 1). Then $2^m - 1$ has a proper factor $2^r - 1$. $\square$

Remark 4. A number of the form $F_n = 2^{2^n} + 1$ is called a **Fermat number**. Example 4 (iii) in Chapter 6 points out an interesting property of Fermat numbers: They are pairwise coprime. (From this, we can immediately deduce that there is an infinite number of prime numbers.)

The prime numbers in Fermat numbers are called Fermat prime numbers. For example, $F_0 = 3$, $F_1 = 5$, $F_2 = 17$, $F_3 = 257$, and $F_4 = 65537$ are all prime numbers. But $F_5 = 641 \times 6700417$ is not prime. At present, it is not known whether there is an infinite number of Fermat prime numbers, and whether there is an infinite number of composite numbers in F_n. (Of course, at least one of these two statements is true.)

A number of the form $M_p = 2^p - 1$ is called a **Mersenne number**, where p is a prime number. For example, M_2, M_5, and M_7 are all prime numbers, but $M_{11} = 23 \times 89$ is not prime. Whether there is an infinite number of prime numbers and whether there is an infinite number of composite numbers using Mersenne numbers are unsolved problems.

Fermat numbers and Mersenne numbers are very important in theory and applications. In mathematics competitions, there are often problems involving these two kinds of numbers.

Example 2. Let a and b be coprime integers. Prove that $(a^2 + b^2, ab) = 1$.

Proof. Let $d = (a^2 + b^2, ab)$. If $d > 1$, then d has a prime factor p (see Remark 1).

From $d \mid ab$, we obtain $p \mid ab$. But p is prime, so at least one of a and b is divisible by p. Let's suppose $p \mid a$. Combining it with $p \mid (a^2 + b^2)$, we deduce $p \mid b^2$, so $p \mid b$ (using the sharp property of prime numbers again). Therefore, p is a common divisor of a and b, which contradicts $(a, b) = 1$.

This problem can also be proved by the following method (without the help of prime numbers): Notice that $a^2 + b^2$ being coprime with ab is equivalent to $a^2 + b^2$ being coprime with both a and b (see (12) in Chapter 6). Since $(a, b) = 1$, clearly $(a^2 + b^2, a) = (b^2, a) = 1$. (See Exercise 10 (i) in Chapter 6.) Similarly, $(a^2 + b^2, b) = 1$, which proves the conclusion. □

Example 3. Let a, b, and c be positive integers. Prove:
$$\frac{[a, b, c]^2}{[a, b][b, c][c, a]} = \frac{(a, b, c)^2}{(a, b)(b, c)(c, a)}.$$

Proof. Suppose the standard factorizations of a, b, and c are
$$a = p_1^{\alpha_1} \cdots p_k^{\alpha_k}, \quad b = p_1^{\beta_1} \cdots p_k^{\beta_k}, \quad c = p_1^{\gamma_1} \cdots p_k^{\gamma_k},$$
where α_i, β_i, and γ_i are all non-negative integers. From the formula in (9), the problem is equivalent to proving that for non-negative integers α, β, and γ, there is
$$2\max\{\alpha, \beta, \gamma\} - \max\{\alpha, \beta\} - \max\{\beta, \gamma\} - \max\{\gamma, \alpha\}$$
$$= 2\min\{\alpha, \beta, \gamma\} - \min\{\alpha, \beta\} - \min\{\beta, \gamma\} - \min\{\gamma, \alpha\}. \qquad (8)$$
It is not difficult to directly check the formula (8) (it may be assumed that if $\alpha \le \beta \le \gamma$, then the left and right sides are both equal to $-\beta$), but a more general proof is to use the inclusion-exclusion principle:

Let $N = \max\{\alpha, \beta, \gamma\}$. If $N = 0$, then there is nothing to prove. We may suppose $N \geq 1$. We introduce three properties into the set $S = \{1, 2, \ldots, N\}$: $\mathbf{P}_\alpha$ is "not greater than α," $\mathbf{P}_\beta$ is "not greater than β," and $\mathbf{P}_\gamma$ is "not greater than γ." Then there is 0 element in S that does not have any properties of $\mathbf{P}_\alpha$, $\mathbf{P}_\beta$, and $\mathbf{P}_\gamma$; there are α elements with property $\mathbf{P}_\alpha$; there are $\min\{\alpha, \beta\}$ elements with properties $\mathbf{P}_\alpha$ and $\mathbf{P}_\beta$, etc. From Theorem 1 in Chapter 4,

$$\max\{\alpha, \beta, \gamma\} - \alpha - \beta - \gamma + \min\{\alpha, \beta\} + \min\{\beta, \gamma\}$$
$$+ \min\{\gamma, \alpha\} - \min\{\alpha, \beta, \gamma = 0\}. \tag{9}$$

We also have

$$\left. \begin{aligned} \max\{\alpha, \beta\} + \min\{\alpha, \beta\} &= \alpha + \beta, \\ \max\{\beta, \gamma\} + \min\{\beta, \gamma\} &= \beta + \gamma, \\ \max\{\gamma, \alpha\} + \min\{\gamma, \alpha\} &= \gamma + \alpha. \end{aligned} \right\} \tag{10}$$

From (9) and (10), we obtain (8). □

Example 4. Let n, a, and b be integers such that $n > 0$ and $a \neq b$. Prove that if $n \mid (a^n - b^n)$, then $n \left| \frac{a^n - b^n}{a - b} \right.$.

Proof. Let p be prime and $p^\alpha \parallel n$. We only need to prove $p^\alpha \left| \frac{a^n - b^n}{a - b} \right.$.

Let $t = a - b$. If $p \nmid t$, then since p is prime, $(p^\alpha, t) = 1$, so from

$$p^\alpha \mid a^n - b^n = t \cdot \frac{a^n - b^n}{t},$$

we can deduce that $p^\alpha \left| \frac{a^n - b^n}{a - b} \right.$.

If $p \mid t$, then by the binomial theorem,

$$\frac{a^n - b^n}{t} = \frac{(b + t)^n - b^n}{t} = \sum_{i=1}^{n} \binom{n}{i} b^{n-i} t^{i-1}. \tag{11}$$

Suppose $p^\beta \parallel i$. Then clearly $\beta \leq i - 1$. Therefore, the power of p contained in $\binom{n}{i} t^{i-1} = \frac{n}{i} \binom{n-1}{i-1} t^{i-1}$ is at least α, that is, each term on the right side of (11) is divisible by p^α, so $p^\alpha \left| \frac{a^n - b^n}{a - b} \right.$. (Refer to Remark 2 in Chapter 6.) □

Remark 5. In order to prove $n \mid m$, usually we make the standard factorization of n, i.e., $n = p_1^{\alpha_1} \cdots p_k^{\alpha_k}$, and then the problem is decomposed into proving $p_i^{\alpha_i} \mid m (i = 1, \ldots, k)$ (refer to Remark 8 in Chapter 6). The advantage of doing so is that it allows us to apply the sharp property of prime numbers. Please check the role of prime numbers in the above argument.

Example 5. Let m and n be non-negative integers. Prove: $\frac{(2m)!(2n)!}{m!n!(m+n)!}$ is an integer.

Proof. We just need to prove that for each prime p, the power of p in the denominator does not exceed its power in the numerator (see Remark 5). From the formula in (11), this is equivalent to proving that

$$\sum_{l=1}^{\infty}\left(\left[\frac{2m}{p^l}\right]+\left[\frac{2n}{p^l}\right]\right) \geq \sum_{l=1}^{\infty}\left(\left[\frac{m}{p^l}\right]+\left[\frac{n}{p^l}\right]+\left[\frac{m+n}{p^l}\right]\right). \tag{12}$$

We hope that for $l = 1, 2, \ldots,$

$$\left[\frac{2m}{p^l}\right]+\left[\frac{2n}{p^l}\right] \geq \left[\frac{m}{p^l}\right]+\left[\frac{n}{p^l}\right]+\left[\frac{m+n}{p^l}\right],$$

thus, it proves (12). (Refer to Remark 2 in Chapter 6.)

In fact, the following stronger inequality is valid: For any real numbers x and y,

$$[2x] + [2y] \geq [x] + [y] + [x+y]. \tag{13}$$

To prove (13), first, $[k + \alpha] = k + [\alpha]$ for any integer k and real number α. Clearly, if we subtract an integer from x (or y), then the inequality (13) will not change. Therefore, for $0 \leq x < 1$ and $0 \leq y < 1$, it only needs to prove (13). Hence, the problem is reduced to proving that

$$[2x] + [2y] \geq [x+y].$$

Note that now $[x + y] = 0$ or 1. If $[x + y] = 0$, then the above inequality is obviously true. If $[x + y] = 1$, then $x + y \geq 1$, so one of x and y is $\geq \frac{1}{2}$. Suppose $x \geq \frac{1}{2}$. Then $[2x] + [2y] \geq [2x] \geq 1$. Therefore, the result is correct.

The following method can also be used for this problem. Let

$$f(m, n) = \frac{(2m)!(2n)!}{m!n!(m + n)!}.$$

Prove by induction on m: For any integer $n \geq 0$, the number $f(m, n)$ is always an integer.

When $m = 0$, obviously $f(m, n) = \binom{2n}{n}$ is an integer. It's not hard to verify that there exists the recurrence formula

$$f(m + 1,\, n) = 4f(m, n) - f(m, n + 1).$$

Therefore, from the inductive hypothesis that $f(m, n)$ is an integer for all n, we derive that $f(m+1, n)$ is an integer for all n. (Refer to Remark 9 in Chapter 3.) □

Example 6. Let m and n be positive integer. Prove: $n!(m!)^n \mid (mn)!$.

Proof. Using the first solution of the above example, we prove that for any prime number p,

$$\sum_{l=1}^{\infty} \left[\frac{mn}{p^l} \right] \geq \sum_{l=1}^{\infty} \left(\left[\frac{n}{p^l} \right] + n \left[\frac{m}{p^l} \right] \right). \tag{14}$$

However, unlike the case of Example 5, the stronger inequality

$$\left[\frac{mn}{p^l} \right] \geq \left[\frac{n}{p^l} \right] + n \left[\frac{m}{p^l} \right]. \tag{15}$$

isn't valid generally (the readers can give a counterexample). To prove (14), we distinguish two situations.

(i) If $p \nmid m$, then by the division with remainder, $m = p^l q + r$ with $1 \leq r < p^l$, where $q = \left[\frac{m}{p^l} \right]$ (see Remark 5 in Chapter 6). Therefore, $m \geq \left[\frac{m}{p^l} \right] \cdot p^l + 1$, and multiplying it by $\frac{n}{p^l}$ to both sides of it, we get

$$\left[\frac{mn}{p^l} \right] \geq \left[n \left[\frac{m}{p^l} \right] + \frac{n}{p^l} \right] = n \left[\frac{m}{p^l} \right] + \left[\frac{n}{p^l} \right],$$

that is, (15) is established and thus (14) is proved.

(ii) If $p \mid m$, let $m = p^\alpha m'$ and $p \nmid m'$. Clearly (14) becomes

$$\sum_{l=1}^{\infty} \left[\frac{m'n}{p^l} \right] \geq \sum_{l=1}^{\infty} \left(\left[\frac{n}{p^l} \right] + n \left[\frac{m'}{p^l} \right] \right),$$

which reduces to case (i), so (14) is proved.

There is also a combinatorial proof of this problem: There are $(mn)!$ methods to make a full permutation of mn different elements. Consider the following three steps. Step 1: We divide mn elements into n groups with m elements in each group, and suppose there are k methods (of course, k is an integer). Step 2: We treat a group as an element and make a permutation of these groups. Then there are $n!$ arrangements. Step 3: The elements in each group have $m!$ arrangements. Therfore there are $k \cdot n! \cdot (m!)^n$ arrangements in total. Hence $k \cdot n! \cdot (m!)^n = (mn)!$, which obviously leads to the conclusion.

A more direct argument of Example 6 is based on recursion. We induct on n. When $n = 1$, the conclusion is obviously true. Assume that the conclusion is true for n, that is, $\frac{(mn)!}{n!(m!)^n}$ is an integer for any positive integer m. We have

$$\frac{(m(n+1))!}{(n+1)!(m!)^{n+1}} \div \frac{(mn)!}{n!(m!)^n}$$

$$= \frac{(m(n+1))!}{(n+1)!(m!)^{n+1}} \times \frac{n!(m!)^n}{(mn)!}$$

$$= \frac{(mn+m)(mn+m-1)\cdots(mn+1)}{(n+1)\cdot m!}$$

$$= \frac{(mn+m-1)\cdots(mn+1)}{(m-1)!}$$

$$= \binom{mn+m-1}{m-1}, \text{ an integer.}$$

Combining the above with the inductive hypothesis, we deduce that $\frac{(m(n+1))!}{(n+1)!(m!)^{n+1}}$ is an integer for all positive integers m. $\qquad\square$

Example 7. (i) Let p be a prime number. Prove that $p \,\big|\, \binom{p}{k}$ for $k = 1, 2, \ldots, p-1$.

(ii) Let n be a composite number. Prove: $\binom{n}{1}, \binom{n}{2}, \ldots, \binom{n}{n-1}$ cannot be all divisible by n.

Proof. (i) Because $k\binom{p}{k} = p\binom{p-1}{k-1}$ (the formula (11) in Chapter 2), $p \,\big|\, k\binom{p}{k}$. But $1 \leq k \leq p-1$, so $p \nmid k$, and p is prime, thus $p \,\big|\, \binom{p}{k}$.

(ii) Let p be a prime factor of the composite number n, and let

$$p^k \leq n < p^{k+1} \ (k \text{ is a positive integer}). \tag{16}$$

If $n \neq p^k$, then $p^k \leq n-1$. We now prove $p \nmid \binom{n}{p^k}$, and then combining it with $p \,|\, n$, we know that $n \nmid \binom{n}{p^k}$.

Because $\binom{n}{p^k} = \frac{n!}{p^k!(n-p^k)!}$, it can be seen from the result in Remark 3 that the power of p in $\binom{n}{p^k}$ is

$$\frac{S_p(p^k) + S_p(n-p^k) - S_p(n)}{p-1}. \tag{17}$$

From (16), the p-ary representation of n has $k+1$ digits (that is, it's in the form of $a_k p^k + \cdots + a_1 p + a_0$), so the numerator of (17) is

$$1 + ((a_k - 1) + a_{k-1} + \cdots + a_1 + a_0) - (a_k + a_{k-1} + \cdots + a_1 + a_0) = 0,$$

thus $p \nmid \binom{n}{p^k}$.

Now let $n = p^k$. Notice that the power of p in $\binom{p^k}{p^{k-1}}$ is $\frac{S_p(p^{k-1}) + S_p(p^k - p^{k-1}) - S_p(p^k)}{p-1} = \frac{1 + (p-1) - 1}{p-1} = 1$.

Since n is a composite number, $k \geq 2$, thus, $p^k \nmid \binom{p^k}{p^{k-1}}$, i.e., $n \nmid \binom{p^k}{p^{k-1}}$. (Notice that $p^{k-1} < n - 1$.) □

Question (ii) of this example can also be solved by formula (7), but it is more convenient to use the result in Remark 3.

Our purpose is to select a proper $t \leq n - 1$ for a prime factor p of n, so that the power of p in n is greater than that in $\binom{n}{t}$, leading to $n \nmid \binom{n}{t}$.

The inequality (16) gives some information about the base-p representations of n, so we choose t, which makes the base-p representations of t and $n - t$ have applicable information. In this way, we can determine $S_p(t) + S_p(n - t) - S_p(n)$, so as to find the power of p in $\binom{n}{t}$.

Exercises

Group A

1. Prove that a positive integer n can be uniquely expressed in the form $n = q^2 r$, where q and r are both positive integers and r has no square factor (that is, it's not divisible by a square number greater than 1).

2. Let $\tau(n)$ be the number of positive divisors of n. Prove:
   (i) $\tau(n) < 2\sqrt{n}$;
   (ii) the product of all positive divisors of n is $n^{\frac{\tau(n)}{2}}$.

3. Let $M_p = 2^p - 1$ and p be a prime number. Prove: If $p \neq q$, then $(M_p, M_q) = 1$.

4. Let n be a positive integer. Prove: $2^{2^n} + 2^{2^{n-1}} + 1$ has at least n distinct prime factors.

5. Prove that there is an infinite number of prime numbers in the form of $4k - 1$ and an infinite number of prime numbers in the form of $6k - 1$ (k is a positive integer).

6. Let n be a given positive integer. Prove that there exist n consecutive positive integers, each of which is not a prime number.

7. Prove that all integers greater than 11 can be expressed as the sums of two composite numbers.

Group B

8. If a positive integer n satisfies $\sigma(n) = 2n$, then n is called a perfect number. Prove: An even number n is a perfect number if and only if $n = 2^{k-1}(2^k - 1)$ and $2^k - 1$ is a prime number.
9. Let $n > 1$. Prove: $1 + \frac{1}{2} + \cdots + \frac{1}{n}$ is not an integer.
10. Let $n > 1$. Prove: $1 + \frac{1}{3} + \cdots + \frac{1}{2n-1}$ is not an integer.
11. Let $k > 1$ be an integer. Prove that there exist prime numbers p and q such that $p \mid q^k + 1$ and $q \mid p^k + 1$.
12. Let $a_1, a_2, \ldots, a_n$ be positive integers. Prove:

$$[a_1, \ldots, a_n] \geq \frac{a_1 \cdots a_n}{\prod_{1 \leq i < j \leq n} (a_i, a_j)}.$$

13. Prove: If positive integers m and n satisfy

$$(m, n) + [m, n] = m + n,$$

then one of m and n is a multiple of the other.
14. Let positive integers a, b, x, and y satisfy $(a^2 + b^2) \mid (ax + by)$. Prove: $x^2 + y^2$ and $a^2 + b^2$ are not coprime.

Chapter 8

Congruence (1)

Congruence is an important concept in number theory, which is widely used. This chapter briefly introduces congruence, including the basic properties of congruence, the complete residue system and reduced residue system, Euler's theorem, and other famous theorems in number theory.

Congruence and Congruence Class

Let m be a positive integer, and a and b be integers. If $m \mid (a - b)$, then we say that a and b are **congruent modulo m**, written as
$$a \equiv b \pmod{m}.$$
If $m \nmid (a - b)$, then we say that a and b are **not congruent modulo m**, written as
$$a \not\equiv b \pmod{m}.$$
It is obvious that a and b are congruent modulo m if and only if a and b have the same remainder on division by m.

For a fixed modulo m, the congruence has many properties similar to ordinary equations.

(1) (Reflexivity) $a \equiv a \pmod{m}$.
(2) (Symmetry) If $a \equiv b \pmod{m}$, then $b \equiv a \pmod{m}$.
(3) (Transitivity) If $a \equiv b \pmod{m}$ and $b \equiv c \pmod{m}$, then $a \equiv c \pmod{m}$.
(4) If $a \equiv b \pmod{m}$ and $c \equiv d \pmod{m}$, then $a \pm c \equiv b \pm d \pmod{m}$.
(5) If $a \equiv b \pmod{m}$ and $c \equiv d \pmod{m}$, then $ac \equiv bd \pmod{m}$.
(6) If $a \equiv b \pmod{m}$ and k and c are integers with $k > 0$, then $a^k c \equiv b^k c \pmod{m}$.

These properties are not difficult to prove by definition. For example, (5) can be derived from $ac - bd = a(c - d) + d(a - b)$ and the known conditions. It's not hard to see that (4) or (5) can be used repeatedly to establish the formulae of addition, subtraction, and multiplication for more than two congruences (with the same modulo).

Notice that the cancellation law of congruence does not hold, that is,

$$ac \equiv bc(\mod m)$$

may not imply $a \equiv b(\mod m)$ (the readers can give a counterexample). However, we have the following results:

(7) If $ac \equiv bc(\mod m)$, then $a \equiv b\left(\mod \frac{m}{(c,m)}\right)$. In particular, when $(c, m) = 1$, there is $a \equiv b(\mod m)$, that is, when c and m are coprime, c can be canceled from both sides of the congruence without changing the modulo.

It's not hard to prove (7). Because $m \mid c(a - b)$, which is equivalent to

$$\frac{m}{(c,m)} \left| \frac{c}{(c,m)}(a - b),\right.$$

according to the properties (11) and (13) in Chapter 6, it follows that $\frac{m}{(c,m)} \left| (a - b).\right.$

Now we mention several simple but useful properties involving congruence:

(8) If $a \equiv b(\mod m)$ and $d \mid m$, then $a \equiv b(\mod d)$.

(9) If $a \equiv b(\mod m)$ with $d \neq 0$, then $da \equiv db(\mod dm)$.

(10) If $a \equiv b(\mod m_i)(i = 1, \ldots, k)$, then $a \equiv b(\mod[m_1, \ldots, m_k])$.

In fact, (8) and (9) are obvious, while (10) is an equivalent expression of property (14) in Chapter 6 (refer to Remark 9 in Chapter 6).

The above properties (1), (2), and (3) show that congruence is an equivalent relation in the set of integers, so all integers can be classified by modulo m. Specifically, if a and b are congruent modulo m, then a and b belong to the same class; otherwise, they do not belong to the same class. Every such class is called a **congruence class modulo m**.

By the division with remainder, any integer must be congruent modulo m with a number in $0, 1, \ldots, m - 1$, and the m numbers $0, 1, \ldots, m - 1$ are not congruent modulo m with each other, so modulo m has m distinct congruence classes, that is

$$M_i = \{n \mid n \in \mathbb{Z}, \, n \equiv i(\mathrm{mod}\, m)\}, \, i = 0, 1, \ldots, m-1.$$

For example, there are two congruence classes modulo 2, and one is the odd class and the other is even. The numbers in the two classes have the forms $2k + 1$ and $2k$, respectively (k is an integer).

Take a representative in each of the m residue classes. Then such m numbers are called a **complete residue system** modulo m, or a **complete system** for short. In other words, m numbers $c_1, \ldots, c_m$ are called a complete system modulo m if they are not congruent to each other modulo m.

For example, $0, 1, \ldots, m-1$ form a complete system modulo m, which is called the **minimum nonnegative complete system** (modulo m).

If i and m are coprime, then we know clearly that all numbers in the congruence class M_i are all coprime with m (Chapter 6, property (6)). Such a congruence class is called a **reduced congruence class** modulo m. We use $\varphi(m)$ to denote the number of reduced congruence classes modulo m. This is called the **Euler function**, which is an important function in number theory. Obviously, $\varphi(1) = 1$, while for $m > 1$, $\varphi(m)$ is the number of the integers in $1, 2, \ldots, m-1$ that are coprime with m. For example, if p is a prime, then $\varphi(p) = p - 1$.

Take a number as a representative in each of the $\varphi(m)$ reduced congruence classes, such $\varphi(m)$ numbers are called a **reduced residue system** modulo m, or a **reduced system** for short. That is, $\varphi(m)$ numbers $r_1, \ldots, r_{\varphi(m)}$ are called a reduced system modulo m if they are all coprime with m and are not congruent to each other modulo m.

The $\varphi(m)$ positive integers which are not greater than m and coprime with m are called the **least positive reduced system** modulo m.

Remark 1. According to the definition, for any two complete (reduced) systems modulo m, we may adjust the order of the numbers in one of them appropriately, so that the numbers in the two systems are congruent modulo m correspondingly. Therefore, from the properties (4) and (5), the sum, product, sum of squares, and other quantities of the numbers in the two complete (reduced) systems must be congruent modulo m.

It is a quite basic technique in number theory to generate a congruence relation (i.e., the exact division relation) by two complete (reduced) systems. Many problems in this chapter and this book do so.

The following results, which generate another complete (reduced) system from one complete (reduced) system, are very useful.

(11) Suppose $(a, m) = 1$ and b is an arbitrary integer.

    (i) If $c_1, \ldots, c_m$ form a complete system modulo m, then $ac_1 + b, \ldots, ac_m + b$ also form a complete system modulo m.

    (ii) If $r_1, \ldots, r_{\varphi(m)}$ form a reduced system modulo m, then $ar_1, \ldots, ar_{\varphi(m)}$ also form a reduced system modulo m.

In fact, the m numbers $ac_1 + b, \ldots, ac_m + b$ are not congruent to each other modulo m. If $ac_i + b \equiv ac_j + b \pmod{m} \, (i \neq j)$, then $ac_i \equiv ac_j \pmod{m}$. Since $(a, m) = 1$, we have $c_i \equiv c_j \pmod{m}$ from (7), which is impossible.

To prove (ii), notice that the $\varphi(m)$ numbers $ar_1, \ldots, ar_{\varphi(m)}$ are all coprime with m (property (12) in Chapter 6). From (i) we deduce that they are not congruent to each other modulo m, which proves (ii). It is not difficult to derive the result by (11)(i).

(12) Suppose $(a, m) = 1$ and b is an arbitrary integer. Then there exists an integer x such that $ax \equiv b \pmod{m}$; all such x form a congruence class modulo m.

In fact, take any complete system $c_1, \ldots, c_m$ modulo m. Then from 11(i), $ac_1 - b, \ldots, ac_m - b$ form a complete system modulo m, so there exists i such that $ac_i - b \equiv 0 \pmod{m}$, that is, $ac_i \equiv b \pmod{m}$. If two numbers x and x' both satisfy $ax \equiv b \pmod{m}$ and $ax' \equiv b \pmod{m}$, then $ax \equiv ax' \pmod{m}$. Because $(a, m) = 1$, we have $x \equiv x' \pmod{m}$, that is, x and x' belong to the same congruence class modulo m.

Remark 2. If $(a, m) = 1$, then from (12), there exists x such that

$$ax \equiv 1 \pmod{m}. \tag{1}$$

The congruence (1) is similar to the equation $ax = 1$ in complex numbers. Therefore, we call x that satisfies (1) the **inverse of a modulo m**, written as a^{-1} or $\frac{1}{a} \pmod{m}$, and they form a congruence class modulo m (that is, they are uniquely determined in the sense of modulo m). In particular, there is one a^{-1} satisfying $1 \leq a^{-1} < m$. (Notice that if (1) has a solution, then a must be coprime with m, so the notation a^{-1} is only meaningful when $(a, m) = 1$.)

Remark 3. The inverse modulo m has the following simple properties that are very convenient for applications.

Suppose a and b are both coprime with m.

(i) If $a \equiv b(\bmod m)$. Then $\frac{1}{a} \equiv \frac{1}{b}(\bmod m)$, and vice versa;

(ii) $\frac{1}{ab} \equiv \frac{1}{a} \cdot \frac{1}{b}(\bmod m)$;

(iii) $\frac{1}{a} + \frac{1}{b} \equiv \frac{a+b}{ab}(\bmod m)$.

All the proofs are simple. It is not difficult to derive (i) and (ii) from the definition and (7). Also from (ii) and the definition, we see that

$$\frac{a+b}{ab} = \frac{1}{a} \cdot a \cdot \frac{1}{b} + \frac{1}{a} \cdot b \cdot \frac{1}{b} \equiv \frac{1}{b} + \frac{1}{a}(\bmod m),$$

which implies (iii).

Note that from (i) we can deduce: If $a_1, \ldots, a_{\varphi(m)}$ form a reduced system modulo m, then $\frac{1}{a_1}, \ldots, \frac{1}{a_{\varphi(m)}}$ also form a reduced system modulo m.

Some Famous Theorems in Number Theory

In this section, we will prove Euler's theorem, Fermat's little theorem, and Wilson's theorem.

(13) (**Euler's theorem**) If $(a, m) = 1$, then $a^{\varphi(m)} \equiv 1(\bmod m)$.

(14) (**Fermat's little theorem**) If p is a prime number and $p \nmid a$, then $a^{p-1} \equiv 1(\bmod p)$.

(15) (**Wilson's theorem**) If p is a prime number, then $(p - 1)! \equiv -1(\bmod p)$.

To prove (13), let's take any reduced system $r_1, \ldots, r_{\varphi(m)}$ modulo m. Since $(a, m) = 1$, we see that $ar_1, \ldots, ar_{\varphi(m)}$ also form a reduced system modulo m (see (11) (ii)). Therefore (refer to Remark 1)

$$r_1 \cdots r_{\varphi(m)} \equiv (ar_1) \cdots (ar_{\varphi(m)}) \equiv a^{\varphi(m)} r_1 \cdots r_{\varphi(m)}(\bmod m).$$

Because $r_1, \ldots, r_{\varphi(m)}$ are coprime with m, there is $(r_1 \cdots r_{\varphi(m)}, m) = 1$. We may remove $r_1 \cdots r_{\varphi(m)}$ on both sides of the above formula (see (7)) and get the result.

(14) is a corollary of (13): Take $m = p$, and note that $\varphi(p) = p - 1$. Proof of (15): When $p = 2$, the conclusion is obvious. If $p \geq 3$, then from Remark 2, for each a with $1 \leq a \leq p - 1$, there exists a unique a^{-1} with $1 \leq a^{-1} \leq p - 1$ such that $aa^{-1} \equiv 1(\bmod p)$. And $a \equiv a^{-1}(\bmod p)$ is

equivalent to $a^2 \equiv 1 \pmod{p}$, that is, $a \equiv a^{-1} \equiv \pm 1 \pmod{p}$. Hence, the $p-3$ numbers $2, 3, \ldots, p-2$ can be matched into $\frac{p-3}{2}$ pairs, and each pair $\{a, a^{-1}\}$ satisfies $aa^{-1} \equiv 1 \pmod{p}$. Therefore

$$(p-1)! \equiv 1 \cdot 2 \cdot \cdots \cdot (p-2)(p-1)$$

$$\equiv 1 \cdot (p-1) \equiv -1 \pmod{p}.$$

Remark 4. Fermat's little theorem can be expressed as (sometimes more convenient): If p is prime, then $a^p \equiv a \pmod{p}$ for any integer a. (This is obviously true when $p \,|\, a$; it is equivalent to (14) when $p \nmid a$.)

Remark 5. In (15), $p > 1$ is a prime number, which is also a necessary condition for $(p-1)! \equiv -1 \pmod{p}$. In fact, if p has a prime factor $q \le p-1$, then $q \,|\, (p-1)!$, so it is easy to know $q \,|\, 1$, which is impossible.

Although the congruence $(p-1)! \equiv -1 \pmod{p}$ is a characterization of $p > 1$ being a prime number, it is not a practical method to distinguish prime numbers because $(p-1)!$ is too large to calculate (refer to (10) in Chapter 7).

Basic Properties of Euler's Function

This section introduces some basic results of Euler's function.

(16) If $m > 2$, then $\varphi(m)$ is an even number.

(17) Let $m = p_1^{\alpha_1} \cdots p_k^{\alpha_k}$ be the standard factorization of m. Then

$$\varphi(m) = m\left(1 - \frac{1}{p_1}\right) \cdots \left(1 - \frac{1}{p_k}\right). \tag{2}$$

Therefore, if $(m, n) = 1$, then $\varphi(mn) = \varphi(m)\varphi(n)$.

(18) $\sum_{d \,|\, m} \varphi(d) = m$, where the summation $\sum_{d \,|\, m}$ represents that d takes over all the positive divisors of m. (e.g., $\sum_{d \,|\, 6} \varphi(d) = \varphi(1) + \varphi(2) + \varphi(3) + \varphi(6)$.)

To prove (16), take $a = -1$ in Euler's theorem (13), and we obtain $(-1)^{\varphi(m)} \equiv 1 \pmod{m}$. Because $m > 2$, it must be that $(-1)^{\varphi(m)} = 1$, namely, $\varphi(m)$ is even.

Another proof: For any positive integer i that does not exceed m and is coprime with m, clearly $m - i$ and m are coprime, and $1 \le m - i < m$. If there exists i such that $m - i = i$, then $m = 2i$. Because $m > 2$, we have $i > 1$, so $(m, i) = i > 1$, which is contrary to $(i, m) = 1$. Therefore, all positive integers that do not exceed m and are coprime with m can be paired by sum m, which means that $\varphi(m)$ is even.

Proof of (17): We use the inclusion-exclusion principle. Let $S = \{1, \ldots, m\}$ and $S_i = \{x \in S | \ p_i \text{ exactly divides } x\}$ $(i = 1, \ldots, k)$. Then $\varphi(m) = |\overline{S}_1 \cap \cdots \cap \overline{S}_k|$. Also we can find out that $|S| = m$, $|S_i| = \frac{m}{p_i}$, $|S_i \cap S_j| = \frac{m}{p_i p_j} (1 \leq i < j \leq k)$, and etc. Hence from Theorem 1 in Chapter 4,

$$\varphi(m) = m - \sum_{i=1}^{k} \frac{m}{p_i} + \sum_{1 \leq i < j \leq k} \frac{m}{p_i p_j} - \cdots + (-1)^k \frac{m}{p_1 \cdots p_k}$$

$$= m \left(1 - \frac{1}{p_1}\right) \cdots \left(1 - \frac{1}{p_k}\right).$$

In addition, if $(m, n) = 1$, then m and n do not have common prime factors. From formula (2), we derive that $\varphi(mn) = \varphi(m)\varphi(n)$.

Formula (2) can also be demonstrated by the following method: For positive integers n and k, define

$$f_k(n) = \begin{cases} 1, & \text{if } n \text{ is divisible by } k, \\ 0, & \text{if } n \text{ isn't divisible by } k. \end{cases}$$

For prime p, because $p \nmid n$ is equivalent to $(p, n) = 1$, from the definition of Euler's function, we see that (notice that $p_1, \ldots, p_k$ are all the distinct prime factors of m):

$$\varphi(m) = \sum_{n=1}^{m} (1 - f_{p_1}(n)) \cdots (1 - f_{p_k}(n))$$

$$= m - \sum_{n=1}^{m} f_{p_1}(n) - \sum_{n=1}^{m} f_{p_2}(n) - \cdots + \sum_{n=1}^{m} f_{p_1}(n) f_{p_2}(n) + \cdots$$

$$+ (-1)^k \sum_{n=1}^{m} f_{p_1}(n) f_{p_2}(n) \cdots f_{p_k}(n). \tag{3}$$

From the definition of $f_k(n)$, the sum $\sum_{n=1}^{m} f_k(n)$ is equal to the number of integers divisible by k in $1, 2, \ldots, m$. When $(k_1, k_2) = 1$, the relation $k_1 k_2 | n$ is equivalent to $k_1 | n$ and $k_2 | n$, so $f_{k_1 k_2}(n) = f_{k_1}(n) f_{k_2}(n)$. Thus

$$\text{RHS of (3)} = m - \frac{m}{p_1} - \frac{m}{p_2} - \cdots + \frac{m}{p_1 p_2} + \cdots + (-1)^k \frac{m}{p_1 p_2 \cdots p_k}$$

$$= m \left(1 - \frac{1}{p_1}\right) \cdots \left(1 - \frac{1}{p_k}\right).$$

(The $f_k(n)$ defined above is the characteristic function of the set of integers divisible by k. For the characteristic functions of a set, see Remark 2 in Chapter 17.)

We refer to equation (2) in (17), sometimes rewritten as

$$\varphi(m) = \prod_{i=1}^{k} p_i^{\alpha_i - 1}(p_i - 1), \text{ or } \varphi(m) = \prod_{p^\alpha \| m} p^{\alpha - 1}(p - 1),$$

which is easier for practical uses.

Proof of (18): Let $S = \{1, 2, \ldots, m\}$ and $S_d = \{x \in S | (x, m) = d\}$. It is obvious that d takes over all positive divisors of m, and these S_d form a partition of S (that is, each number in S must belong to a certain S_d, and there is no common element between different S_d). Therefore, by the addition principle (Remark 1 in Chapter 1), we obtain that

$$\sum_{d \mid m} |S_d| = |S| = m. \tag{4}$$

On the other hand, for a fixed $d \mid m$, if $x \in S_d$, then $x = dx'$, and $\left(x', \frac{m}{d}\right) = 1$. Because $1 \le x' = \frac{x}{d} \le \frac{m}{d}$, the value $|S_d|$ is equal to the number of x' satisfying $\left(x', \frac{m}{d}\right) = 1$ in $1, \ldots, \frac{m}{d}$, that is, $\varphi\left(\frac{m}{d}\right)$. Hence from (4),

$$\sum_{d \mid m} \varphi\left(\frac{m}{d}\right) = m.$$

But when d runs over all positive divisors of m, the quotient $\frac{m}{d}$ also runs over all positive divisors of m. Let $\frac{m}{d} = d'$. Then the above formula becomes

$$\sum_{d' \mid m} \varphi(d') = m.$$

Illustrative Examples

Example 1. Prove: Perfect squares are congruent to 0 or 1 modulo 4.

Proof. If n is any integer, then the conclusion can be proved by classifying n modulo 4. But in fact, by classifying n modulo 2, we can determine the remainder of n^2 on division by 4: Let k be an integer. If $n = 2k$, then $n^2 = 4k^2$ is congruent to 0 modulo 4; if $n = 2k + 1$, then $n^2 = 4k(k+1) + 1$ is congruent to 1 modulo 4. □

Remark 6. The conclusion of (and similar to) this question is quite simple but useful. It gives a necessary condition for an integer to be a perfect square through the remainder: If an integer is not congruent to 0 or 1 modulo 4, then it must not be a perfect square.

If m is a certain positive integer, then there are m possible values for the remainders of any integer modulo m, but the number of remainders modulo m of the square (cube, etc.) of an integer may be greatly reduced. This fact shows a basic idea of how congruence can be used to solve many problems.

Example 2. Let a and b be integers, p be prime, and $a^p \equiv b^p \pmod{p}$. Prove: $a^p \equiv b^p \pmod{p^2}$.

Proof. From the known conditions and Fermat's little theorem, $a \equiv b \pmod{p}$ (see Remark 4), that is

$$a = b + mp \ (m \text{ is an integer}).$$

From the above formula and the binomial theorem

$$a^p = (b + mp)^p = b^p + \binom{p}{1} mp \cdot b^{p-1} + \cdots + \binom{p}{p} (mp)^p$$

$$\equiv b^p \pmod{p^2}.$$

(Note that in the above sum, except for b^p, each term is divisible by p^2.)

$\square$

Example 3. Let $(a, m) = 1$. Then there exists an integer k with $1 \leq k < m$, such that $a^k \equiv 1 \pmod{m}$.

Proof. Because $(a, m) = 1$, the m integers $a, a^2, \ldots, a^m$ are all coprime with m, so there are at most $m - 1$ possibilities for their remainders on division by m (i.e., $1, 2, \ldots, m-1$). Therefore, there must exist i and $j (1 \leq i < j \leq m)$ satisfying $a^i \equiv a^j \pmod{m}$, that is, $a^i(a^{j-i} - 1) \equiv 0 \pmod{m}$. Because $(a^i, m) = 1$, we have $m \mid (a^{j-i} - 1)$. And $1 \leq j - i < m$, so $k = j - i$ meets the requirement. $\square$

Remark 7. Euler's theorem shows that the positive integer k in Example 3 can be (quantitatively) taken as $\varphi(m)$.

Clearly we know from Example 3 that if $(a, m) = 1$, then the numbers in the sequence $\{a^n\} (n \geq 1)$ modulo m must be periodic; the minimum positive period is less than m. However, it is a very complicated problem to determine the minimum positive period, i.e., the minimum positive integer k satisfying the requirement of Example 3. We will introduce it briefly in Chapter 19.

Example 4. Let p be a prime number. Prove:

(i) $\binom{p-1}{k} \equiv (-1)^k \pmod p$ for $k = 0, 1, \ldots, p-1$.

(ii) For any positive integer k, there holds that $\binom{k}{p} \equiv \left[\frac{k}{p}\right] \pmod p$.

Proof. (i) When $k = 0$, the conclusion is obvious. For $k = 1, \ldots, p-1$, because $p \mid \binom{p}{k}$ (see Example 7 in Chapter 7), combining the recurrence relation of binomial coefficients

$$\binom{p}{k} = \binom{p-1}{k} + \binom{p-1}{k-1},$$

we see that

$$\binom{p-1}{k} + \binom{p-1}{k-1} \equiv 0 \pmod p.$$

The result can be obtained by induction on k from the above formula.

(ii) The p consecutive integers $k-1, \ldots, k-p+1$ form a complete system modulo p, in which there is exactly one number divisible by p. Suppose this number is $k - i$ for some $0 \le i \le p-1$. Then

$$\left[\frac{k}{p}\right] = \left[\frac{k-i+i}{p}\right] = \frac{k-i}{p} + \left[\frac{i}{p}\right] = \frac{k-i}{p}.$$

On the other hand, the above p numbers form a reduced system modulo p after removing $k - i$.

Thus, if we let $Q = \frac{k(k-1)\cdots(k-p+1)}{k-i}$, then

$$Q \equiv (p-1)! \pmod p. \tag{5}$$

Obviously,

$$Q\left[\frac{k}{p}\right] = \frac{(k-i)}{p} \cdot Q = (p-1)! \binom{k}{p},$$

and then from (5)

$$(p-1)! \left[\frac{k}{p}\right] \equiv Q\left[\frac{k}{p}\right] \equiv (p-1)! \binom{k}{p} \pmod p.$$

Since p is prime, $p \nmid (p-1)!$. Thus from the above formula,

$$\binom{k}{p} \equiv \left[\frac{k}{p}\right] \pmod{p}.$$

$\square$

Example 5. Suppose $p \geq 5$ is prime. For $p \nmid k$, we define $\frac{1}{k}$ as a solution x satisfying $kx \equiv 1 \pmod{p}$ (refer to Remark 2). Prove:

$$\frac{1}{1^2} + \frac{1}{2^2} + \cdots + \frac{1}{(p-1)^2} \equiv 0 \pmod{p}.$$

Proof. Since p is prime, $1, 2, \ldots, p-1$ compose a reduced system modulo p. It follows that $\frac{1}{1}, \frac{1}{2}, \ldots, \frac{1}{p-1}$ also compose a reduced system modulo p (see Remark 3), so for $p \geq 5$, (refer to Remark 1)

$$\sum_{k=1}^{p-1} \left(\frac{1}{k}\right)^2 \equiv \sum_{k=1}^{p-1} k^2 = \frac{1}{6}p(p-1)(2p-1) \equiv 0 \pmod{p}.$$

In addition, due to

$$\left(\frac{1}{k}\right)^2 \equiv \frac{1}{k^2} \pmod{p} (k = 1, 2, \ldots, p-1),$$

by combining the above formula, the conclusion is proved.

The second solution is a little complicated, but more general: Because $\frac{1}{1}, \frac{1}{2}, \ldots, \frac{1}{p-1}$ form a reduced system modulo p, so for any integer a (here $p \nmid a$), we see that $a \cdot \frac{1}{1}, a \cdot \frac{1}{2}, \ldots, a \cdot \frac{1}{p-1}$ also form a reduced system modulo p (see (11)(ii)). Thus

$$\sum_{k=1}^{p-1} \left(a \cdot \frac{1}{k}\right)^2 \equiv \sum_{k=1}^{p-1} \left(\frac{1}{k}\right)^2 \pmod{p},$$

that is

$$(a^2 - 1) \sum_{k=1}^{p-1} \frac{1}{k^2} \equiv 0 \pmod{p}. \tag{6}$$

On the other hand, because p is prime, $a^2 - 1 \equiv 0 \pmod{p}$ implies $a \equiv \pm 1 \pmod{p}$. Hence, for $p \geq 5$, we can choose (parameter) a to satisfy $p \nmid a$ and $p \nmid (a^2 - 1)$. For such a, the result can be obtained from (6). $\square$

Example 6. Let a and b be positive integers, and n be a positive integer. Prove:

$$n! \mid b^{n-1}a(a+b)(a+2b)\cdots(a+(n-1)b).$$

Proof. We just need to prove that for any prime number p with $1 < p \le n$, if $p^\alpha \parallel n!$, then $p^\alpha \mid b^{n-1}a(a+b)\cdots(a+(n-1)b)$. (Refer to Remark 5 in Chapter 7.)

If $p \mid b$, then since (see (11) in Chapter 7)

$$\alpha = \sum_{l=1}^{\infty}\left[\frac{n}{p^l}\right] < \sum_{l=1}^{\infty}\frac{n}{p^l} = \frac{n}{p-1} \le n,$$

which means that $\alpha \le n-1$, we have $p^\alpha \mid b^{n-1}$, and the conclusion follows.

If $p \nmid b$, then there exists b_1 such that $bb_1 \equiv 1 \pmod{p^\alpha}$ (see Remark 2), thus

$$b_1^n a(a+b)\cdots(a+(n-1)b) \equiv ab_1(ab_1+1)\cdots(ab_1+n-1)\pmod{p^\alpha}.$$

The right side of the congruence is the product of n consecutive integers, so it is divisible by $n!$, and it's also divisible by p^α. That is

$$p^\alpha \mid b_1^n a(a+b)\ldots(a+(n-1)b),$$

but $p \nmid b_1$, so $(p^\alpha, b_1^n) = 1$, thus

$$p^\alpha \mid a(a+b)\cdots(a+(n-1)b).$$

$\square$

This method converts integers that form an arithmetic sequence into consecutive integers, in the sense of congruence.

Example 7. Let m and n be positive integers. Prove:

$$n! \mid (m^n-1)(m^n-m)\cdots(m^n-m^{n-1}).$$

Proof. We have

$$(m^n-1)(m^n-m)\cdots(m^n-m^{n-1})$$
$$= m \cdot m^2 \cdots\cdots m^{n-1} \cdot (m-1)\cdots\cdots(m^n-1) \tag{7}$$
$$= m^{\frac{n(n-1)}{2}}(m-1)\cdots(m^n-1).$$

When $n = 1$, the conclusion is obvious. When $n = 2$, because $m(m - 1)$ is even, the conclusion is also true. Let $n \geq 3$, and suppose p is a prime number. Then the power of p that appears in $n!$ is $\alpha = \sum_{l=1}^{\infty} \left[\frac{n}{p^l}\right]$. Let β denote the power of p on the right side of (7). We prove that $\alpha \leq \beta$ is always true, and then the conclusion is proved. Distinguish two situations: If $p \mid m$, then $\beta \geq \frac{n(n-1)}{2}$, thus (notice that $n \geq 3$)

$$\alpha < \sum_{l=1}^{\infty} \frac{n}{p^l} = \frac{n}{p-1} \leq n \leq \frac{n(n-1)}{2} \leq \beta.$$

If $p \nmid m$, then $(p, m) = 1$. From Fermat's little theorem, $m^{p-1} \equiv 1 \pmod{p}$, thus, for any positive integer k divisible by $p-1$, we have $p \mid m^k - 1$. And in $1, 2, \ldots, n$, there are $\left[\frac{n}{p-1}\right]$ numbers divisible by $p-1$, so there are at least $\left[\frac{n}{p-1}\right]$ numbers divisible by p in $m - 1, m^2 - 1, \ldots, m^n - 1$, so $\beta \geq \left[\frac{n}{p-1}\right]$. And $\alpha < \frac{n}{p-1}$, thus the integer $\alpha \leq \left[\frac{n}{p-1}\right]$, hence $\alpha \leq \beta$. □

Exercises

Group A

1. Prove: A perfect square is congruent to 0 or 1 modulo 3; it is congruent to 0, 1, or 4 modulo 5.
2. Prove: A perfect square is congruent to 0, 1, or 4 modulo 8.
3. Prove: A perfect square is congruent to 0 or ± 1 modulo 9.
4. Prove: The fourth power of an integer is congruent to 0 or 1 modulo 16.
5. Let a be an odd number and n be a positive integer. Prove: $a^{2^n} \equiv 1 \pmod{2^n + 2}$.
6. Let p be an odd prime number. Prove:

   (i) $1^2 \cdot 3^2 \cdots \cdots (p - 2)^2 \equiv (-1)^{\frac{p+1}{2}} \pmod{p}$;

   (ii) $2^2 \cdot 4^2 \cdots \cdots (p - 1)^2 \equiv (-1)^{\frac{p+1}{2}} \pmod{p}$.

7. If p and q are distinct odd prime numbers such that $pq \mid 2^{pq-1} - 1$, prove that $p \mid 2^{q-1} - 1$ and $q \mid 2^{p-1} - 1$; also prove the converse.
8. For a positive integer n, $S(n)$ denotes the sum of all digits in its decimal representation; $T(n)$ denotes the alternating sum of its digits (starting from the rightest digit). Prove: $S(n) \equiv n \pmod{9}$ and $T(n) \equiv n \pmod{11}$.

9. Prove: The sum of positive integers that are not greater than m and coprime with m is $\frac{1}{2}m\varphi(m)$, where $m \geq 2$.

10. Prove Wilson's theorem by Fermat's little theorem and Euler's identity (formula (6) in Chapter 4).

11. Suppose a positive integers a is coprime with 10. Find the last two digits of a^{20} (in the decimal system).

12. Let n be an even number, and $a_1, \ldots, a_n$, and $b_1, \ldots, b_n$ are both complete systems modulo n. Prove: $a_1 + b_1, \ldots, a_n + b_n$ do not form a complete system modulo n.

13. Let $p \geq 3$ be a prime number, and $a_1, \ldots, a_{p-1}$ and $b_1, \ldots, b_{p-1}$ both are reduced systems modulo p. Prove: $a_1 b_1, \ldots, a_{p-1} b_{p-1}$ do not form a reduced system modulo p.

Group B

14. Prove: Euler's theorem can be derived from Fermat's little theorem.

15. Give a proof of Fermat's little theorem (independent of Euler's theorem).

16. Let p be a prime number. Prove that there exists an infinite number of positive integers n such that $p \mid 2^n - n$.

17. Prove: For any given integer m, there always exists an infinite number of positive integers n such that

$$2^n + 3^n - 1, \ 2^n + 3^n - 2, \ldots, 2^n + 3^n - m$$

are all composite numbers.

Chapter 9

Indeterminate (Diophantine) Equations (1)

Indeterminate (Diophantine) equations, generally speaking, refer to the equations in which the number of unknowns is greater than that of equations, and the range of unknowns is restricted (e.g., integers, positive integers, and rational numbers).

The indeterminate equation is an important topic in number theory, and also a very difficult and complicated one.

In mathematics competitions, indeterminate equations are often involved. The purpose of this chapter is to briefly introduce some basic methods of dealing with indeterminate equations, as well as several kinds of widely used equations.

Basic Methods of Solving Indeterminate Equations

In elementary mathematics, there are three basic methods to deal with indeterminate equations:

(i) Factorization method.
(ii) Congruence method.
(iii) Inequality estimation method.

Let's give a few examples to illustrate the spirit of these methods.

Example 1.

(i) Prove: The equation $x(x+1)+1 = y^2$ has no positive integer solutions.
(ii) Let k be a positive integer and $k \geq 2$. Prove: The equation $x(x+1) = y^k$ has no positive integer solutions.

Proof. Multiply both sides of the equation in (i) by 4, and rearrange into $(2x+1)^2 + 3 = 4y^2$, that is

$$(2x + 2y + 1)(-2x + 2y - 1) = 3. \tag{1}$$

Now assume that the equation has positive integer solutions x and y. Then $2x + 2y + 1 > -2x + 2y - 1$ are both integers (in fact, they are both positive integers), and since 3 is prime, so it is deduced from (1) and the unique factorization theorem ((5) in Chapter 7) that

$$-2x + 2y - 1 = 1, \ 2x + 2y + 1 = 3. \tag{2}$$

The integer solution to the equations (2) is $x = 0$ and $y = 1$, a contradiction.

Proof of (ii): The above factorization method is not easy to work now. Let's use another decomposition method: If the equation has positive integer solutions x and y, then because x and $x + 1$ are coprime, and their product is the kth power of an integer, so x and $x + 1$ are both the kth power of positive integers (refer to Exercise 11 in Chapter 6), that is

$$x = u^k, \ x + 1 = v^k, \ y = uv, \tag{3}$$

where u and v are positive integers. From (3), we get $v^k - u^k = 1$, that is

$$(v - u)(v^{k-1} + v^{k-2}u + \cdots + vu^{k-2} + u^{k-1}) = 1. \tag{4}$$

Because $k \geq 2$, it is obviously impossible. $\square$

Remark 1. Factorizing is the most basic method for indeterminate equations, and its proof is based on the unique factorization theorem of integers.

The main function of factorizing is to convert the original equation into several equations that are easier to deal with. (For example, the binary quadratic equation (1) was converted into a system of binary linear equations (2), and the equation in (ii) was converted into a system of equations (3), which is easier for a further demonstration.)

In general, there are two ways to factorize. One is to use the factorization of polynomials to generate the factorization of integers (refer to (i) and (4)). The other is to use the properties of integers (coprime, divisibility, etc.) to derive specific properties in the factorization (refer to question (ii)).

Of course, there is no fixed procedure to follow for the decomposition method. Sometimes, factorizing is quite difficult, and sometimes it's hard to demonstrate further properties. Some examples and exercises are selected in this book, which will be discussed later. The factorization method is often used in combination with other methods (such as congruence, etc.); please refer to Example 3 below.

Congruence is also a powerful tool for dealing with indeterminate equations.

Example 2. Prove that the equation

$$x^2 + y^2 - 4z^2 = 7 \tag{5}$$

has no integer solutions.

Proof. If integers x, y, and z satisfy (5), then because the square of an integer is congruent to 0 or 1 modulo 4, the left side of (5) $\equiv x^2 + y^2 \equiv$ 0, 1, 2(mod 4), but the right side $\equiv 3$(mod 4), which is contradictory. (We actually have proved that there are no solutions to the equation (5) modulo 4; refer to Remark 5 in Chapter 8.) □

Remark 2. The solution of Example 2 shows a basic spirit of applying congruence to solve indeterminate equation problems: If integer $A = 0$, then the remainder of A on division by any positive integer n $(n > 1)$ is certainly 0. Therefore, if a certain $n > 1$ can be found so that A is not congruent to 0 modulo n then the integer A must not be 0. The essence of this idea is to convert a problem in the (infinite) set of integers into a problem in the residue class a (finite set) modulo n. The latter can be greatly simplified, so it is easy to produce a contradiction or get applicable information.

The congruence method deals with indeterminate equations, which is mainly used to prove that there is no solution to the equation or derive some necessary conditions of the equation having solutions, so as to prepare for a further solution (demonstration).

The key of the congruence method is, of course, to choose an appropriate modulo, but how to choose the modulo depends on the specific problem, and it may take several attempts. (e.g., equation (5) has an integer solution modulo 3, so modulo 3 cannot solve Example 2. Similarly, it is easy to verify neither modulo 5 nor modulo 7 works.)

On the other hand, congruence is not a "universal method" to prove that the equation has no solution. That is, an equation has no integer solution does not necessarily imply that there exists $n > 1$, making the equation have no solution modulo n.

For a simple example of this, please refer to Exercise 15.

Example 3. Find all positive integer solutions (m, n, x) to the equation

$$2^m + 3^n = x^2. \tag{6}$$

Solution   Obviously $3 \nmid x$, so $x^2 \equiv 1 (\text{mod } 3)$. Taking equation (6) modulo 3 gives that

$$1 \equiv 2^m + 3^n \equiv (-1)^m \ (\text{mod } 3). \tag{7}$$

Thus, m is even. Suppose $m = 2k$. Convert (6) into

$$(x - 2^k)(x + 2^k) = 3^n. \tag{8}$$

It is deduced from (8) and the unique factorization theorem that both positive integers $x - 2^k$ and $x + 2^k$ are powers of (prime) 3, but the sum of these two numbers is $2x$ (notice that $3 \nmid x$), so $(x - 2^k, x + 2^k) = 1$, hence

$$x - 2^k = 1, \ x + 2^k = 3^n.$$

By eliminating x from the above two formulas, we get

$$3^n - 1 = 2^{k+1}. \tag{9}$$

If n is odd, then 3^n is three times the square of an odd number, so the left side of (9) $\equiv 3 - 1 \equiv 2 (\text{mod } 4)$, but the right side $\equiv 0 (\text{mod } 4)$, which is impossible. Hence n is even (this can also be derived from equation (6) modulo 4). Let $n = 2l$, and decompose the left side of (9) by using the difference of two squares formula. It is not difficult to find its solution, but we prefer to use the following method: The number 3^l is odd. Supposed it is $2u + 1$. Then (9) becomes

$$u(u + 1) = 2^{k-1}. \tag{10}$$

If $u > 1$, at least one of u and $u + 1$ has an odd prime factor, and (10) is obviously not true. Hence, $u = 1$, thus $k = 2$. Therefore, all the solutions are $(m, n, x) = (4, 2, 5)$.

Remark 3. The purpose of proving that m is even is to create conditions for the decomposition of equation (6) (refer to Remark 1), which can be seen clearly from the above solution.

The fact that m is even is not easy to derive from (6) itself. Taking the equation (6) modulo 3 (see (7)), we can simplify the equation (two unknowns are eliminated), and then some (weaker but applicable) results are easily produced.

By the way, we mention the difference between modulo 4 and modulo other integers (3, 5, 7, etc.) in Example 2: Taking modulo 5 as an example, we see that the left side of equation (5) modulo 5 still has three terms, while a square number modulo 5 has three possible values 0, 1, and 4. The combination of these three terms can produce all the residue classes modulo 5, so it can't solve the problem.

Finally, we briefly talk about the estimation method of dealing with indeterminate equations.

Example 4.

(i) Find all the integer solutions to the equation $5x^2 - 6xy + 7y^2 = 130$.
(ii) Prove that the equation $x(x+1) + 1 = y^2$ has no positive integer solutions.

Solution Suppose the equation in (i) has an integer solution, it should have a real solution first, so its discriminant as a quadratic equation of x should be nonnegative, i.e.,

$$(-6y)^2 - 4 \times 5 \times 7y^2 + 4 \times 650 \geq 0.$$

It can be found that $y^2 \leq 25$, i.e., $|y| \leq 5$. Therefore, the integer y can take the values $0, \pm 1, \pm 2, \ldots, \pm 5$. Substitute them into the original equation one by one to verify (we can first check whether the above discriminant is a perfect square), so all the solutions are $(x, y) = (3, 5)$ and $(-3, -5)$.

(ii) is (i) of Example 1, and we use the estimation method to solve: If the equation has a positive integer solution, then from $x^2 < x(x+1) + 1 = y^2$, we see that $y > x$. Since x and y are integers, $y \geq x + 1$. Hence

$$y^2 > (x+1)^2 > x(x+1) + 1,$$

a contradiction.

The argument can also be expressed in another way: If the equation has positive integer solutions x and y, then

$$x^2 < x(x+1) + 1 < (x+1)^2,$$

which indicates that $x(x+1)+1$ is between two consecutive square numbers, so it cannot be a square number.

Remark 4. The estimation method generally includes two aspects. First, if an indeterminate equation has an integer solution, it certainly has a real solution. When the real solution set of the equation is a bounded set, we can use this necessary condition to determine a bound of the integer solution (there is at most a finite number of integers in a bounded range), and then verify one by one to determine all the solutions. (The graph of the equation in (i) is an ellipse, and the range of its real solutions is bounded.)

If the range of the real solutions of the equation is unbounded then the above method can't work. In this case, we should focus on integers, and use the properties of integers to generate applicable inequalities. (The argument in (ii) applies the most basic property of integers: If integers x and y satisfy $y > x$, then $y \geq x + 1$.)

The estimation method, of course, is not limited to indeterminate equations. Many problems can be solved by this method. Please pay attention to the related examples and exercises in this book.

Example 5. Find all positive integer solutions to the indeterminate equation

$$(a^2 - b)(a + b^2) = (a + b)^2. \tag{11}$$

Solution   Obviously, if $b = 1$, then $a = 2$. In the following we consider the case of $b > 1$. Suppose $(a, b) = d$. Then $a = a_1 d$, and $b = b_1 d$, where the positive integers a_1 and b_1 are coprime. Then (11) is converted into:

$$a_1[a_1 b_1^2 d^2 + a_1(da_1 - 1) - 3b_1] = b_1^2(db_1 + 1).$$

Because b_1 exactly divides the right side of the formula above, it exactly divides the left side as well, so $b_1 \mid (da_1 - 1)$, which means that $b_1 \mid (a - 1)$. Because $a > 1$, in particular $a - 1 \geq b_1$ (refer to Remark 3 in Chapter 6). From (11) (notice that $b > 1$),

$$(a + b)^2 > (a^2 - b)(a + b),$$

so $a + b > a^2 - b$, that is, $2b > a(a - 1)$. Thus

$$2 > \frac{a(a - 1)}{b} = a_1 \cdot \frac{a - 1}{b_1} \geq a_1,$$

so $a_1 = 1$, and $a - 1 = b_1$. Hence $b = b_1 d = a(a - 1)$. Substitute it into (11), and then

$$a + a^2 (a - 1)^2 = a^3.$$

The left side of the above formula should be divisible by a^2, i.e., $a^2 \mid a$, so $a = 1$, which is impossible. Therefore, the only positive integer solution of the equation is $a = 2$ and $b = 1$.

Some Kinds of Indeterminate Equations

In this section, we briefly introduce some basic and widely used indeterminate equations.

(1) **Linear indeterminate equations** This is the most thoroughly understood kind of indeterminate equations. If such an equation has one integer solution, it has an infinite number of integer solutions. And people can effectively judge whether there is a solution to a linear equation and can give all its solutions. For the purpose of this book, we will not discuss the general situation. For the results of the most commonly used binary linear indeterminate equations, please refer to Exercise 7 and Exercise 9 in Chapter 6.

(2) **Pell equation** The binary quadratic indeterminate equations are more complicated, which are essentially reduced to the study of (hyperbolic) equations

$$x^2 - dy^2 = c \tag{12}$$

where d and c are integers, $d > 0$ and d is not a square number, and $c \neq 0$. A special form

$$x^2 - dy^2 = 1 \tag{13}$$

of equation (12) is the most fundamental and important, which is called the **Pell equation**. It can be proved (but not discussed in this book) that the equation (13) must have an infinite number of positive integer solutions. Assume that (x_1, y_1) is the solution in the positive integer solutions (x, y) of (13) that minimizes $x + y\sqrt{d}$ (called the basic solution of (13)). Then all the positive integer solutions of (13) are given by

$$
\begin{cases}
x_n = \dfrac{1}{2}[(x_1 + \sqrt{d}y_1)^n + (x_1 - \sqrt{d}y_1)^n], \\
y_n = \dfrac{1}{2\sqrt{d}}[(x_1 + \sqrt{d}y_1)^n - (x_1 - \sqrt{d}y_1)^n]
\end{cases}
\tag{14}
$$

$(n = 1, 2, \cdots)$.

It is not difficult for the readers to derive the linear recurrence relation satisfied by x_n and y_n from (14):

$$
\begin{cases}
x_n = 2x_1 x_{n-1} - x_{n-2}, \\
y_n = 2x_1 y_{n-1} - y_{n-2}.
\end{cases}
\tag{15}
$$

Remark 5. The Pell equation is mainly used to prove that there is an infinite number of integer solutions to a problem in mathematics competitions, so in fact, it is not necessary to apply the above profound theorem, since for a specific d, a group of positive integer solutions (x_1, y_1) of (13) can be found by using the trial-and-error method (we actually know that the equation must have solutions!). No matter whether (x_1, y_1) is the basic solution or not, an infinite number of positive integer solutions of the equation is given from (14).

We also briefly talk about the difference between equations (12) and (13). The equation (12) does not always have solutions, but if it has a solution, it is not difficult to prove that (12) also has an infinite number of positive integer solutions.

In fact, if (u, v) is a of positive integer solution of (12), and (x_n, y_n) is any positive integer solution of (13) (there is an infinite number of such solutions), then

$$x + \sqrt{d}y = \left(u + v\sqrt{d}\right)\left(x_n + y_n\sqrt{d}\right)$$
$$= ux_n + dvy_n + (uy_n + vx_n)\sqrt{d} \qquad (16)$$

leads to an infinite number of positive integer solutions (x, y) of (12) (please verify). (However, positive integer solutions of (12) may not be all given by (16), which will not be discussed in this book.)

(3) **Pythagorean equation** $x^2 + y^2 = z^2$. This is a very special quadratic indeterminate equation with three unknowns, which has a distinguished geometric significance and is widely used.

We only discuss positive integer solutions of Pythagorean equation. Because of the homogeneity of the equation, we can assume that $(x, y, z) = 1$. Such a solution is called a primitive solution of the equation, also called primitive Pythagorean numbers.

Let's note that if $(x, y, z) = 1$, then suppose that $(x, y) = d$ and we see that $d^2 | z^2$ from the equation, so $d | z$ (see Exercise 12 in Chapter 6), thus $d | (x, y, z)$. Hence, $d = 1$. Similarly, $(x, z) = 1$ and $(y, z) = 1$. Thus, x, y, and z are pairwise coprime! Therefore, we only need to discuss positive integer solutions satisfying $(x, y) = 1$.

Taking the equation modulo 4 and since $(x, y) = 1$, we can see that one of x and y is odd and the other is even. Assume that is even. The following result (sometimes called the Pythagorean number theorem) gives all the primitive solutions of Pythagorean equation.

Theorem All positive integer solutions $(x,\,y,\,z)$ of the equation $x^2 + y^2 = z^2$ satisfying $(x,\,y) = 1$ and $2\,|\,y$ can be expressed as $x = a^2 - b^2$, $y = 2ab$ and $z = a^2 + b^2$, where a and b are any integers satisfying $a > b > 0$, one of a and b is odd and the other is even, and $(a,\,b) = 1$.

It's easy to prove the theorem (in both aspects). We leave it for practice.

(4) **Indeterminate equation $xy = zt$.** This quadratic equation with four unknowns also has many uses. All its positive integer solutions are very easy to find:

Let $(x,\,z) = a$. Then $x = ac$ and $z = ad$, where $(c,\,d) = 1$. Hence, $acy = adt$, i.e., $cy = dt$. Because $(c,\,d) = 1$, then $d\,|\,y$. Let $y = bd$. Then $t = bc$. Therefore, the positive integer solutions to the equation $xy = zt$ can be expressed as

$$x = ac,\ y = bd,\ z = ad,\ t = bc,$$

where a, b, c, and d are all positive integers, and $(c,\,d) = 1$. Conversely, we easily know that x, y, z and t given above are solutions.

The following derivation can also be used (it may be easier to remember):

Suppose $\frac{x}{z} = \frac{t}{y} = \frac{c}{d}$, where $\frac{c}{d}$ is an irreducible fraction, i.e., $(c,\,d) = 1$. Since $\frac{c}{d}$ is obtained after the reduction of $\frac{x}{z}$, then $x = ac$ and $z = ad$ (this can also be derived by divisibility as above); similarly, $t = cb$ and $y = db$.

Example 6. Prove: There is an infinite number of triangle numbers that are perfect squares (a triangle number is a number of the form $\frac{n(n+1)}{2}$, where n is a natural number).

Proof. The problem is to prove that the indeterminate equation

$$\frac{n(n+1)}{2} = k^2,$$

i.e.,

$$(2n + 1)^2 - 2 \cdot (2k)^2 = 1, \tag{17}$$

has an infinite number of positive integer solutions.

We know (see (2) and Remark 5) that the Pell equation

$$x^2 - 2y^2 = 1 \tag{18}$$

has an infinite number of positive integer solutions (x, y), and x must be odd. Take modulo 4 on (18). Then y must be an even number. Thus, let $n = \frac{x-1}{2}$ and $k = \frac{y}{2}$, and we can get a positive integer solution of (17) from a positive integer solution of (18). Hence there is an infinite number of positive integer solutions of (17). (In fact, we can find all the solutions of this problem.)   □

Example 7. Let a, b, c, and d be all positive integers, such that $ab = cd$. Prove: $a^4 + b^4 + c^4 + d^4$ is not prime.

Proof. From $ab = cd$, we obtain that $a = us$, $b = vt$, $c = vs$ and $d = ut$, where u, v, s, and t are all positive integers (see (4)). Thus

$$a^4 + b^4 + c^4 + d^4 = (u^4 + v^4)(s^4 + t^4)$$

is not prime.

The argument applied the factorization method. Please compare it with Example 1 in Chapter 7.   □

Example 8. Let $n > 1$ be an odd number. Suppose n has a decomposition $n = uv$, where $0 < u - v \le 4\sqrt[4]{n}$. Prove: Such a decomposition of n is unique.

Proof. Prove by contradiction. If there are two different decompositions of n, that is $n = uv = u_1 v_1$, then $u = ad$, $v = bc$, $u_1 = bd$, and $v_1 = ac$ (see (4)). Obviously, $u - v \ne u' - v'$ (otherwise, from $uv = u'v'$, it is easy to obtain $u = u'$, and $v = v'$). Assume that $u - v > u' - v'$. Then

$$16\sqrt{n} > (u - v)^2 - (u' - v')^2$$
$$= (ad - bc)^2 - (bd - ac)^2$$
$$= (a^2 - b^2)(d^2 - c^2)$$
$$= (a - b)(d - c)(a + b)(c + d). \tag{19}$$

Since $u - v > u' - v'$, we have $(a - b)(d - c) > 0$. And because n is an odd number, a, b, c, and d are all odd numbers. Therefore, it is derived from (19) that

$$16\sqrt{n} > 2 \cdot 2 \cdot 2\sqrt{ab} \cdot 2\sqrt{cd} = 16\sqrt{abcd} = 16\sqrt{n},$$

a contradiction.   □

Exercises

Group A

1. Let x and y be integers greater than 1. Find all positive integer solutions of the equation $x^y = 2^z - 1$.
2. Let $y > 1$. Find all positive integer solutions of $x^y = 2^z + 1$.
3. Prove that the product of three consecutive positive integers cannot be the kth power of an integer, where $k \geq 2$ is a given positive integer.
4. Try to determine all integers that can be expressed as the difference of squares of two integers.
5. Prove: If $n \equiv 4 \pmod 9$, then n cannot be expressed as the cubic sum of three integers.
6. Prove: The equation $x_1^4 + \cdots + x_{14}^4 = 1599$ doesn't have integer solutions.
7. Find all integer solutions to the equation $x^2 + x = y^4 + y^3 + y^2 + y$.
8. Let a, b, and c be three distinct positive integers. Prove: $a + b$, $b + c$, and $c + a$ can't all be powers of 2.
9. Find all the triangles whose side lengths are integers and the (value of) perimeter is twice the area.
10. Find all positive integer solutions of $3^x + 4^y = 5^z$.
11. Prove the Pythagorean number theorem in (3) of this chapter.

Group B

12. Find all positive integers m and n, such that $1! + 2! + \cdots + m! = n^2$.
13. Find all finite sets of distinct positive integers such that their product is equal to their sum.
14. Prove: The equation $x^2 + y^2 + z^2 = 6xyz$ has the only solution $x = y = z = 0$.
15. Prove: For any positive integer m, the congruence equation

$$6xy - 2x - 3y + 1 \equiv 0 \pmod m$$

always has integer solutions, but the indeterminate equation $6xy - 2x - 3y + 1 = 0$ has no integer solutions.
16. Let p be prime. Prove: If $2^p - 1$ is a power of a prime number, then it must be prime.

Chapter 10

Problems in Number Theory

This chapter introduces some number theory problems in mathematics competitions.

Example 1. Let a and b be positive integers with $1 \leq a < b$. Prove that there must exist two numbers in b consecutive integers such that the product of the two numbers is divisible by ab.

Proof. Note that for any $n > 1$, any two numbers in n consecutive integers must not be congruent modulo n, that is, they form a complete system modulo n, so there is one (and only one) number divisible by n. From this and $1 \leq a < b$, it can be seen that one of the given consecutive b integers is a multiple of b, denoted as x; There's also one number that's a multiple of a, denoted as y. If $x \neq y$, then xy is a multiple of ab, and the conclusion is proved. If $x = y$, then from $a|x$ and $b|x$, we know that $[a, b] \mid x$.

If $(a, b) = d$, then $d|a$ and $d|b$. And because $a < b$, we have $b \geq 2d$. Therefore, in the given $b (\geq 2d)$ consecutive integers, either at least d integers are less than x, or at least d integers are greater than x. In either case, from the above fact, we can always choose a number from the given integers, which is different from x and is a multiple of d. Suppose this number is z. Then xz is a multiple of $[a, b]d = ab$. $\qquad\square$

Example 2. Let $S = 1!2! \cdots 100!$. Prove that there is an integer k with $1 \leq k \leq 100$ such that $\frac{S}{k!}$ is a perfect square and k is unique.

Proof. Using $(2i-1)!(2i)! = [(2i-1)!]^2 \cdot 2i$, we know that

$$S = \prod_{i=1}^{50} [(2i-1)!]^2 (2i) = 2^{50} \cdot 50! \left(\prod_{i=1}^{50} (2i-1) \right)^2.$$

Thus, take $k = 50$, and then $\frac{S}{k!}$ is a perfect square.

In order to prove that k must be equal to 50, notice that if $53 \leq k \leq 100$, then the prime number 53 only appears once in (the standard factorization of) $\frac{k!}{50!}$, so $\frac{k!}{50!}$ is not a perfect square, and the above has proved that $\frac{S}{50!}$ is a perfect square, thus

$$\frac{S}{k!} = \frac{S}{50!} \Big/ \frac{k!}{50!}$$

is not a perfect square. If $k = 51$ or 52, obviously $\frac{k!}{50!}$ is not a perfect square, so $\frac{S}{k!}$ is not a perfect square.

If $k \leq 46$, then the prime number 47 appears once in $\frac{50!}{k!}$, so $\frac{50!}{k!}$ is not a perfect square, and $\frac{S}{k!} = \frac{S}{50!} \cdot \frac{50!}{k!}$ is not a perfect square. If $k = 47$, 48, or 49, we easily know that $\frac{50!}{k!}$ is not a perfect square, and it is also deduced that $\frac{S}{k!}$ is not a perfect square. Therefore, k can only be 50. Combining it with the previous result, we find that $k = 50$ is the only solution.   □

Example 3. Prove: If integers a and b satisfy $2a^2 + a = 3b^2 + b$, then both $a - b$ and $2a + 2b + 1$ are perfect squares.

Proof The known equation can be rearranged into

$$(a-b)(2a+2b+1) = b^2. \tag{1}$$

Note that $2a+2b+1$ is odd, so it is not 0, thus the greatest common divisor of $a - b$ and $2a + 2b + 1$ exists, which we denote as d. We first prove $d = 1$. If $d > 1$, then d has a prime factor p. We know $p \mid b^2$ from $p \mid a - b$ and (1), so the prime number p exactly divides b, but $p \mid 2a + 2b + 1$, thus we deduce that $p \mid 1$, a contradiction. Hence $d = 1$, that is, $a - b$ and $2a + 2b + 1$ are coprime. Because the right side of (1) is a square number, from Exercise 11 in Chapter 6, $|a - b|$ and $|2a + 2b + 1|$ are both perfect squares.

Next we prove that $a - b \geq 0$, so that in combination with (1), we know $2a + 2b + 1 > 0$, and the conclusion then follows.

Suppose $a - b < 0$. Since $|a - b|$ is a square number, $b - a = r^2$ with $r > 0$. From (1), r exactly divides b, and then from $b - a = r^2$, there is $r \mid a$. Suppose $b = b_1 r$ and $a = a_1 r$, and substitute into the equation in this problem. Then it can be reduced to

$$a_1^2 + 6a_1 r + 3r^2 + 1 = 0. \tag{2}$$

Taking the above equation modulo 3, we obtain $a_1^2 + 1 \equiv 0 \pmod 3$, which is impossible (since squares of integers are congruent to or 1 modulo 3). Therefore, the assumption $a - b < 0$ leads to a contradiction, so our assertion is proved.

Remark 1. It can also be proved that $a - b$ and $2a + 2b + 1$ are coprime by other methods, but it is more direct to prove it by prime numbers and the equation (1).

Remark 2. The proof of $a - b \geq 0$ is another important point of the above argument, which seems not easy from the algebraic point of view. We derive the equation (2) by the assumption and proven result, and then obtain a contradiction from arithmetic point of view (compare the remainders modulo 3 on both sides of (2)).

By the way, the above integer r is actually equal to $(a - b, b) = (a, b)$ (see Exercise 11 in Chapter 6), but we do not need this conclusion.

Remark 3. The equation in the problem can also be rearranged into

$$(a - b)(3a + 3b + 1) = a^2.$$

Therefore, it can also be proved that $3a + 3b + 1$ is a perfect square in a similar way.

Example 4. Prove: For any prime number p, there exists a positive integer n such that

$$p \mid 2^n + 3^n + 6^n - 1.$$

Proof. When $p = 2$ or 3, just take $n = 2$.

Let $p > 3$. We want to select n that can make 2^n, 3^n and 6^n modulo p greatly simplified. Fermat's little theorem has such an effect: From this theorem we have

$$2^{p-1}, \ 3^{p-1}, \ 6^{p-1} \equiv 1 \pmod p.$$

Thus

$$3 \cdot 2^{p-1} + 2 \cdot 3^{p-1} + 6^{p-1} \equiv 3 + 2 + 1 = 6 \pmod p,$$

that is $6 \cdot 2^{p-2} + 6 \cdot 3^{p-2} + 6 \cdot 6^{p-2} \equiv 6 \pmod p$.

Therefore, $p \mid 2^{p-2} + 3^{p-2} + 6^{p-2} - 1$, that is for $p > 3$, we can take $n = p - 2$.

Using the inverse of $\bmod\, p$, the above argument can also be expressed more directly as follows:

From Fermat's little theorem,

$$2^{p-2} \equiv \frac{1}{2},\ 3^{p-2} \equiv \frac{1}{3},\ 6^{p-2} \equiv \frac{1}{6}\ (\bmod\, p),$$

where $\frac{1}{2}$, $\frac{1}{3}$, and $\frac{1}{6}$ are the inverses of 2, 3, and 6 modulo p respectively. Therefore, according to the operational rule of inverses (see Remark 3 in Chapter 8) we could know:

$$2^{p-2} + 3^{p-2} + 6^{p-2} - 1 \equiv \frac{1}{2} + \frac{1}{3} + \frac{1}{6} - 1$$

$$= \frac{3+2+1}{6} - 1 = 0\ (\bmod\, p).$$

Notice that from the above proof, for any nonnegative integer k, the number $n = (p-1)k + p - 2$ always meets the requirement, so there is actually an infinite number of such n.

Example 5. Let $a > 1$ be an integer and set $S = \{a^2 + a - 1,\ a^3 + a^2 - 1,\ a^4 + a^3 - 1,\ \cdots\}$. Prove that there exists an infinite subset in S, in which the numbers are pairwise coprime.

Proof. We inductively construct an infinite subset $T = \{b_1,\ b_2,\ \cdots\}$ of S, which has the properties in the problem.

Let $b_1 = a^2 + a - 1$. Now we prove that for all $n \geq 1$, if we have already found n numbers $b_1, \ldots, b_n$ in S, then we can always choose $b_{n+1} = a^{m+1} + a^m - 1$ in S, which is coprime with all of $b_1, \ldots, b_n$ (that is coprime with $N = b_1 \cdots \cdot b_n$), where m is a (to be determined) parameter, depending on $b_1, \ldots, b_n$. Therefore, $b_1, \ldots, b_n, b_{n+1}$ are pairwise coprime. By mathematical induction, we construct such an infinite subset T of S.

In fact, since the numbers in S are all coprime with a, we have $(N, a) = 1$. Therefore, for $m = \varphi(N)$, from Euler's theorem, we obtain $a^{\varphi(N)} \equiv 1 (\bmod\, N)$, thus

$$b_{n+1} = a^{\varphi(N)+1} + a^{\varphi(N)} - 1 \equiv a + 1 - 1 = a\ (\bmod\, N).$$

Because $(a, N) = 1$, it is known from the above formula that $(b_{n+1}, N) = 1$, that is, the number b_{n+1} that we selected meets the above requirements.

$\square$

Remark 4. To find (or prove) that a certain number is coprime with N, we can consider the decomposition of its prime factors. However, for the number in the form of $a^{m+1} + a^m - 1$ involved in this problem, it seems that this idea is not easy to realize.

The above solution is based on the consideration of congruence: A number being coprime with N is equivalent to that it belongs to a reduced congruence class modulo N, that is, it is $\equiv r(\mathrm{mod}\, N)$, where r is an integer coprime with N. For this problem, in order to realize this idea, we should first select an appropriate m, which can make a^m modulo N greatly simplified. For this purpose, Euler's theorem can be used (see the above proof).

Example 6. Prove: For any positive integer $n \geq 2$, there exists a set S consisting of n positive integers such that for any $a \in S$ and $b \in S\, (a \neq b)$, there is $(a - b)^2 \mid ab$.

Proof. We inductively construct a set of n positive integers $S_n = \{a_1, \ldots, a_n\}$, so that for any a_i and $a_j\, (i \neq j)$, there are $a_i - a_j \mid a_i$ and $a_i - a_j \mid a_j$. Therefore, there is $(a_i - a_j)^2 \mid a_i a_j$, so S_n meets the requirement of the problem.

When $n = 2$, take $S_2 = \{1, 2\}$. Assuming that the existing n-element set S_n satisfies the requirement, we will construct S_{n+1} based on it. However, unlike the case in Example 5, we cannot expect that S_{n+1} here is generated by adding a certain positive integer to S_n (for all $n \geq 2$). (See Remark 5 below.)

Let $S_n' = \{a_1 + x, \ldots, a_n + x\}$, where $x > 0$ is any common multiple of the numbers in S_n. It can be seen from the inductive hypothesis of S_n and the selection of x that for any $i \neq j$, the difference $(a_i + x) - (a_j + x) = a_i - a_j$ exactly divides a_i and a_j. Since it also exactly divides x, the difference exactly divides $a_i + x$ and $a_j + x$. This means that S_n' has the same properties as S_n.

Now let $S_{n+1} = S_n' \bigcup x$. Because $(a_i + x) - x = a_i$ exactly divides x, it also exactly divides $a_i + x$. Since it has been proved that S_n' has the said properties, the set S_{n+1} also meets our requirement. This completes the inductive construction. $\qquad\square$

Remark 5. Notice that if for each $n \geq 1$, there exists an n-element set with a certain property (the elements may be all related to n), it may not follow that there exists an infinite set with this property.

As far as this problem is concerned, it is easy to prove that there cannot be an infinite set consisting of positive integers to meet the requirement of the problem. Suppose $\{a_1, a_2, \ldots\}$ is such a set, where $a_1 < a_2 < \ldots$. Then particularly there is $(a_n - a_1)^2 \mid a_n a_1$ (for all $n \geq 2$). But a_n is a positive integer, so $a_n \to \infty$. For a_n that is large enough, obviously $(a_n - a_1)^2 > a_n a_1$, a contradiction!

Therefore, the following proposition cannot be true:

For all $n \geq 2$, if S_n satisfies the requirement of the problem, then there exists a positive integer $a_{n+1} \notin S_n$, which makes $S_{n+1} = S_n \bigcup a_{n+1}$ also satisfy the requirement of the problem.

(Otherwise, from this proposition and the induction principle, we can recursively generate an infinite set consisting of positive integers from S_2 to meet the requirement, which is inconsistent with the fact mentioned above. Please compare it with the solution of Example 5.)

Remark 6. In consideration of Remark 5, suppose there already exists an n-element set S_n satisfying the requirement. To construct S_{n+1}, we first construct an n-element set S_n' (its elements directly or indirectly depend on n) based on S_n that still meets the requirement. Then add an appropriate number to S_n' to generate S_{n+1}.

There are two basic methods to construct (above) S_n' from S_n. One is "translation", and the other is "scalar multiplication".

(1) A translation of S_n refers to the set $S_n + x = \{a + x | a \in S_n\}$, where $x \neq 0$ is a (undetermined) parameter.

For this problem, using the inductive hypothesis of S_n, it is not difficult to see that if $x > 0$ is a common multiple of all numbers in S_n, then the n-element set $S_n + x$ has the required properties.

(2) A scalar multiplication of S_n refers to the set $t S_n = \{at | a \in S_n\}$, where $t \neq 0$ and 1 is a (undetermined) parameter.

We notice that the properties of S_n in this problem maintain unchanged under scalar multiplications, that is, for any positive integer $t \geq 2$, the n-element set $t S_n$ always has the properties mentioned above. But this fact does not seem convenient for us to construct S_{n+1} by adding a positive integer y to $t S_n$. (The readers can try to choose positive integers t and y, so that $ta_i - y \mid y$ and $ta_i - y \mid ta_i$ for $i = 1, \ldots, n$.) In contrast, in order to add a positive integer y to the translation set $x + S_n$ to generate S_{n+1} (where $x > 0$ is a common multiple of all the numbers in S_n), we need to select y such that $x + a_i - y \mid y$ and $x + a_i - y \mid (x + a_i)$ for $i = 1, \ldots, n$. It is easy to see that taking $y = x$ will meet the requirement (see the solution of Example 6).

Remark 7. The first step in the solution of Example 6 is to decompose the original quadratic problem into two (stronger) linear problems, which are often easier to deal with. But for this problem, this procedure doesn't have special benefits. By the way, our proof actually gives the following stronger (and perhaps more interesting) result: For any $n \geq 2$, there exist n positive integers $a_1 < a_2 < \cdots < a_n$ such that $a_i - a_j = (a_i, a_j)$ (for all $1 \leq j < i \leq n$).

Example 7. Let p be an odd prime number. Prove:

$$\sum_{j=0}^{p} \binom{p}{j} \binom{p+j}{j} \equiv 2^p + 1 \pmod{p^2}.$$

Proof. Note that for $j = 0, 1, \cdots, p-1$,

$$\binom{p+j}{j} = \frac{(p+j)(p+j-1)\cdots(p+1)}{j!}$$

$$\equiv \frac{j(j-1)\cdots 1}{j!}$$

$$\equiv 1 \pmod{p}.$$

Because $\binom{p}{j}$ is a multiple of p when $0 < j < p$ (Example 7 in Chapter 7), from the above formula,

$$\binom{p+j}{j}\binom{p}{j} \equiv \binom{p}{j} \pmod{p^2},$$

which also holds obviously when $j = 0$. Thus

$$\sum_{j=0}^{p} \binom{p}{j}\binom{p+j}{j} \equiv \sum_{j=0}^{p-1}\binom{p}{j} + \binom{2p}{p}$$

$$\equiv 2^p + \binom{2p}{p} - 1 \pmod{p^2}. \tag{3}$$

Finally, from Vandermonde's, identity i.e. (10) in Chapter 2,

$$\binom{2p}{p} = \sum_{j=0}^{p}\binom{p}{j}^2 = 2 + \sum_{j=1}^{p-1}\binom{p}{j}^2 \equiv 2 \pmod{p^2},$$

Combining the above with formula (3), we obtain the result. $\qquad\square$

Remark 8. For a prime number $p \geq 3$, the congruence

$$\binom{2p}{p} \equiv 2 \pmod{p^2}$$

has other proof methods. Let's introduce two more here by the way.

One method is to use: For $1 \leq j \leq \frac{p-1}{2}$,

$$(2p - j)(p + j) \equiv j(p - j) \pmod{p^2},$$

thus

$$\binom{2p}{p} = 2 \cdot \frac{1}{(p-1)!} \prod_{j=1}^{\frac{p-1}{2}} (2p - j)(p + j)$$

$$\equiv 2 \cdot \frac{1}{(p-1)!} \prod_{j=1}^{\frac{p-1}{2}} j(p - j)$$

$$\equiv 2 \cdot \frac{1}{(p-1)!} (p - 1)!$$

$$\equiv 2 \pmod{p^2}.$$

Another method is to note: For $1 \leq k \leq p - 1$, if k^{-1} is the inverse of k modulo p, then $\frac{1}{1}, \frac{1}{2}, \ldots, \frac{1}{p-1}$ form a reduced system modulo p, so in the sense of modulo p, they are an arrangement of $1, 2, \ldots, p - 1$. Hence

$$\binom{2p}{p} = \frac{p+1}{1} \cdot \frac{p+2}{2} \cdot \ldots \cdot \frac{p+(p-1)}{p-1} \cdot \frac{2p}{p}$$

$$= 2 \left(1^{-1} \cdot p + 1 \right) \left(2^{-1} \cdot p + 1 \right) \cdots \left((p - 1)^{-1} p + 1 \right)$$

$$\equiv 2(1 \cdot p + 1)(2 \cdot p + 1) \cdots ((p - 1)p + 1)$$

$$\equiv 2 \left(\left(\sum_{k=1}^{p-1} k \right) p + 1 \right) \equiv 2 \pmod{p^2}.$$

Example 8. For a prime number $p \geq 5$, prove that $\sum_{0 < k < \frac{2p}{3}} \binom{p}{k} \equiv 0 \pmod{p^2}$.

Proof. Because $p \left| \binom{p}{k} \right.$ for $0 < k < \frac{2p}{3}$, so we just need to prove that

$$\sum_{0<k<\frac{2p}{3}} \frac{1}{p} \binom{p}{k} \equiv 0 \,(\mathrm{mod}\,p^2).$$

For $0 < k < p$,

$$\begin{aligned}
\frac{1}{p} \binom{p}{k} &= \frac{(p-1)\cdots(p-k+1)}{k!} \\
&\equiv \frac{(-1)(-2)\cdots(-k+1)}{k!} \\
&\equiv (-1)^{k-1} \frac{1}{k} \quad (\mathrm{mod}\,p).
\end{aligned}$$

Thus (notice that $p \geq 5$),

$$\begin{aligned}
\sum_{0<k<\frac{2p}{3}} \frac{1}{p} \binom{p}{k} &\equiv \sum_{0<k<\frac{2p}{3}} (-1)^{k-1} \frac{1}{k} \\
&\equiv \sum_{0<k<\frac{2p}{3}} \frac{1}{k} - 2 \sum_{0<k<\frac{p}{3}} \frac{1}{2k} \\
&\equiv \sum_{0<k<\frac{2p}{3}} \frac{1}{k} - \sum_{0<k<\frac{p}{3}} \frac{1}{k} \\
&= \sum_{\frac{p}{3}<k<\frac{2p}{3}} \frac{1}{k} \quad (\mathrm{mod}\,p). \quad (4)
\end{aligned}$$

Let $S = \sum_{\frac{p}{3}<k<\frac{2p}{3}} \frac{1}{k}$. Then (pairing!)

$$2S = \sum_{\frac{p}{3}<k<\frac{2p}{3}} \left(\frac{1}{k} + \frac{1}{p-k} \right) = \sum_{\frac{p}{3}<k<\frac{2p}{3}} \frac{p}{k(p-k)} \equiv 0 \,(\mathrm{mod}\,p).$$

The result can be obtained from the above formula and (4).

Example 9. Find all odd integers $n > 1$ such that for any positive integers a and b that satisfy $a \mid n$, $b \mid n$, and $(a, b) = 1$, there is $a + b - 1 \mid n$.

Solution Obviously, if n is a power of an odd prime number, then it meets the requirement of the problem. We prove in the following that such n are all the solutions.

Let n have at least two distinct odd prime factors, and take any one of the prime factors p of n. Then n can be expressed as $n = p^k s$, where $k \geq 1$, odd number $s > 1$, and $p \nmid s$.

Because $p \mid n$, $s \mid n$, and $(p, s) = 1$, it can be seen from the condition that $p + s - 1 \mid n$. We hope that p can be selected to make $p + s - 1$ coprime with s. Note that $(p + s - 1, s) = (p - 1, s)$. If $(p - 1, s) > 1$, then there exists a prime number q such that $q \mid s$ and $q \mid p - 1$. Particularly, $q < p$. But $q \mid s$ and $s \mid n$, so $q \mid n$. Therefore, if (parameter) p is the minimum prime factor of n, then $(p + s - 1, s) = 1$ (otherwise, from the above argument, n will have a smaller prime factor q, which contradicts the selection of p). Because $p + s - 1$ exactly divides n, that is, exactly divides $p^k s$, we have $p + s - 1 \mid p^k$. Since p is a prime number, $p + s - 1$ is also a power of p. Note that $s > 1$, so $p + s - 1 > p$, and there exists $\alpha > 1$ such that

$$p + s - 1 = p^\alpha. \tag{5}$$

Further, from $s \mid n$, and $p + s - 1 \mid n$ and $(p + s - 1, s) = 1$, we know that $(p + s - 1) + s - 1 = p + 2(s - 1)$ exactly divides $n (= p^k s)$.

Note that $p - 2 > 0$, so from the minimality of p,

$$(p + 2(s - 1), s) = (p - 2, s) = 1.$$

Hence there exists $\beta > 1$ such that

$$p + 2(s - 1) = p^\beta. \tag{6}$$

From 5 and 6, we get $p + p^\beta = 2p^\alpha$, but this is impossible, because p^α and p^β are both divisible by p^2, but $p^2 \nmid p$. (That is to say: the equation doesn't hold modulo p^2.)

It can also be seen from the perspective of algebra: Since $\beta > \alpha$, we have $\beta \geq \alpha + 1$, so

$$p + p^\beta > p \cdot p^{\beta - 1} \geq p \cdot p^\alpha > 2p^\alpha.$$

Example 10. Find all the positive integer solutions (x, y, z) to the equation

$$x^3 + y + 1 = xyz.$$

Solution Rearrange the equation into

$$x(yz - x^2) = y + 1.$$

Let $yz - x^2 = n$. Then the above equation becomes $xn = y + 1$, so

$$x^2 = yz - n = (xn - 1)z - n = xzn - z - n. \tag{7}$$

Thus $zn > x$. Let

$$zn = x + k \ (k > 0). \tag{8}$$

Then (7) gives

$$xk = z + n. \tag{9}$$

From (8) and (9), we get (note that at this point we do not need k, n, x, and z to be positive integers)

$$(z - 1)(n - 1) + (x - 1)(k - 1) = 2. \tag{10}$$

Obviously, $(z - 1)(n - 1)$ and $(x - 1)(k - 1)$ are all nonnegative integers, so the following three possible cases can be obtained from (10):

$$(z - 1)(n - 1) = 0, \ (x - 1)(k - 1) = 2;$$
$$(z - 1)(n - 1) = 2, \ (x - 1)(k - 1) = 0;$$
$$(z - 1)(n - 1) = 1, \ (x - 1)(k - 1) = 1.$$

By solving the above three equations in turn, we can get the solution of this problem as follows, $(x, y, z) = (2, 1, 5)$, $(2, 9, 1)$, $(3, 2, 5)$, $(3, 14, 1)$; $(1, 1, 3)$, $(1, 2, 2)$, $(5, 9, 3)$, $(5, 14, 2)$; $(2, 3, 2)$. There are nine solutions total.

The above solution of Example 10 mainly relies on an algebraic deformation to get the equation (10) that is easy to deal with. However, in Example 11 below, a property of integers is needed to generate an applicable decomposition.

Example 11. Let n be a given odd integer. Then the equation

$$x^2 + n^a = 4^b$$

has at most a finite number of positive integer solutions (a, b, x).

Proof. Obviously x is an odd number, and a is also an odd number, otherwise n^a is the square of an odd number, and take the original equation modulo 4: its left side $\equiv 1 + 1 \equiv 2 (\bmod \, 4)$, while the right side $\equiv 0 (\bmod \, 4)$, a contradiction.

Decompose the equation into

$$(2^b - x)(2^b + x) = n^a. \tag{11}$$

Let $(2^b - x, 2^b + x) = d$. Then $d \mid (2^b - x) + (2^b + x)$, i.e., $d \mid 2^{b+1}$, so d is a power of 2. Because $d \mid (2^b + x) - (2^b - x)$, that is, $d \mid 2x$, and x is odd, we get $d = 1$ or 2. Since $2^b - x$ is odd, it follows that $d = 1$, that is, $2^b - x$ and $2^b + x$ are coprime.

Because the right side of (11) is the ath power of an integer, $2^b - x$ and $2^b + x$ are both the ath power of positive integers, so there exist positive integers u and v such that

$$\begin{cases} 2^b - x = u^a, \\ 2^b + x = v^a, \\ uv = n. \end{cases} \tag{12}$$

From (12), we obtain that

$$u^a + v^a = 2^{b+1}. \tag{13}$$

Equivalently,

$$(u + v)(u^{a-1} - u^{a-2}v + \cdots - uv^{a-2} + v^{a-1}) = 2^{b+1}. \tag{14}$$

Because n is odd, from $uv = n$, we know that u and v are both odd numbers, and also a is an odd number, so the second factor of the left side of (14) is the sum of an odd number of odd integers, which is an odd integer, so it must be 1 (since the right side of (14) is a power of 2). Combining the above with 14, we know that

$$u + v = 2^{b+1}. \tag{15}$$

Comparing (13) and (15), we see that $a = 1$.

Because $uv = n$, we see that u and v have only a finite number of solutions, so we know from (15) that b has at most a finite number of solutions, and from 12, x has at most a finite number of solutions. Therefore, the equation in the problem has at most a finite number of positive integer solutions (a, b, x).

Example 12. Let a, b, and c be given integers. Prove that there is an infinite number of positive integers n such that $n^3 + an^2 + bn + c$ is not a perfect square.

Proof. Let $a_n = n^3 + an^2 + bn + c$. Then for any positive integer k,
$$a_{4k+1} \equiv 1 + a + b + c \pmod 4,$$
$$a_{4k+2} \equiv 2b + c \pmod 4,$$
$$a_{4k+3} \equiv -1 + a - b + c \pmod 4,$$
$$a_{4k} \equiv c \pmod 4,$$
so (eliminating a and c)
$$a_{4k+2} - a_{4k} \equiv 2b \pmod 4,$$
$$a_{4k+1} - a_{4k+3} \equiv 2b + 2 \pmod 4.$$
Hence, either $a_{4k+2} - a_{4k} \equiv 2 \pmod 4$ or $a_{4k+1} - a_{4k+3} \equiv 2 \pmod 4$. If the former is true, then since a square number modulo 4 is 0 or 1, a_{4k+2} and a_{4k} cannot both be perfect squares. (Otherwise, their difference modulo 4 can only be 0, -1, or 1, but not 2.) This proves that for each k, at least one of the four numbers a_{4k+1}, a_{4k+2}, a_{4k+3}, and a_{4k} is not a square number, so there is an infinite number of terms in a_n that are not squares.

Remark 9. It can be proved by a deeper method that if the three (complex) roots of the equation $x^3 + ax^2 + bx + c = 0$ are different from each other, then there is at most a finite number of positive integers n which make $n^3 + an^2 + bn + c$ a perfect square.

Exercises

Group A

1. Let k be a given positive integer. Suppose $a_1 = k + 1$, and $a_{n+1} = a_n^2 - ka_n + k$ ($n \geq 1$). Prove: Any two terms in the sequence $\{a_n\}$ are pairwise coprime.

2. For $n > 1$, let $1 = d_1 < \cdots < d_k = n$ be all positive divisors of n. Let $S = d_1 d_2 + d_2 d_3 + \cdots + d_{k-1} d_k$. Prove: $S < n^2$.

3. Let p be an odd prime number. Suppose integers $a_1, \cdots, a_p$ form an arithmetic sequence, the common difference of which is not divisible by p. Prove: There exists i such that $a_1 a_2 \cdots a_p + a_i \equiv 0 \pmod{p^2}$.

4. Let positive integers a, b, and c satisfy $(a, b, c) = 1$, and $c = \frac{ab}{a-b}$. Prove that $a - b$ is a perfect square.

5. Does there exist two powers of 2 (in decimal representation), such that the digits of one number is an arrangement of those of the other?

6. Find all positive integer solutions (x, y, z) of the equation $2^x \cdot 3^y = 5^z + 1$.

Group B

7. Let a and b be integers. If for all positive integers n, the number $2^n a + b$ is always a perfect square, prove that $a = 0$.

8. Let a sequence $\{a_n\}$ $(n \geq 1)$ be defined as: $a_1 = 2$, $a_2 = 5$, and $a_{n+1} = \left(2 - n^2\right) a_n + \left(2 + n^2\right) a_{n-1}$ for $n \geq 2$. Prove that for any positive integers r and s, the number $a_r \cdot a_s$ is not a term in $\{a_n\}$.

9. Let $a_1, a_2, \cdots, a_n$ be pairwise coprime positive integers and they are all composite numbers. Prove:

$$\frac{1}{a_1} + \cdots + \frac{1}{a_n} < 1.$$

10. Let n and k be positive integers with $1 < k < n - 1$. Prove that $\binom{n}{k}$ cannot be a power of a prime number.

Chapter 11

Operations and Exact Division of Polynomials

Polynomial is a fundamental research object in algebra. This chapter mainly introduces the basic knowledge of univariate polynomials whose coefficients belong to the rational number set, real number set, and complex number set, including divisibility, the division with remainder, greatest common factors and the unique factorization theorem. The readers will see that these concepts, results and methods are very similar to those for integers.

Basic Concepts of Univariate Polynomials

In this chapter, for convenience, we use D to denote $\mathbb{Z}$ (the set of integers), $\mathbb{Q}$ (the set of rational numbers), $\mathbb{R}$ (the set of real numbers), or $\mathbb{C}$ (the set of complex numbers).

Let n be a non-negative integer. The expression

$$a_n x^n + a_{n-1} x^{n-1} + \cdots + a_0, \tag{1}$$

where $a_i \in D\,(i = 0, 1, \ldots, n)$, is called a **univariate polynomial whose coefficients belong to** D, and also called a univariate polynomial over D. (For example, we say a polynomial with "coefficients being rational" or "rational coefficients".) The set of all polynomials in the form of (1) is denoted as $D[x]$. We often use $f(x), g(x), \ldots$ or $f, g, \ldots$ to represent specific polynomials.

In general we call (1) a polynomial function, that is, we treat x as a variable in D (refer to Chapter 12), but in this chapter, we do not concern the value of the polynomial, but treat x in (1) as an indeterminate — not an undetermined element in D, but only a notation that does not belong to D.

In the polynomial (1), $a_i x^i$ is called the ith **term**, and a_i is called the **coefficient** of the ith term. If two polynomials $f(x)$ and $g(x)$ are **equal** (or **identically equal**), then it means that the coefficients of the same term in $f(x)$ and $g(x)$ are equal, denoted as $f(x) = g(x)$ (or $f(x) \equiv g(x)$). The polynomial whose coefficients are all 0 is called the **zero polynomial**, which is called the zero for short, denoted as 0.

In (1), if $a_n \neq 0$, then $a_n x^n$ is the **leading term** of the polynomial (1), a_n is the **leading coefficient**, and n is called the degree of the polynomial (1), denoted as $n = \deg f(x)$, or $n = \deg f$ for short. Note that zero-degree polynomials are non-zero elements in D, and the zero polynomial is the only polynomial whose degree is undefined.

The polynomials in $D[x]$ can be added, subtracted, and multiplied. Suppose

$$f(x) = a_n x^n + a_{n-1} x^{n-1} + \cdots + a_0, \tag{2}$$

$$g(x) = b_m x^m + b_{m-1} x^{m-1} + \cdots + b_0 \tag{3}$$

are two polynomials in $D[x]$. When we express the sum of $f(x)$ and $g(x)$, if $n \geq m$, then for convenience we assume that $b_n = \cdots = b_{m+1} = 0$. Define

$$f(x) + g(x) = (a_n + b_n) x^n + (a_{n-1} + b_{n-1})x^{n-1} + \cdots + (a_0 + b_0).$$

It's easy to verify that the addition of polynomials satisfies the associative law and commutative law (because they are all attributed to the addition of the coefficients). In addition, for any $f(x)$, the polynomial obtained by changing the sign of each coefficient is denoted as $-f(x)$, and we define $f(x) - g(x) = f(x) + (-g(x))$. Then we clearly see that one can perform the subtraction operation in $D[x]$.

Multiplication can also be implemented in $D[x]$. Let $f(x)$ and $g(x)$ be (2) and (3), respectively, and define the product of $f(x)$ and $g(x)$ as

$$f(x) \cdot g(x) = c_{n+m} x^{n+m} + c_{n+m-1} x^{n+m-1} + \cdots + c_1 x + c_0,$$

where the coefficients $c_i (i = 0, 1, \ldots, n + m)$ are

$$c_i = \sum_{r+s=i} a_r b_s = a_i b_0 + a_{i-1} b_1 + \cdots + a_1 b_{i-1} + a_0 b_i. \tag{4}$$

It is also easy to verify that the multiplication in $D[x]$ satisfies the associative law, the commutative law, and the distribution law (because these are also attributed to the addition and multiplication of elements in D).

The simple fact that every nonzero polynomial corresponds to a non-negative integer, i.e., its degree, is very important. (Refer to the proofs of (5) and (6) below).

Here are some basic results about the degree of polynomials.

(1) Let $f(x)$ and $g(x)$ be nonzero polynomials in $D[x]$.

   (i) If $f(x) + g(x)$ is not zero, then

$$\deg(f+g) \leq \max(\deg f, \deg g).$$

   (ii) $\deg fg = \deg f + \deg g$, and in particular, $f(x)g(x) \neq 0$.

The property (i) is a trivial corollary of the definition of additions and degrees of polynomials.

To prove (ii), let $f(x)$ and $g(x)$ be in the forms of (2) and (3), in which $a_n \neq 0$ and $b_m \neq 0$. We know from the multiplication rules of polynomials (see (4)) that

$$f(x)g(x) = a_n b_m x^{n+m} + \text{lower terms}.$$

Because $a_n b_m \neq 0$, clearly $\deg fg = n + m = \deg f + \deg g$.

Division with Remainder and Greatest Common Factors

There is a big difference between the three sets $\mathbb{Q}$, $\mathbb{R}$, and $\mathbb{C}$, and the integer set $\mathbb{Z}$ in operations: Division can be performed in the first three number sets, but in general, it cannot be performed in $\mathbb{Z}$ (refer to Chapter 6). Therefore, there are many differences between polynomials with coefficients in $\mathbb{Q}$, $\mathbb{R}$, and $\mathbb{C}$ and polynomials with integer coefficients in property and treatment. In this chapter and the next chapter, we mainly consider polynomials with coefficients in $\mathbb{Q}$, $\mathbb{R}$, and $\mathbb{C}$. For convenience, let F denote $\mathbb{Q}$, $\mathbb{R}$, or $\mathbb{C}$. When some concepts and results are also applicable to polynomials with integer coefficients, we will propose them in particular.

The concept of exact division in $F[x]$ is similar to that for integers. Suppose $f(x)$ and $g(x) \in F[x]$, and $g(x)$ is not zero. If there exists $h(x) \in F[x]$ such that $f(x) = g(x)h(x)$, then $g(x)$ (in $F[x]$) **exactly divides** $f(x)$, denoted as $g(x) \mid f(x)$. We call $g(x)$ a **factor** of $f(x)$ and $f(x)$ a **multiple** of $g(x)$. If such $h(x)$ does not exist, then $g(x)$ **doesn't exactly divide** $f(x)$, denoted as $g(x) \nmid f(x)$.

From the definition of exact division, it is not difficult to prove the following properties (let's suppose $f(x)$, $g(x)$, $h(x) \in F[x]$):

(2) If $g(x) \mid h(x)$ and $h(x) \mid f(x)$, then $g(x) \mid f(x)$.

(3) If $g(x) \mid f(x)$ and $g(x) \mid h(x)$, then for any $\alpha(x)$, $\beta(x) \in F[x]$,

$$g(x) \mid \alpha(x)f(x) + \beta(x)h(x).$$

(4) If $g(x) \mid f(x)$, then either $f(x) = 0$ or $\deg g \leq \deg f$. In addition, if $g(x) \mid f(x)$ and $f(x) \mid g(x)$, then $f(x) = cg(x)$, where c is a non-zero element of F.

The properties (2) and (3) are obvious. To prove (4), we suppose $f(x) = g(x)h(x)$. If $f(x) \neq 0$, then $h(x) \neq 0$. Therefore, from (ii) in (1), we know that $\deg f \geq \deg g$.

If there is also $f(x) \mid g(x)$, then $g(x) = f(x)h_1(x)$, so

$$f(x)(h(x)h_1(x) - 1) = 0.$$

From (ii) in (1), we know that $h(x)h_1(x) - 1 = 0$, so $h(x)$ and $h_1(x)$ are non-zero numbers in F.

Remark 1. The exact division of polynomials with integer coefficients can be defined the same as above, and the properties (2) and (3) still hold (just replace F with $\mathbb{Z}$). But the second part of property (4) is somewhat different:

(4′) Let $f(x), g(x) \in \mathbb{Z}[x]$ such that $f(x) \mid g(x)$ and $g(x) \mid f(x)$. Then $f(x) = \pm g(x)$.

The fundamental cause for this difference is that division can't be performed in $\mathbb{Z}$: If the product of two integers is 1, then the two numbers can only be ± 1.

The following (5) shows that the division with remainder can be performed in $F[x]$, which is similar to the division with remainder for integers.

(5) (**Division with remainder**) Let $f(x)$, $g(x) \in F[x]$ and $g(x) \neq 0$. Then there exist polynomials $q(x)$ and $r(x)$ in $F[x]$ such that

$$f(x) = g(x)q(x) + r(x), \tag{5}$$

where $r(x)$ is either zero or $\deg r < \deg g$; the **quotient** $q(x)$ and **remainder** $r(x)$ are uniquely determined by these conditions.

Proof. Note that if $f(x) = 0$ or $\deg f < \deg g$, then we can take $q(x) = 0$ and $r(x) = f(x)$. If $g(x)$ is a constant c, then we can take $q(x) = \frac{1}{c}f(x)$ and $r(x) = 0$. Hence, in the following we suppose $\deg f \geq \deg g > 0$.

Let $n = \deg f$ and $m = \deg g$, and let a_n and b_m be the coefficients of the leading terms of $f(x)$ and $g(x)$ respectively. Denote $q_1(x) = a_n b_m^{-1} x^{n-m}$ and suppose

$$f_1(x) = f(x) - q_1(x)g(x).$$

Then it is easy to know that either $f_1(x) = 0$ or $\deg f_1 < \deg f$. If $f_1 \neq 0$ and $\deg f_1 \geq \deg g$, repeat the above process for $f_1(x)$. In this way, we can get a sequence of polynomials $f_1(x), \ldots, f_i(x), \ldots$, and $q_1(x), \ldots, q_i(x), \ldots$, such that

$$f_{i+1}(x) = f_i(x) - q_{i+1}(x)g(x),$$

and $\deg f > \deg f_1 > \cdots \geq \deg g(>0)$. Because $\deg f_i$ are all positive integers, this process must stop after a finite number of steps, that is, there exists k that makes $f_k(x) = 0$ or $\deg f_k < \deg g$. Therefore

$$q(x) = q_1(x) + \cdots + q_k(x), \quad r(x) = f_k(x)$$

gives the said representation.

To prove the uniqueness, suppose $f(x)$ has another satisfactory representation: $f(x) = g(x)q_1(x) + r_1(x)$. Combining it with (5), we get

$$(q(x) - q_1(x))g(x) = r_1(x) - r(x).$$

The right side of the above formula is either zero or the degree is $< \deg g$. Therefore, both sides must be zero, so $q(x) = q_1(x)$, and $r(x) = r_1(x)$ (refer to (1)).

Remark 2. Since division cannot be performed in $\mathbb{Z}$, the division with remainder cannot be implemented in $\mathbb{Z}[x]$ as in (5). For example, for polynomials $f(x) = x^2$ and $g(x) = 2x$ in $\mathbb{Z}[x]$, it is easy to know that there doesn't exist $q(x), r(x) \in \mathbb{Z}[x]$ such that $x^2 = 2x \cdot q(x) + r(x)$, where $r(x) = 0$ or $\deg r < 1$ (i.e., $r(x)$ is an integer).

However, if $f(x), g(x) \in \mathbb{Z}[x]$, and the leading coefficient of $g(x)$ is ± 1, then the same argument as (5) shows that there exist unique $q(x), r(x) \in \mathbb{Z}[x]$ that have the properties mentioned in (5).

Remark 3. Uniqueness in the division with remainder is very important. It can be seen that the formula of the division with remainder does not change with the expansion of the set to which the polynomial coefficients belong.

For example, let $f(x)$, $g(x) \in \mathbb{Q}[x]$ and $g(x) \neq 0$. The quotient $q(x)$ and the remainder $r(x)$ can be obtained in $\mathbb{Q}[x]$, so that

$$f(x) = g(x)q(x) + r(x), \ r(x) = 0 \quad \text{or} \quad \deg r < \deg g.$$

Because $\mathbb{Q} \subset \mathbb{R} \subset \mathbb{C}$, the above formula can naturally be treated as an equality in $\mathbb{R}[x]$ or $\mathbb{C}[x]$. Hence from the uniqueness in (5), we see that in $\mathbb{R}[x]$ or $\mathbb{C}[x]$, the quotient and remainder generated from dividing $f(x)$ by $g(x)$ are the same as $q(x)$ and $r(x)$ above, respectively.

In particular, to say that $g(x)$ exactly divides $f(x)$ in $\mathbb{Q}[x]$ is the same thing as saying that $g(x)$ exactly divides $f(x)$ in $\mathbb{R}[x]$ or $\mathbb{C}[x]$.

Remark 4. There are some differences in the case of polynomials with integer coefficients, because the division with remainder can't be performed in $\mathbb{Z}[x]$ in general. However, if $f(x)$, $g(x) \in \mathbb{Z}[x]$ and the coefficient of the first term of $g(x)$ is ± 1, then it can be seen from Remark 2 and Remark 3 that in $\mathbb{Z}[x]$, the equality (of the division with remainder) generated from dividing $f(x)$ by $g(x)$ is the same as that in $\mathbb{Q}[x]$, $\mathbb{R}[x]$, or $\mathbb{C}[x]$.

Proof. In particular, if we can prove that $g(x)$ exactly divides $f(x)$ in $\mathbb{C}[x]$, then there is also $g(x) \mid f(x)$ in $\mathbb{Z}[x]$. (However, if the leading coefficient of $g(x)$ is not ± 1, then it doesn't have to be true. For example, if $f(x) = x + 1$ and $g(x) = 2x + 2$, then $g(x) \mid f(x)$ in $\mathbb{Q}[x]$, but $g(x) \nmid f(x)$ in $\mathbb{Z}[x]$.)

The following results are the basis of defining the greatest common factor of polynomials.

(6) Let $f(x)$ and $g(x)$ be two polynomials in $F[x]$ that are not both zero. Denote

$$S = \{f(x)\alpha(x) + g(x)\beta(x) \mid \alpha(x), \ \beta(x) \in F[x]\}.$$

Then there exists a unique polynomial $d(x) \in S$ with leading coefficient 1 such that $d(x)$ exactly divides all polynomials in S.

Obviously, there are nonzero polynomials in S. Let $d(x)$ be (one of) the polynomial with the lowest degree, and we can assume that the leading coefficient of $d(x)$ is 1. Next we prove that any polynomial in S is a multiple of $d(x)$.

In fact, for any $f(x) \in S$, from (5), there exist $q(x), r(x) \in F[x]$ such that $f(x) = d(x)q(x) + r(x)$, where $r(x) = 0$ or $\deg r < \deg d$. If $r(x) = 0$, then there is no need to prove it. Otherwise, because $d(x) \in S$ and $d(x)q(x) \in S$, it is easy to know that

$$r(x) = f(x) - d(x)q(x) \in S,$$

but $\deg r < \deg d$, which contradicts the selection of $d(x)$.

To prove the uniqueness, assume that $d_1(x) \in S$ also has the said properties. Then $d(x) \mid d_1(x)$ and $d_1(x) \mid d(x)$.

Since the coefficients of the first term of $d(x)$ and $d_1(x)$ are all 1, we see that $d(x) = d_1(x)$ (refer to (4). The selection of the first term coefficient $d(x)$ to be 1 is only to ensure the uniqueness). $\qquad\square$

From (6), $d(x) \mid f(x)$ (take $\alpha(x) = 1$ and $\beta(x) = 0$) in S), and similarly, $d(x) \mid g(x)$. Hence $d(x)$ is a common factor of $f(x)$ and $g(x)$. On the other hand, because $d(x) \in S$, there exist $\alpha(x), \beta(x) \in F[x]$ such that

$$f(x)\alpha(x) + g(x)\beta(x) = d(x).$$

Therefore, if $d_1(x)$ is any common factor of $f(x)$ and $g(x)$, then $d_1(x) \mid d(x)$ from the above formula (refer to property (3)), thus $\deg d_1 \le \deg d$. Therefore $d(x)$ is the unique common factor of $f(x)$ and $g(x)$ with the highest degree and the leading coefficient 1. We define it as the **greatest common factor** of $f(x)$ and $g(x)$, denoted as $(f(x), g(x))$.

When $(f(x), g(x)) = 1$, then $f(x)$ and $g(x)$ are **coprime**. Therefore, $f(x)$ and $g(x)$ are coprime means that their common factors are only non-zero elements of F.

From (6) we get

(7) If $f(x), g(x) \in F[x]$ and not both of them are zero, then there exist $\alpha(x), \beta(x) \in F[x]$ such that

$$f(x)\alpha(x) + g(x)\beta(x) = (f(x), g(x)). \tag{6}$$

In particular, if $f(x)$ and $g(x)$ are coprime, then there exist $\alpha(x), \beta(x) \in F[x]$ such that

$$f(x)\alpha(x) + g(x)\beta(x) = 1. \tag{7}$$

Conversely, if there exist $\alpha(x), \beta(x) \in F[x]$ such that (7) holds, then $f(x)$ and $g(x)$ are coprime.

The formula (7) is also known as **Bézout's identity**, which is very useful.

Similar to the proofs in (12) and (13) in Chapter 6, the following basic properties of the greatest common factor can be easily established by (7). (We suppose $f(x)$, $g(x)$, $h(x) \in F[x]$)

   (8) Any common divisor of $f(x)$ and $g(x)$ must exactly divide their greatest common factor.

   (9) If $(f(x), h(x)) = 1$ and $(g(x), h(x)) = 1$, then $(f(x)g(x), h(x)) = 1$.

  (10) Suppose $g(x) \mid f(x)h(x)$ and $(g(x), h(x)) = 1$. Then $g(x) \mid f(x)$.

Remark 5. In the case of integers, the existence and uniqueness of the greatest common factor of some integers (not all zero) are self-evident, but not in the case of polynomials.

   The property (6) is our foundation, upon which we established the definition of the greatest common factor and its basic properties (please note that (6) is dependent on division with remainder (5)).

Remark 6. The greatest common factor does not change with the expansion of the set to which the coefficients belong.

   For example, (not both zero) polynomials $f(x)$ and $g(x)$ with rational coefficients have the greatest common factor $d(x) \in \mathbb{Q}[x]$. Because $\mathbb{Q} \subset \mathbb{R} \subset \mathbb{C}$, we see that $f(x)$ and $g(x)$ can also be treated as polynomials in $\mathbb{R}[x]$ or $\mathbb{C}[x]$, and let $d_1(x)$ and $d_2(x)$ be their greatest common factor in $\mathbb{R}[x]$ and $\mathbb{C}[x]$, respectively. It is not difficult for us to prove that $d(x) = d_1(x) = d_2(x)$. In fact, there exist $\alpha(x)$, $\beta(x) \in \mathbb{Q}[x]$ such that

$$f(x)\alpha(x) + g(x)\beta(x) = d(x).$$

This equality also holds in $\mathbb{R}[x]$, so $d_1(x) \mid d(x)$. But on the other hand, $d(x)$ is a common factor of $f(x)$ and $g(x)$ in $\mathbb{R}[x]$, so $d(x) \mid d_1(x)$. Since the leading coefficients of $d(x)$ and $d_1(x)$ are both 1, we get that $d(x) = d_1(x)$. Similarly, $d(x) = d_2(x)$.

Remark 7. For two polynomials in $F[x]$, Euclidean algorithm similar to that for integers can also be established to actually find the greatest common factor, but this book will not discuss it.

Remark 8. For polynomials with integer coefficients, we can't define the greatest common factor as we did above, because (6) doesn't hold for polynomials with integer coefficients (see Exercise 4 in this chapter).

   In Chapter 13, we will briefly discuss the greatest common factor of polynomials with integer coefficients. We will only mention that they are different from the case in $F[x]$. For example, we can easily know that polynomials $x + 2$ and x are coprime in $\mathbb{Z}[x]$, but obviously there is no $\alpha(x)$, $\beta(x) \in \mathbb{Z}[x]$, which makes $(x + 2)\alpha(x) + x\beta(x) = 1$, that is, Bézout's identity in $\mathbb{Z}[x]$ does not hold in general.

Irreducible Polynomials and Unique Factorization Theorem

Let $p(x) \in F[x]$ and $\deg p \geq 1$. If $p(x)$ cannot be factored into the product of two polynomials of positive degrees in F, then $p(x)$ is said to be an **irreducible polynomial in $F[x]$**, or **irreducible over F**; otherwise, $p(x)$ is **reducible over F**.

Note that unlike the greatest common factor, irreducibility of polynomials may change with the expansion of the set to which the coefficients belong. For example, $x^2 + 1$ is an irreducible polynomial in $\mathbb{R}[x]$, but it is reducible in $\mathbb{C}[x]$.

Similar to the case of integers, it is very easy to prove (refer to Chapter 7) the following facts:

(11) There must be an irreducible polynomial factor for any non-constant polynomial in $F[x]$.

(12) Let $p(x) \in F[x]$ be an irreducible polynomial. Then for any $f(x) \in F[x]$, there is either $p(x) \mid f(x)$ or $(p(x), f(x)) = 1$.

For any $a \in F$, clearly $x - a$ is an irreducible polynomial in $F[x]$, and since F contains an infinite number of elements, we have

(13) There is an infinite number of irreducible polynomials with leading coefficient 1 in $F[x]$.

In addition, from (9) and (12) we derive

(14) Let $p(x)$ be an irreducible polynomial in $F[x]$. If the product of two polynomials $f(x)$ and $g(x)$ in $F[x]$ is divisible by $p(x)$, then at least one of $f(x)$ and $g(x)$ is divisible by $p(x)$.

With (14) as a tool, it is not difficult to prove the following conclusion.

(15) (**Unique factorization theorem**) A polynomial $f(x)$ of any positive degree in $F[x]$ can be decomposed into the product of a constant in F and a finite number of irreducible polynomials with leading coefficient 1 in $F[x]$. Moreover, if the order of these irreducible polynomials in the product is ignored, then they are uniquely determined by $f(x)$.

Theorem (15) is similar to the unique factorization theorem for positive integers in (5) of Chapter 7, and the proof can also be carried out similarly. The details are omitted here.

Polynomials Modulo a Prime

The polynomials discussed in this section are all polynomials with integer coefficients, and let p be a fixed prime number.

If the coefficients of the corresponding terms of two polynomials $f(x)$ and $g(x)$ are all congruent modulo p, then $f(x)$ and $g(x)$ are **congruent modulo p**, or are identical modulo p, denoted as

$$f(x) \equiv g(x)(\bmod p).$$

If the coefficients of $f(x)$ are not all divisible by p, then the highest power, in which the coefficients are not divisible by p, is called the **degree of $f(x)$ modulo p**. For example, the degree of the polynomial $30x^4 + 6x^3 + 9$ modulo 5 is 3, the degree modulo 2 is 0, and the degree modulo 3 is not defined.

The integer coefficient polynomial congruence modulo p has similar properties to the congruence of integers in Chapter 8:

(16) $f(x) \equiv f(x)(\bmod p)$.

(17) If $f(x) \equiv g(x)(\bmod p)$, then $g(x) \equiv f(x)(\bmod p)$.

(18) If $f(x) \equiv g(x)(\bmod p)$ and $g(x) \equiv h(x)(\bmod p)$, then $f(x) \equiv h(x)$ $(\bmod p)$.

(19) If $f(x) \equiv g(x)(\bmod p)$ and $f_1(x) \equiv g_1(x)(\bmod p)$, then

$$f(x) \pm f_1(x) \equiv g(x) \pm g_1(x)(\bmod p),$$

and

$$f(x)f_1(x) \equiv g(x)g_1(x)(\bmod p).$$

(20) $f^p(x) \equiv f(x^p)(\bmod p)$.

Properties (16) $\sim$ (19) can be easily derived from the definition and the congruence properties of integers. To prove (20), suppose

$$f(x) = a_n x^n + a_{n-1} x^{n-1} + \cdots + a_1 x + a_0 \in \mathbb{Z}[x].$$

From the binomial theorem, Example 7 of Chapter 7, and Fermat's little theorem (Remark 3 of Chapter 8),

$$f^p(x) = [a_n x^n + (a_{n-1} x^{n-1} + \cdots + a_0)]^p$$
$$\equiv a_n^p x^{np} + \left(a_{n-1} x^{n-1} + \cdots + a_0\right)^p$$
$$\equiv a_n x^{np} + \left(a_{n-1} x^{n-1} + \cdots + a_0\right)^p$$
$$\equiv \cdots$$
$$\equiv a_n x^{np} + a_{n-1} x^{(n-1)p} + \cdots + a_1 x^p + a_0$$
$$= f(x^p) \pmod{p}.$$

The integer coefficient polynomial congruence modulo p is very important in number theory, but it is not discussed in this book.

By the way, for a non-prime positive integer m, we can define the congruence relation of integer coefficient polynomials modulo m similarly, and the properties (16) $\sim$ (19) are still valid, but the property (20) (replacing p with m) is not correct. In addition, polynomials modulo m have other bad properties (see Exercise 5 in this chapter), so they are rarely used.

Illustrative Examples

Example 1. Express the polynomial $x^4 + 3x^2 - 2x + 3$ as the difference of squares of two integer coefficient polynomials of unequal degrees.

Solution   Suppose $x^4 + 3x^2 - 2x + 3 = f^2(x) - g^2(x)$, where $\deg f \neq \deg g$. Because the leading coefficient of the given polynomial is positive, $\deg f = 2$ and $\deg g \leq 1$. Suppose $f(x) = x^2 + ax + b$ and $g(x) = cx + d$. Then $x^4 + 3x^2 - 2x + 3 = x^4 + 2ax^3 + (a^2 + 2b - c^2)x^2 + (2ab - 2cd)x + b^2 - d^2$.

From the definition of equality of polynomials,

$$2a = 0,\ a^2 + 2b - c^2 = 3,\ 2ab - 2cd = -2,\ b^2 - d^2 = 3.$$

Hence $a = 0$, $b = 2$, $c = 1$ and $d = 1$, i.e., $f(x) = x^2 + 2$ and $g(x) = x + 1$ are solutions that meet the requirement.

Remark 9. The method of Example 1 is called the undetermined coefficient method, which is based on the definition of equality of polynomials. In this method, we make equations of the coefficients of the like terms, and then determine the polynomial or derive a contradiction.

Example 2. Find all real coefficient polynomials $f(x)$ satisfying $f(x^2) = f^2(x)$.

Solution   Let such a polynomial be

$$f(x) = a_n x^n + a_{n-1} x^{n-1} + \cdots + a_1 x + a_0,$$

where $a_n \neq 0$. We prove that $a_{n-1} = \cdots = a_1 = a_0 = 0$. Suppose a_i ($i = 0, 1, \ldots, n-1$) are not all zero, and let $k < n$ be the largest subscript such that $a_k \neq 0$. From $f(x^2) = f^2(x)$, we obtain that

$$a_n x^{2n} + a_k x^{2k} + \cdots + a_1 x^2 + a_0 = \left(a_n x^n + a_k x^k + \cdots + a_1 x + a_0\right)^2.$$

By comparing the coefficients of x^{n+k}, we have $0 = 2a_n a_k$, which contradicts $a_n \neq 0$ and $a_k \neq 0$. Therefore, the above assertion is correct, that is, $f(x) = a_n x^n$. In addition, from $f(x^2) = f^2(x)$, we know that $a_n = 1$, so $f(x) = x^n$.

Example 3. Let $n \geq 0$ be an integer. Prove that the polynomial $(x+1)^{2n+1} + x^{n+2}$ is divisible by $x^2 + x + 1$.

Proof.   Prove by induction on n. When $n = 0$, the conclusion is obviously true. Assume that the conclusion holds for $n-1$, that is, $(x+1)^{2n-1} + x^{n+1}$ is divisible by $x^2 + x + 1$. Then (using the inductive hypothesis)

$$(x+1)^{2n+1} + x^{n+2}$$
$$= (x+1)^2 (x+1)^{2n-1} + x \cdot x^{n+1}$$
$$= \left(x^2 + x + 1\right)(x+1)^{2n-1} + x((x+1)^{2n-1} + x^{n+1})$$

is divisible by $x^2 + x + 1$, so the conclusion is also true for n. $\square$

Example 4. Let n be a positive integer. Prove:

$$(1+x)\left(1+x^2\right)\left(1+x^{2^2}\right)\cdots(1+x^{2^n}) = 1 + x + x^2 + x^3 + \cdots + x^{2^{n+1}-1}.$$

Proof 1 This is a direct method: Let $f(x)$ be the left side. Then

$$f(x) = \frac{1}{1-x}(1-x)f(x) = \frac{1}{1-x}(1-x^2)(1+x^2)\cdots(1+x^{2^n})$$

$$= \cdots = \frac{1-x^{2^{n+1}}}{1-x} = \frac{(1-x)}{1-x}(1+x+x^2+\cdots+x^{2^{n+1}-1})$$

$$= 1 + x + x^2 + x^3 + \cdots + x^{2^{n+1}-1}.$$

Here we used the decomposition (5) in Chapter 6.

Proof 2 If the left side is expanded to $a_0 + a_1 x + \cdots + a_N x^N$, then obviously $N = 1 + 2 + \cdots + 2^n = 2^{n+1} - 1$. Since the coefficients of non-zero terms in each factor $1 + x^{2^i}$ on the left side of the equality are all 1, it can be seen from the definition of the polynomial multiplication that a_i is exactly the number of subsets of the set $S = \{1, 2, 2^2, \ldots, 2^n\}$ whose sum of elements is i $(0 \le i \le 2^{n+1} - 1)$.

Obviously, $a_0 = 1$. For $1 \le i \le 2^{n+1} - 1$, clearly we know that the highest power of 2 in the binary representation of i is, not greater than n, so the set S must have a subset, whose sum of elements is i. From the uniqueness of the binary representation of positive integers (see Exercise 14 in Chapter 6), there is exactly one such subset, that is, $a_i = 1$ $(1 \le i \le 2^{n+1} - 1)$, which proves the conclusion.

Example 5. Let $f(x) = x^4 + x^3 - 3x^2 + x + 2$. Prove: For any positive integer n, the polynomial $f^n(x)$ has at least one negative coefficient.

Proof. We calculate the coefficient of x^3 in $f^n(x)$. Consider

$$f^n(x) = \{(x^4 + x^3 - 3x^2 + x + 2)(x^4 + x^3 - 3x^2 + x + 2)$$
$$\cdots (x^4 + x^3 - 3x^2 + x + 2)\}$$

From the definition of the polynomial multiplication, the x^3 terms from the expansion can be generated in two ways: one is that the x^3 term in one factor on the right side of the above formula multiplies the constant terms of the other factors, which results in n numbers of x^3 in total; the other is that the x^2 term in one factor on the right multiplies the x term an other factor and the constant terms in the remaining factors, which results in $n(n-1)$ numbers of x^3. Clearly we know that the coefficient of x^3 in $f^n(x)$ is $n \cdot 2^{n-1} - 3n(n-1) \cdot 2^{n-2} = n \cdot 2^{n-2}(-3n + 5)$, which is negative when $n \ge 2$, while the conclusion is obviously true when $n = 1$. $\square$

Note that the reason to consider the coefficient of x^3 is that the constant term and the coefficient of the linear term in $f^n(x)$ are obviously positive, while the coefficient of x^2 is $n \cdot 2^{n-2}(n - 7)$, which is positive when $n > 7$.

Example 6. Given a polynomial $f(x) = ax^2 + bx + c$ with $a \ne 0$, prove that for each positive integer n, there exists at most one polynomial $g(x)$ of degree n, which makes

$$f(g(x)) = g(f(x)).$$

Proof. If there is a polynomial $g(x) = b_n x^n + \cdots + b_1 x + b_0$, of degree n that satisfies the condition, then compare the coefficients of x^{2n} terms of $f(g(x))$ and $g(f(x))$ to obtain $b_n a^n = ab_n^2$, that is, $b_n = a^{n-1}$.

Therefore, if for a certain n, there are two distinct polynomials $g_1(x)$ and $g_2(x)$ of degree n satisfying the conditions of the problem, then since the leading coefficients of $g_1(x)$ and $g_2(x)$ are both a^{n-1}, the degree k of the polynomial $h(x) = g_1(x) - g_2(x)$ is less than n, and

$$
\begin{aligned}
h(f(x)) &= g_1(f(x)) - g_2(f(x)) \\
&= f(g_1(x)) - f(g_2(x)) \\
&= a(g_1^2(x) - g_2^2(x)) + b(g_1(x) - g_2(x)) \\
&= (g_1(x) - g_2(x))(a(g_1(x) + g_2(x)) + b) \\
&= h(x) \cdot T(x),
\end{aligned}
$$

where $T(x) = a(g_1(x) + g_2(x)) + b$.

Therefore, comparing the degrees of polynomials on both sides of the above equality, we see that $2k = k + n$, that is, $k = n$, contradicting the previous result $k < n$. (Note that since the leading coefficients of $g_1(x)$ and $g_2(x)$ are the same, the degree of $T(x)$ is also the degree of $g_1(x)$ and $g_2(x)$, which is n.) $\qquad\square$

Exercises

Group A

1. Let $f(x) = x^7 - 1$ and $g(x) = x^3 + x + 1$. Find the quotient and remainder of $f(x)$ on division by $g(x)$.
2. In $\mathbb{R}[x]$, express $x^4 + x^3 + x + 1$ as the difference of squares of two polynomials of different degrees.
3. Let $f(x) \in \mathbb{Z}[x]$, p be prime, and k be a positive integer. Prove that

$$
f^{p^k}(x) \equiv f(x^{p^k}) \pmod{p}.
$$

Group B

4. Let $S = \{(x+2)\alpha(x) + x\beta(x) \mid \alpha(x), \beta(x) \in \mathbb{Z}[x]\}$. Prove: There doesn't exist $d(x) \in S$ such that all the polynomials in S are divisible by $d(x)$.
5. (i) Let $f(x), g(x) \in \mathbb{Z}[x]$ and p be prime. Prove: If $f(x)g(x) \equiv 0 \pmod{p}$, then $f(x) \equiv 0 \pmod{p}$ or $g(x) \equiv 0 \pmod{p}$.

(ii) Prove: If m is a composite number, then there exist $f(x)$, $g(x) \in \mathbb{Z}[x]$ such that $f(x) \not\equiv 0 \pmod m$ and $g(x) \not\equiv 0 \pmod m$, but $f(x)g(x) \equiv 0 \pmod m$.

6. Determine which of the two polynomials $\left(1 + x^2 - x^3\right)^{100}$ and $\left(1 - x^2 + x^3\right)^{100}$ has a greater coefficient of x^{20}.

7. Find all positive integers k such that the polynomial $x^{2k+1} + x + 1$ is divisible by $x^k + x + 1$.

Chapter 12

Zeros of Polynomials

This chapter introduces the zeros of polynomials and their related contents, including the identity theorem of polynomials, Lagrange's theorem of the congruence equation modulo a prime, the rational roots of polynomials with integer coefficients, and the Fundamental Theorem of Algebra.

Zeros of Polynomials and the Identity Theorem

In this section, we use D to denote $\mathbb{Z}$, $\mathbb{Q}$, $\mathbb{R}$, or $\mathbb{C}$.

Let $f(x) = a_n x^n + \cdots + a_1 x + a_0 \in D[x]$, For any $\alpha \in D$, suppose

$$f(\alpha) = a_n \alpha^n + \cdots + a_1 \alpha + a_0. \tag{1}$$

Then $f(\alpha) \in D$, that is, that is each $\alpha \in D$ corresponds to a value in D via $f(x)$, so we also call $f(x)$ a polynomial function in D, and the indeterminate α can be called a variable in D. The expression (1) is called the value of $f(x)$ when $x = \alpha$, also called the value of $f(x)$ at α.

Clearly we see that the relationship represented by the addition and multiplication of polynomials remains true after the indeterminate is substituted by any number in D, that is, if

$$f(x) + g(x) = h(x), \ f(x)g(x) = h_1(x),$$

then

$$f(\alpha) + g(\alpha) = h(\alpha), \ f(\alpha)g(\alpha) = h_1(\alpha).$$

If a number α in D makes $f(\alpha) = 0$, then α is a **zero** of $f(x)$, or a root of $f(x)$, and this situation is particularly important. The following theorem is called the factor theorem, and its effect is to infer the properties of polynomials from zeros.

(1) (**Factor theorem**) Let $f(x) \in D[x]$. Then $x = \alpha$ is a zero of $f(x)$ if and only if $f(x)$ is divisible by $x - \alpha$.

In fact, divide $f(x)$ by $x - \alpha$ to get (refer to (5) and Remark 2 in Chapter 11)

$$f(x) = (x - \alpha)q(x) + r(x),$$

where $q(x)$, $r(x) \in D[x]$, and $r(x) = 0$ or $\deg r < 1$, thus $r(x)$ is a constant. Let $r(x) = r$, and substitute $x = \alpha$ into the above formula. Then we get $r = f(\alpha)$. Therefore, both two aspects of the theorem are proved.

From (1) and by mathematical induction, we can immediately obtain:

(2) Let $f(x) \in D[x]$ and $\alpha_1, \ldots, \alpha_k \in D$ are distinct zeros of $f(x)$. Then $f(x)$ is divisible by $(x - \alpha_1) \cdots (x - \alpha_k)$.

The following results are obtained from (2):

(3) An nth degree polynomial in $D[x]$ has at most n distinct zeros.
(4) If a polynomial $f(x) = a_n x^n + \cdots + a_1 x + a_0$ in $D[x]$ has at least $n + 1$ (distinct) zeros in D, then $f(x)$ is the zero polynomial (that is, all coefficients are 0).

To prove (3), let an nth degree polynomial $f(x)$ have $n+1$ distinct zeros $\alpha_1, \ldots, \alpha_{n+1}$. From (2), $f(x) = (x - \alpha_1) \cdots (x - \alpha_{n+1})q(x)$, and compare the degrees on both sides of the equality (Chapter 11, (1)). Then we can derive a contradiction.

Property (4) is a corollary of (3): If $f(x)$ is not the zero polynomial, then its degree is at most n, resulting in a contradiction from (3).

Now we can prove a very useful theorem:

(5) (**Identity theorem**) Let $f(x)$, $g(x) \in D[x]$. If there is an infinite number of $\alpha \in D$ to make $f(\alpha) = g(\alpha)$, then $f(x) = g(x)$ (that is, the coefficients of the same powers in f and g are equal).

Since, if we let $h(x) = f(x) - g(x)$, then $h(x)$ has an infinite number of zeros, and it is known from (4) that $h(x)$ is identically equal to zero.

Remark 1. Let $f(x)$ and $g(x)$ be two polynomials in D. From the identity theorem, if $f(\alpha) = g(\alpha)$ for all $\alpha \in D$, then $f(x) = g(x)$. The opposite conclusion is also obviously true. Therefore, two polynomials in D are equal as functions in D if and only if they are equal as polynomials.

At the end of this section, we briefly talk about multiple roots of a polynomial. Let $f(x) \in D[x]$, and $\alpha \in D$ is a zero of $f(x)$. If $f(x)$ is divisible by $(x - \alpha)^m$, but not by $(x - \alpha)^{m+1}$, that is

$$f(x) = (x - \alpha)^m q(x),$$

where $q(x) \in D[x]$, and $q(\alpha) \neq 0$, then we call the positive integer m the **multiplicity** of zero α. When $m = 1$, the number α is a **single root** (or single zero) of $f(x)$. If $m > 1$, then α is a **multiple root** of $f(x)$ with multiplicity m. It is easy to see that the number of zeros (multiple roots are still counted by their multiplicity) of polynomials of degree n in $D[x]$ can not exceed n.

Congruence Equation Modulo a Prime Number

The content of this section is similar to that of the previous section. In fact, from a higher point of view, the two can be expressed in a unified way.

Let $f(x)$ be a polynomial with integer coefficients and p be a prime number. If an integer a satisfies $f(a) \equiv 0 \pmod{p}$, then a is **a zero of $f(x)$ modulo p**, or a solution to the congruence equation

$$f(x) \equiv \pmod{p}. \tag{2}$$

(Please note that (2) is called a conditional congruence, or a **congruence equation**, which is the same as the notation of being identically zero modulo p in Chapter 11, but the meanings are different.)

Clearly we know that if $a \equiv b \pmod{p}$, then $f(a) \equiv f(b) \pmod{p}$. Therefore, all numbers in the congruence class $x \equiv a \pmod{p}$ are solutions of (2). We say that a solution of (2) actually refers to a congruence class modulo p. If two integers satisfy (2) and are not congruent modulo p, then they are called different solutions of (2) (i.e., two different congruence classes). Therefore, the number of distinct solutions of (2) does not exceed p.

The following (6), which is a similar result to (3), is called Lagrange's theorem.

(6) (**Lagrange's theorem**) If p is a prime number and $f(x)$ is a polynomial with integer coefficients such that its degree modulo p is n, then the congruence equation (2) has at most n distinct (i.e., are not congruent to each other modulo p) solutions.

When $n = 0$ and 1, the conclusion is obvious (refer to (12) in Chapter 8). Prove by induction on n. We assume that the theorem is true for $n - 1$.

Now we consider the polynomial $f(x)$ of degree n (modulo p). If there is no solution to the congruence equation, then there is no need to prove it. If $x = a$ is one of its solutions, divide the polynomial $f(x)$ by $x - a$, and we obtain (refer to the proof of (1))

$$f(x) = g(x)(x - a) + A, \quad A \text{ is an integer.}$$

If the congruence equation (2) has no solution except for $x \equiv a \pmod p$, then the proof is completed (because $n \geq 1$). Otherwise, let $x = b$ be another solution of (2), i.e.,

$$0 \equiv f(b) = g(b)(b - a) + A \pmod p.$$

But obviously

$$0 \equiv f(a) = g(a)(a - a) + A \pmod p,$$

so $A \equiv 0 \pmod p$. Thus $g(b)(b - a) \equiv 0 \pmod p$. Since p is prime and $p \nmid (b - a)$, we have $g(b) \equiv 0 \pmod p$. It shows that the rest of the solutions of (2) are the solutions of the congruence equation $g(x) \equiv 0 \pmod p$ except for the solution $x \equiv a \pmod p$. But the degree of $g(x)$ modulo p is obviously $n - 1$. According to the inductive hypothesis, $g(x) \equiv 0 \pmod p$, has at most $n - 1$ solutions that are not congruent to each other, so the congruence equation (2) has at most $(n - 1) + 1 = n$ solutions that are not congruent to each other.

Remark 2. The conclusion in (6) is nontrivial only when $n < p$.

In addition, we remind the readers that if the prime number in (6) is replaced by a composite number, then the proposition is no longer correct. For example, $x^2 \equiv 1 \pmod 8$ has four solutions $x \equiv \pm 1$ and $\pm 3 \pmod 8$; $x^2 \equiv 1 \pmod{15}$ has four solutions ± 1 and $\pm 4 \pmod{15}$. The readers should see the role of prime numbers in the above proof.

It is not difficult to deduce the following (7) from (6), which is similar to (4).

(7) Let $f(x) = a_n x^n + a_{n-1} x^{n-1} + \cdots + a_1 x + a_0 \in \mathbb{Z}[x]$, p be a prime number and $n < p$. If the congruence equation (2) has at least $n + 1$ solutions that are not congruent to each other, then $f(x)$ modulo p is the zero polynomial, that is, all the coefficients a_i $(i = 0, 1, \ldots, n)$ are divisible by p.

This is because if $f(x)$ modulo p is a nonzero polynomial, then its degree modulo p is not greater than n, so from (6), the congruence equation (2) has at most n distinct solutions, which contradicts the assumption.

Remark 3. Property (7) provides a way to prove that a prime p exactly divides an integer: Construct a polynomial in the form of (7) such that the integer under consideration is a coefficient of it, and the number of different solutions of the polynomial modulo p exceeds n, so (7) gives the desired result. (Refer to Example 9 below.)

Rational Roots of Polynomials with Integer Coefficients

(8) Let $f(x) = a_n x^n + \cdots + a_1 x + a_0 \in \mathbb{Z}[x]$, $n \geq 1$, $a_n a_0 \neq 0$, and $(a_n, \ldots, a_0) = 1$. If $\frac{b}{c}$ is a rational root of $f(x)$, where the integers b and c are coprime, then $c \mid a_n$ and $b \mid a_0$.

In particular, a rational root of a polynomial with integer coefficients, whose coefficient of the first term is ± 1, must be an integer.
To prove it, from $f\left(\frac{b}{c}\right) = 0$, we obtain

$$a_n b^n + \cdots + a_1 bc^{n-1} + a_0 c^n = 0.$$

Since $a_{n-1}b^{n-1}c, \ldots, a_1 bc^{n-1}, a_0 c^n$ are all divisible by c, we see that $c \mid a_n b^n$, but $(b, c) = 1$, so $(c, b^n) = 1$, therefore $c \mid a_n$. Similarly, $b \mid a_0$.

Remark 4. Since there is only a finite number of divisors of a nonzero integer, in principle, all rational roots of a polynomial with integer coefficients can be determined by (8). This is especially useful when the number of nonzero terms of a polynomial and the number of divisors of a_n and a_0 are not too large.

Remark 5. Property (8) provides a basic way of transition from rational numbers to integers: To prove that a complex number is actually an integer, if you can prove that it is a rational number first, then by proving that it is a zero of the polynomial with integer coefficients and the coefficient ± 1 of its first term, the result is derived from (8).

Fundamental Theorem of Algebra

The famous **Fundamental Theorem of Algebra** asserts that:

(9) Any nonzero polynomial in $\mathbb{C}[x]$ has at least one complex root.

The proof of this theorem will not be discussed in this book. We only deal with a few common corollaries:

(10) Irreducible polynomials in $\mathbb{C}[x]$ are all linear polynomials.
(11) Any nth degree polynomial $f(x)(n \geq 1)$ in $\mathbb{C}[x]$ can be uniquely decomposed into

$$f(x) = a_n (x - \alpha_1) \cdots (x - \alpha_n) \tag{3}$$

in $\mathbb{C}[x]$, where a_n is the leading coefficient of $f(x)$, so an nth degree polynomial with complex coefficients has exactly n complex roots (counted with their multiplicity).

The assertion (10) is a trivial corollary of (9), and (11) can be derived from (10) and the unique factorization theorem in $\mathbb{C}[x]$ (Chapter 11, (15)).

In addition, let the distinct roots of $f(x)$ be $\alpha_1, \ldots, \alpha_k$ and their multiplicities be $m_1, \ldots, m_k$, respectively. Then $f(x)$ can be uniquely decomposed into

$$f(x) = a_n (x - \alpha_1)^{m_1} \cdots (x - \alpha_k)^{m_k},$$

which is **the standard factorization of a polynomial with complex coefficients**.

Remark 6. Let $f(x) = a_n x^n + a_{n-1} x^{n-1} + \cdots + a_1 x + a_0$ and expand the right side of equation (3). Then we obtain that

$$\alpha_1 + \cdots + \alpha_n = -\frac{a_{n-1}}{a_n},$$

$$\alpha_1 \alpha_2 + \cdots + \alpha_{n-1} \alpha_n = \frac{a_{n-2}}{a_n},$$

$$\cdots$$

$$\alpha_1 \cdots \alpha_n = (-1)^n \frac{a_0}{a_n}.$$

This is the famous **Vieta's theorem**, which gives the relation between the roots and the coefficients of an algebraic equation.

From the Fundamental Theorem of Algebra, all irreducible polynomials in $\mathbb{R}[x]$ can also be determined.

(12) Let $f(x) \in \mathbb{R}[x]$ be an irreducible polynomial. Then $f(x)$ is a linear polynomial or $f(x) = ax^2 + bx + c$, where $b^2 - 4ac < 0$.

Such said polynomial is obviously irreducible in $\mathbb{R}[x]$. Conversely, if $f(x)$ is an irreducible polynomial in $\mathbb{R}[x]$, and $\deg f > 1$, then $f(x)$ has no real roots. It can be seen from (9) that $f(x)$ has a nonreal root α, and $\overline{\alpha}$ is also a root of $f(x)$ (Exercise 7 in this chapter), and $\alpha \neq \overline{\alpha}$. Therefore, $f(x)$ is divisible by $(x - \alpha)(x - \overline{\alpha})$ (using (2)) in $\mathbb{C}[x]$. But clearly we know that

$$g(x) = (x - \alpha)(x - \overline{\alpha}) = x^2 - (\alpha + \overline{\alpha})x + \alpha\overline{\alpha} \in \mathbb{R}[x],$$

therefore, $g(x)$ also exactly divides $f(x)$ (refer to Remark 3 in Chapter 11) in $\mathbb{R}[x]$, so $f(x)$ must be a quadratic polynomial since it is irreducible, and $f(x) = ag(x)(a \in \mathbb{R})$, so $f(x)$ has the form stated in the theorem.

Remark 7. When $b^2 - 4ac < 0$, the quadratic polynomial $ax^2 + bx + c$ can be reduced into $a((x - \alpha)^2 + \beta^2)$ (α and β are real numbers) by completing the square. Therefore, from the unique factorization theorem in $\mathbb{R}[x]$ and (12), a real polynomial $f(x)$ of any positive degree can be uniquely expressed as

$$f(x) = a(x - a_1) \cdots (x - a_k)((x - b_1)^2 + c_1^2) \cdots ((x - b_l)^2 + c_l^2), \quad (4)$$

where a, a_i, b_j, and c_j are all real numbers and $k + 2l = \deg f$.

We can also collect the same factors in (4) and write the standard factorization (in $\mathbb{R}[x]$) of $f(x)$, but (4) is usually more applicable.

Complex Roots and Divisibility of Polynomials with Rational (Integer) Coefficients

By using the complex roots of polynomials with rational (integer) coefficients, we can infer their divisibility.

(13) Let $f(x)$, $g(x) \in \mathbb{Q}[x]$. If any root α of $g(x)$ is also a root of $f(x)$, and the multiplicity of α in $g(x)$ does not exceed its multiplicity in $f(x)$, then $g(x) \mid f(x)$ in $\mathbb{Q}[X]$. The converse is also true.

In addition, if the above $f(x)$ and $g(x)$ are both polynomials with integer coefficients, and the coefficient of the first term of $g(x)$ is ± 1, then $g(x) \mid f(x)$ in $\mathbb{Z}[X]$. In fact, from the known conditions and the standard factorization of $f(x)$ and $g(x)$ (see the previous section), we see that $g(x) \mid f(x)$ in $\mathbb{C}[x]$, that is, there exists $h(x) \in \mathbb{C}[x]$ such that $g(x) = f(x)h(x)$. From the uniqueness of the division with remainder, $h(x)$ is actually a polynomial with rational coefficients, that is, $g(x) \mid f(x)$ in $\mathbb{Q}[x]$ (refer to Remark 3 of Chapter 11). The converse is also true.

If $f(x)$, $g(x) \in \mathbb{Z}[x]$, and the coefficient of the first term of $g(x)$ is ± 1, then in $\mathbb{Z}[x]$, we can do the division with remainder of $f(x)$ on division by $g(x)$ (Remark 2 of Chapter 11). Therefore, from Remark 4 of Chapter 11, we know that $h(x)$ must be a polynomial with integer coefficients, that is, $g(x) \mid f(x)$ in $\mathbb{Z}[x]$.

Further, we have

(14) Let $g(x)$ be an irreducible polynomial in $\mathbb{Q}[x]$. If $f(x) \in \mathbb{Q}[x]$, and $f(x)$ and $g(x)$ have a common complex root, then $g(x) \mid f(x)$ (in $\mathbb{Q}[x]$).

In addition, if the above $f(x)$ and $g(x)$ are both polynomials with integer coefficients, and the coefficient of the first term of $g(x)$ is ± 1, then $g(x) \mid f(x)$ (in $\mathbb{Z}[x]$).

In fact, because $g(x)$ is irreducible, if $g(x) \nmid f(x)$, then $g(x)$ and $f(x)$ are coprime, so there exist $u(x)$, $v(x) \in \mathbb{Q}[x]$ such that

$$f(x)u(x) + g(x)v(x) = 1$$

(see (7) in Chapter 11). By substituting the common root of $f(x)$ and $g(x)$ into the above formula, we get $0 = 1$, a contradiction! Hence, $g(x) \mid f(x)$ (in $\mathbb{Q}[x]$).

If $f(x)$, $g(x) \in \mathbb{Z}[x]$ and the coefficient of the first term of $g(x)$ is ± 1, then $g(x) \mid f(x)$ in $Q[x]$, which means that it is also true in $\mathbb{Z}[x]$ (refer to Remark 4 of Chapter 11).

Remark 8. The claim (14) is a basic result of connecting irreducible polynomials and their complex roots in $\mathbb{Q}[x]$, which is very useful.

By the way, it can be seen from (3) in the following Chapter 13 that a polynomial with integer coefficients is irreducible in $\mathbb{Q}[x]$, is equivalent to that it's irreducible in $\mathbb{Z}[x]$.

Remark 9. The divisibility of polynomials with integer coefficients is related to the divisibility of integers, which is common in some problems of mathematical competitions. We give a brief introduction to this.

In order to prove $b|a$, sometimes a and b can be treated as the values of polynomials $f(x)$ and $g(x)$ with integer coefficients at a certain integer k respectively, and then we expect to prove (stronger result):

$$g(x) \mid f(x) \text{ in } \mathbb{Z}[x]. \tag{5}$$

If the leading coefficient of $g(x)$ is ± 1, from (13) (or (14)), (5) may be easier to prove by considering the complex roots of $g(x)$ (but please note, that if (5) is not true, the conclusion $g(k) \nmid f(k)$ cannot be derived).

On the other hand, for non-constant polynomials with integer coefficients $f(x)$ and $g(x)$, if there is an infinite number of positive integers n, which make $g(n) \mid f(n)$, then there must be $g(x) \mid f(x)$ in $Q[x]$.

This result is to be proved by the readers (Exercise 15 in this chapter).

Illustrative Examples

Example 1. Find the remainder of $x^{100} - 2x^{55} + 1$ on division by $x^2 - 1$.

Solution Because $x^2 - 1$ is a quadratic polynomial,

$$x^{100} - 2x^{55} + 1 = (x^2 - 1)q(x) + ax + b.$$

Take $x = 1$ in the above formula, and then $a + b = 0$; take $x = -1$, and then $b - a = 4$. Hence $a = -2$ and $b = 2$, that is, the remainder is $-2x + 2$.

Example 2. Prove: The polynomial $x^{2n} - 2x^{2n-1} + 3x^{2n-2} - \cdots - 2nx + 2n + 1$ has no real roots.

Proof. Let $f(x)$ denote the polynomial in the problem. From Remark 7, $f(x)$ has no real roots if and only if there is no linear factor in its standard factorization (in $\mathbb{R}[x]$), that is, for all real numbers x, the value $f(x)$ is always positive or always negative. (Hence this problem is actually an inequality problem.)

For $x \leq 0$, obviously $f(x) > 0$. If $x > 0$, then from

$$xf(x) = x^{2n+1} - 2x^{2n} + 3x^{2n-1} - 4x^{2n-2} + \cdots + (2n+1)x,$$

we see that

$$(1+x)f(x) = f(x) + xf(x)$$
$$= x^{2n+1} - x^{2n} + x^{2n-1} - x^{2n-2} + \cdots + x + 2n + 1$$
$$= x\frac{x^{2n+1} + 1}{x + 1} + 2n + 1 > 0.$$

Thus, $f(x) > 0$ for $x > 0$.

Therefore, $f(x) > 0$, for all real numbers x, so $f(x)$ has no real roots.

Example 3. Let a be a real number. Determine the number of real roots of the polynomial $x^4 - 2ax^2 - x + a^2 - a$.

Solution It can be seen from Remark 7 that the number of real roots of a (real coefficient) polynomial is the number of linear divisors (counted by multiplicity) in its standard factorization (in $\mathbb{R}[x]$).

In order to obtain the decomposition of the polynomial in the problem, we treat it as a quadratic polynomial with regard to a (x as a parameter), from which we get

$$x^4 - 2ax^2 - x + a^2 - a$$
$$= a^2 - (2x^2 + 1)a + x^4 - x$$
$$= (a - x^2 + x)(a - x^2 - x - 1)$$
$$= (x^2 - x - a)(x^2 + x - a + 1).$$

The two factors on the right side of the above formula are both quadratic polynomials with real coefficients, and the number of their real roots is easy to determine. Therefore, the number of real roots of the polynomial in the problem can be determined, and the details are left for the readers to complete.

Many problems can be solved by using polynomials, which is sometimes called the polynomial method. The basic spirit of this method is: The problem is converted into a problem about polynomials. By properties of polynomials, we may derive required conclusions or make further arguments (solutions). Here are some examples of this technique.

Example 4. It is known that $a_1, a_2, \ldots, a_n$ are different from each other, and $b_1, b_2, \ldots, b_n$ are also different from each other. Find the values of $x_1, x_2, \ldots, x_n$ satisfying the following system of equations

$$\begin{cases} \dfrac{x_1}{a_1 - b_1} + \dfrac{x_2}{a_1 - b_2} + \cdots + \dfrac{x_n}{a_1 - b_n} = 1, \\[2mm] \dfrac{x_1}{a_2 - b_1} + \dfrac{x_2}{a_2 - b_2} + \cdots + \dfrac{x_n}{a_2 - b_n} = 1, \\[2mm] \qquad\qquad\qquad \cdots \\[2mm] \dfrac{x_1}{a_n - b_1} + \dfrac{x_2}{a_n - b_2} + \cdots + \dfrac{x_n}{a_n - b_n} = 1. \end{cases}$$

Solution Consider the auxiliary equation with regard to t:

$$\frac{x_1}{t - b_1} + \frac{x_2}{t - b_2} + \cdots + \frac{x_n}{t - b_n} = 1 - \frac{(t - a_1)(t - a_2)\cdots(t - a_n)}{(t - b_1)(t - b_2)\cdots(t - b_n)}. \qquad (6)$$

If we remove the denominator of (6), we can either get an identity or get an equation of degree less than n. It can be seen from the condition that there are n distinct roots $t = a_1, a_2, \ldots, a_n$, therefore, after removing the denominator of formula (6), it must be an identity, so that formula (6) itself is an identity.

Multiply both sides of (6) by $t - b_1$, and then let $t = b_1$, we obtain

$$x_1 = -\frac{(b_1 - a_1)(b_1 - a_2) \cdots (b_1 - a_n)}{(b_1 - b_2)(b_1 - b_3) \cdots (b_1 - b_n)}.$$

Similarly,

$$x_2 = -\frac{(b_2 - a_1)(b_2 - a_2) \cdots (b_2 - a_n)}{(b_2 - b_1)(b_2 - b_3) \cdots (b_2 - b_n)},$$

$$\cdots$$

$$x_n = -\frac{(b_n - a_1)(b_n - a_2) \cdots (b_n - a_n)}{(b_n - b_1)(b_n - b_2) \cdots (b_n - b_{n-1})}.$$

Example 5. Let m be a positive integer. Prove that

$$\cot^2 \frac{\pi}{2m+1} + \cot^2 \frac{2\pi}{2m+1} + \cdots + \cot^2 \frac{m\pi}{2m+1} = \frac{m(2m-1)}{3}.$$

Proof. The construction of a polynomial in this problem is a little difficult. First of all, from De Moivre's formula and the binomial theorem, for any positive integer n and real number θ, there is (here $i = \sqrt{-1}$)

$$\cos n\theta + i \sin n\theta = (\cos \theta + i \sin \theta)^n$$

$$= \cos n\theta + \binom{n}{1} i \cos^{n-1} \theta \cdot \sin \theta + \cdots .$$

Comparing the imaginary parts, we get

$$\sin n\theta = \binom{n}{1} \cos^{n-1} \theta \cdot \sin \theta$$

$$- \binom{n}{3} \cos^{n-3} \theta \cdot \sin^3 \theta + \binom{n}{5} \cos^{n-5} \theta \cdot \sin^5 \theta + \cdots .$$

Now take $n = 2m + 1$ and $\theta = \frac{\pi}{2m+1}, \frac{2\pi}{2m+1}, \ldots, \frac{m\pi}{2m+1}$. Then $\sin(2m + 1)\theta = 0$, and neither $\sin \theta$ nor $\cos \theta$ is equal to 0. Dividing both sides of the above formula by $\sin^n \theta$, we get that for $\theta = \frac{\pi}{2m+1}, \frac{2\pi}{2m+1}, \cdots, \frac{m\pi}{2m+1}$,

$$\binom{2m+1}{1}\cot^{2m}\theta - \binom{2m+1}{3}\cot^{2m-2}\theta + \cdots + (-1)^m \binom{2m+1}{2m+1} = 0,$$

i.e., the mth degree polynomial equation

$$\binom{2m+1}{1}x^m - \binom{2m+1}{3}x^{m-1} + \cdots + (-1)^m \binom{2m+1}{2m+1} = 0$$

has m (distinct) roots $x_k = \cot^2 \frac{k\pi}{2m+1}$ with $k = 1, 2 \ldots, m$. Therefore, from Vieta's theorem,

$$\cot^2 \frac{\pi}{2m+1} + \cot^2 \frac{2\pi}{2m+1} + \cdots + \cot^2 \frac{m\pi}{2m+1}$$

$$= \frac{\binom{2m+1}{3}}{\binom{2m+1}{1}} = \frac{m(2m-1)}{3}.$$

Example 6. Given $2n$ distinct complex numbers $a_1, \ldots, a_n, b_1, \ldots, b_n$, fill them into an $n \times n$ grid table according to the following rule: Fill in $a_i + a_j$ $(i, j = 1, \ldots, n)$ in the grid at the intersection of row i and column j. Prove: If the products of the numbers of each column are equal, then the products of the numbers of each row are also equal.

Proof. Let the products of the numbers of each column be equal to c, and consider the polynomial $f(x) = (x + a_1) \cdots (x + a_n) - c$. From the known condition, $f(b_i) = 0$ $(i = 1, \ldots, n)$. Because b_i are not equal to each other, the nth degree polynomial $f(x)$ has n different roots, so from (2) (note that the the leading coefficient of $f(x)$ is 1)

$$f(x) = (x - b_1) \cdots (x - b_n).$$

Hence, we obtain the equality

$$(x + a_1) \cdots (x + a_n) - c = (x - b_1) \cdots (x - b_n).$$

Take $x = -a_i (i = 1, \ldots, n)$. Then $(a_i + b_1) \cdots (a_i + b_n) = (-1)^{n+1}c$, that is, the products of the numbers of each row are all $(-1)^{n+1}c$.

Example 7.

(i) For what positive integers n, the polynomial $x^2 + x + 1$ exactly divides $x^{2n} + x^n + 1$?

(ii) Prove: If n is a positive integer and $4^n + 2^n + 1$ is prime, then n is a power of 3.

Solution (i) Let $f(x) = x^{2n} + x^n + 1$. It is well known that the two (distinct) complex roots of $x^2 + x + 1$ are ω and ω^2, where $\omega = \frac{1}{2}(-1 + \sqrt{3}i)$ and $\omega^3 = 1$.

When $n = 3k + 1$,

$$f(\omega) = \omega^{6k+2} + \omega^{3k+1} + 1 = \omega^2 + \omega + 1 = 0,$$

and $f(\omega^2) = 0$. Thus, in this case, $x^2 + x + 1 \mid f(x)$.

It can also be proved that $x^2 + x + 1 \mid f(x)$ when $n = 3k + 2$. But when $n = 3k$, we have $f(\omega) = 3 \neq 0$, so $x^2 + x + 1 \nmid f(x)$.

To summarize, $x^2 + x + 1$ exactly divides $f(x)$ if and only if $3 \nmid n$. Note that the above proof shows that the divisibility is valid in $\mathbb{C}[x]$, but because the leading coefficient of $x^2 + x + 1$ is 1, the divisibility is also valid in $\mathbb{Z}[x]$ (see (13)).

(ii) Prove by contradiction. Suppose there exists n that is not a power of 3, such that $4^n + 2^n + 1$ is prime.

If the power of 3 contained in n is 3^k, then $n = 3^k(3q + r)$. Here, $r = 1$ and 2 when $q > 0$ and $r = 2$ when $q = 0$. We have

$$4^n + 2^n + 1 = \left(2^{3^k}\right)^{6q+2r} + \left(2^{3^k}\right)^{3q+r} + 1.$$

In (i), taking $n = 3q + r$ and $x = 2^{3^k}$, we deduce that $(2^{3^k})^2 + 2^{3^k} + 1$ exactly divides the right side of the above formula, thus it exactly divides the left side of the above formula (see Remark 9). But $4^n + 2^n + 1$ is prime, so necessarily

$$\left(2^{3^k}\right)^{6q+2r} + \left(2^{3^k}\right)^{3q+r} + 1 = \left(2^{3^k}\right)^2 + 2^{3^k} + 1.$$

Therefore, $3q + r = 1$, thus $q = 0$ and $r = 1$, that is, $n = 3^k$, which contradicts the previous assumption.

Example 8. Prove: If $\frac{\theta}{\pi}$ and $\cos\theta$ are rational numbers, then $\cos\theta$ must be one of the following five numbers: $-1, -\frac{1}{2}, 0, \frac{1}{2}$ and 1.

Proof. Let $\frac{\theta}{\pi} = \frac{m}{n}$, where m and n are integers and $n > 0$. Then

$$\cos n\theta = \cos m\pi = (-1)^m. \tag{7}$$

The key to the demonstration is to prove that for any positive integer n, the expression $2\cos n\theta$ can be expressed as an integer coefficient polynomial of degree n with regard to $2\cos\theta$, whose leading coefficient is 1, i.e.,

$$2\cos n\theta = (2\cos\theta)^n + a_1(2\cos\theta)^{n-1} + \cdots + a_{n-1}(2\cos\theta) + a_n, \tag{8}$$

where $a_1, a_2, \ldots, a_n$ are all integers.

This result is easy to prove by induction. Firstly, for $n = 1$ and 2,

$$2\cos\theta = 2\cos\theta, \ 2\cos 2\theta = (2\cos\theta)^2 - 2.$$

And if the proposition is true for n and $n + 1$, then the equality

$$2\cos(n+2)\theta = (2\cos\theta) \cdot 2\cos(n+1)\theta - 2\cos n\theta$$

shows that the proposition is also true for $n + 2$. This proves the above proposition.

Now, from (7), (8) and the conditions, $x = 2\cos\theta$ is a rational root of the equation

$$x^n + a_1 x^{n-1} + \cdots + a_{n-1}x + a_n - 2(-1)^m = 0.$$

Since this equation has integer coefficients and the coefficient of the first term is 1, its rational root must be an integer (see (8)), that is, $2\cos\theta$ is an integer. Since $-1 \le \cos\theta \le 1$, it must be one of $-2, -1, 0, 1$ and 2, that is, $\cos\theta$ is one of $-1, -\frac{1}{2}, 0, \frac{1}{2}$, and 1.

Example 9. Let p be prime.

(i) Use (7) and Fermat's little theorem to prove Wilson's theorem: $(p - 1)! \equiv -1 \pmod{p}$.
(ii) If $p > 3$, prove

$$p^2 \left| (p-1)! \left(1 + \frac{1}{2} + \cdots + \frac{1}{p-1}\right).\right.$$

Proof. (i) When $p = 2$, the conclusion is obviously true. When $p \ge 3$, consider the polynomial of degree $p - 2$:

$$f(x) = (x-1)(x-2)\cdots(x-p+1) - x^{p-1} + 1.$$

From Fermat's little theorem, the congruence equation $f(x) \equiv 0(\mathrm{mod}\, p)$ has $p - 1$ different solutions $x \equiv 1, 2, \ldots, p - 1(\mathrm{mod}\, p)$, so from (7), the coefficients of $f(x)$ are all divisible by p. In particular, the constant term $(p - 1)! + 1$ is a multiple of p.

(ii) Suppose

$$(x - 1)(x - 2) \cdots (x - p + 1) = x^{p-1} - s_1 x^{p-2} + \cdots + s_{p-1}. \qquad (9)$$

Note that

$$s_{p-1} = (p - 1)!, \quad s_{p-2} = (p - 1)! \left(1 + \frac{1}{2} + \cdots + \frac{1}{p - 1}\right).$$

It has been proved in (i) that the coefficients $s_1, \ldots, s_{p-2}$ are all multiples of p.

Taking $x = p$ in (9), we obtain that

$$p^{p-2} - s_1 p^{p-3} + \cdots + s_{p-3} p - s_{p-2} = 0.$$

If $p > 3$, then $p^2 | s_{p-2}$ can be derived from the above formula and $p \mid s_{p-3}$. (The readers can compare the conclusion of this problem with Example 2 in Chapter 6.)

Example 10. Let p be prime and $p \equiv 3(\mathrm{mod}4)$. Prove that

$$\prod_{k=1}^{p} (1 + k^2) \equiv 4(\mathrm{mod}\, p).$$

Proof Because $(p - k)^2 \equiv k^2(\mathrm{mod}\, p)$, we just need to prove that

$$\prod_{k=1}^{\frac{p-1}{2}} (1 + k^2) \equiv 2(\mathrm{mod}\, p). \qquad (10)$$

There are several different proofs of (10). Here we only introduce a proof that is based on polynomials.

The first step of the following proof is to associate the product on the left side of (10) with the value of a polynomial with integer coefficients. In fact, suppose

$$f(x) = (x - 1^2)(x - 2^2) \cdots \left(x - \left(\frac{p-1}{2}\right)^2\right).$$

Then the product is $(-1)^{\frac{p-1}{2}} f(-1)$.

Since p is prime, the $\frac{p-1}{2}$ numbers

$$1^2, 2^2, \ldots, \left(\frac{p-1}{2}\right)^2 \qquad (11)$$

are not congruent to each other modulo p, so they are all the solutions of $f(x) \equiv 0 \pmod{p}$.

The second step of the proof is to choose another polynomial with integer coefficients of degree $\frac{p-1}{2}$ modulo p, which takes these $\frac{p-1}{2}$ numbers in (11) as its zeros modulo p. According to Fermat's little theorem, $x^{\frac{p-1}{2}} - 1$ meets the requirement.

In the third step, let $g(x) = f(x) - \left(x^{\frac{p-1}{2}} - 1\right)$. Then $g(x)$ is a $\frac{p-3}{2}$th degree polynomial, but the congruence equation $g(x) \equiv 0 \pmod{p}$ has $\frac{p-1}{2}$ solutions $x \equiv k^2 \left(k = 1, 2, \ldots, \frac{p-1}{2}\right)$. Since p is a prime number, $g(x)$ modulo p is the zero polynomial, thus

$$\prod_{k=1}^{\frac{p-1}{2}} (x - k^2) - \left(x^{\frac{p-1}{2}} - 1\right) \equiv 0 \pmod{p}$$

is true for all integers x. In particular, taking $x = -1$ (notice $p \equiv 3 \pmod 4$), we can get (10).

Exercises

Group A

1. Find the remainder of $x^{1959} - 1$ on division by $(x^2 + 1)(x^2 + x + 1)$.
2. Find all positive integers n such that $x^2 + x + 1$ exactly divides $(x + 1)^n - x^n - 1$.
3. Let $f(x) \in \mathbb{Z}[x]$, and let a, b, and c be different integers. Show that if $f(a) = f(b) = f(c) = -1$, then $f(x)$ has no integer root.
4. Let $f(x) \in \mathbb{Z}[x]$. If there exist an odd number k and an even number l such that $f(k)$ and $f(l)$ are both odd numbers, prove that $f(x)$ has no integer root.
5. Let $f(x) = x^n + a_1 x^{n-1} + \cdots + a_{n-1}x + 1$ have n real roots, and the coefficients $a_1, \ldots, a_{n-1}$ are all nonnegative. Prove: $f(2) \geq 3^n$.
6. Find all polynomials $f(x)$ satisfying $f(x^2 + 1) = f^2(x) + 1$ and $f(0) = 0$.
7. Let $f(x) \in \mathbb{R}[x]$ with $\deg f \geq 1$. If α is a complex root of $f(x)$, show that the conjugate $\overline{\alpha}$ of α is also a root of $f(x)$.
8. Let $f(x) \in \mathbb{R}[x]$. Prove that if $f(x) \geq 0$ for any real number x, then $f(x)$ is the sum of the squares of two polynomials with real coefficients.

9. Let p be prime and k be a positive integer, and $(p-1) \nmid k$. Prove: There exists an integer a satisfying $p \nmid a$ and $p \nmid (a^k - 1)$.

Group B

10. Let $f(x) = a_n x^n + \cdots + a_1 x + a_0 \in \mathbb{C}[x]$, $a_n \neq 0$, and α is any root of $f(x)$. Prove:

$$|\alpha| < 1 + \max_{0 \le k \le n-1} \left| \frac{a_k}{a_n} \right|.$$

11. Let $f(x) = a_n x^n + \cdots + a_1 x + a_0 \in \mathbb{Z}[x]$, where $0 \le a_i \le 9$ ($i = 0, 1, \ldots, n$) and $a_n \neq 0$. If α is a complex root of $f(x)$, show that $\mathrm{Re}(\alpha) \le 0$ or $|\alpha| < 4$.

12. Let $f(x) = a_0 x^n + a_1 x^{n-1} + \cdots + a_{n-1} x + a_n$, where $0 < a_0 < a_1 < \cdots < a_{n-1} < a_n$. Prove: The moduli of the roots of $f(x)$ are all greater than 1.

13. Let the product of n quadratic polynomials $x^2 + a_k x + b_k$ ($1 \le k \le n$) with real coefficients be

$$f(x) = x^{2n} + c_1 x^{2n-1} + \cdots + c_{2n-1} x + c_{2n}.$$

If the coefficients $c_1, c_2, \ldots, c_{2n}$ of $f(x)$ are all positive, prove that there must be a certain k ($1 \le k \le n$) such that the corresponding coefficients a_k and b_k are both positive numbers.

14. Is there a finite set S consisting of non-zero real numbers such that for any positive integer n, there is a polynomial whose degree is not less than n, all its coefficients belong to S, and all its roots are in S?

15. Let $f(x)$ and $g(x)$ be non-constant polynomials with integer coefficients. If there exists an infinite number of positive integers n such that $g(n) \mid f(n)$, show that $g(x) \mid f(x)$ in $\mathbb{Q}[x]$.

Chapter 13

Polynomials with Integer Coefficients

This chapter briefly introduces the basic contents of polynomials with integer coefficients, including primitive polynomials and the Gauss lemma, Eisenstein's criterion of irreducible polynomials, etc.

Primitive Polynomials

Let $f(x) = a_n x^n + \cdots + a_1 x + a_0 \in \mathbb{Z}[x]$ and $f(x) \neq 0$. We call the greatest common divisor $(a_0, a_1, \ldots, a_n)$ of $a_0, a_1, \ldots, a_n$ the **capacity** of $f(x)$. A polynomial with a capacity of 1 is called a **primitive polynomial**. The following important result, called the Gauss lemma, is a basis of studying polynomials with integer coefficients.

(1) (**Gauss lemma**) The product of two primitive polynomials in $\mathbb{Z}[x]$ is still a primitive polynomial.

Proof by contradiction. Suppose there are two primitive polynomials

$$f(x) = a_n x^n + \cdots + a_0,$$

$$g(x) = b_m x^m + \cdots + b_0$$

such that $f(x)g(x)$ is not primitive. Then there is a prime p exactly dividing the capacity of $f(x)g(x)$ (refer to Remark 1 in Chapter 7), thus p exactly divides all the coefficients of $f(x)g(x)$. Because $f(x)$ is primitive, p can't exactly divide all a_i. Let r be the minimum subscript such that a_r is not divisible by p; similarly, let s be the minimum subscript such that b_s is not divisible by p. The coefficient of x^{r+s} in $f(x)g(x)$ is

$$a_r b_s + a_{r+1} b_{s-1} + a_{r+2} b_{s-2} + \cdots + a_{r-1} b_{s+1} + a_{r-2} b_{s+2} + \cdots.$$

This sum is, of course, a multiple of p. But on the other hand, since $p \mid a_i$ and $p \mid b_j$ for $i \le r-1$ and $j \le s-1$, every term after the first term in the sum is divisible by p, so also $p \mid a_r b_s$. Since p is prime, $p \mid a_r$ or $p \mid b_s$, which is contradictory to our selections of a_r and b_s.   □

Nonzero polynomials in $\mathbb{Q}[x]$ are closely related to primitive polynomials in $\mathbb{Z}[x]$:

(2) Let $f(x) \in \mathbb{Q}[x]$ and $f(x) \ne 0$. Then there exists a rational number $a \ne 0$ such that $af(x)$ is a primitive polynomial in $\mathbb{Z}[x]$. In addition, if there exists a rational number $b \ne 0$, such that $bf(x)$ is also a primitive polynomial, then $a = \pm b$.

In fact, let $f(x) = a_n x^n + \cdots + a_1 x + a_0$, where $a_0, \ldots, a_n$ are all rational numbers and $a_n \ne 0$. Take an integer c such that $ca_0, \ldots, ca_n$ are all integers, and let $d = (ca_0, \ldots, ca_n)$. Then

$$\frac{c}{d} f(x) = \frac{ca_n}{d} x^n + \cdots + \frac{ca_1}{d} x + \frac{ca_0}{d}$$

is a primitive polynomial in $\mathbb{Z}[x]$ (refer to Chapter 6, (11)).

In addition, if there are rational numbers a and b such that $af(x) = g(x)$ and $bf(x) = h(x)$ are both primitive polynomials, then $g(x) = \frac{a}{b} h(x)$. Because $g(x)$ and $h(x)$ are primitive, $\frac{a}{b}$ must be an integer without prime factors, so $\frac{a}{b} = \pm 1$ that is, $a = \pm b$. This shows that a nonzero polynomial in $\mathbb{Q}[x]$ corresponds to a primitive polynomial uniquely.

Now, let's briefly talk about the factorization of polynomials with integer coefficients.

Let $f(x) \in \mathbb{Z}[x]$, with $\deg f \ge 1$. If $f(x)$ has only a trivial factorization in $\mathbb{Z}[x]$ that is, it cannot be decomposed into the product of two polynomials of positive degree in $\mathbb{Z}[x]$, then $f(x)$ is called an **irreducible polynomial** in $\mathbb{Z}[x]$. Otherwise, $f(x)$ is said to be **reducible** (or **decomposable**) in $\mathbb{Z}[x]$. For example, $2x + 2$ is irreducible and $2x^2 - 2$ is reducible in $\mathbb{Z}[x]$.

In order to study whether $f(x)$ is reducible in $\mathbb{Z}[x]$, we only need to consider the case that $f(x)$ is a primitive polynomial. We note that if $f(x)$

is decomposable in $\mathbb{Z}[x]$, because $\mathbb{Z} \subset \mathbb{Q}$, then it is certainly reducible in $\mathbb{Q}[x]$. The following result shows that the converse is also true. Therefore, a polynomial in $\mathbb{Z}[x]$ is irreducible in $\mathbb{Z}[x]$ if and only if it (regarded as a polynomial in $\mathbb{Q}[x]$) is irreducible in $\mathbb{Q}[x]$.

(3) Let $f(x) \in \mathbb{Z}[x]$ be a primitive polynomial. If $f(x)$ is reducible in $\mathbb{Q}[x]$, then $f(x)$ is also reducible in $\mathbb{Z}[x]$.

To be exact, let $f(x) = g(x)h(x)$, where $g(x), h(x) \in \mathbb{Q}[x]$ and $\deg g$, $\deg h \geq 1$. Then there is a rational number a such that

$$f(x) = ag(x) \cdot \frac{1}{a}h(x), \quad \text{and} \quad ag(x), \frac{1}{a}h(x) \in \mathbb{Z}[x].$$

In fact, it is known from (2) that there are rational numbers a and b such that $ag(x) = g_1(x)$ and $bh(x) = h_1(x)$ are primitive polynomials, so $abf(x) = g_1(x)h_1(x)$.

By the Gauss lemma, $g_1(x)h_1(x)$ is a primitive polynomial, and $f(x)$ is also a primitive polynomial, so ab must be an integer without a prime factor, that is, $ab = \pm 1$.

Therefore, $ag(x)$ and $\frac{1}{a}h(x)$ are (primitive) polynomials with integer coefficients.

Remark 1. From (3) and the unique factorization theorem in $\mathbb{Q}[x]$ ((15) in Chapter 11), it is not difficult to prove the unique factorization theorem in $\mathbb{Z}[x]$:

Any non-constant polynomial in $\mathbb{Z}[x]$ can be decomposed into the product of an integer and a finite number of primitive irreducible polynomials (with a positive coefficient of the first term). Moreover, these irreducible polynomials are unique if the order of the factors in the product is ignored.

From the unique factorization theorem, it is not difficult to define the greatest common factor of two polynomials (refer to (9) in Chapter 7) and establish its basic properties. Since these contents are not needed in this book, we will not discuss them.

Eisenstein's Criterion for Irreducible Polynomials

It is very difficult and complicated to judge whether a polynomial with integer coefficients is irreducible. The following result gives a sufficient condition for the irreducibility of a polynomial.

(4) (**Eisenstein's criterion**) Let $f(x) = a_n x^n + \cdots + a_0$ be a polynomial with integer coefficients, where $n \geq 1$. If there is a prime number p such that $p \nmid a_n$ and $p \mid a_i$ $(i = 0, 1, \ldots, n - 1)$, but $p^2 \nmid a_0$, then $f(x)$ is irreducible in $\mathbb{Z}[x]$ (thus irreducible in $\mathbb{Q}[x]$).

Proof. The proof here is similar to that in (1). Suppose there are two (non-constant) polynomials with integer coefficients

$$g(x) = b_k x^k + \cdots + b_0 \quad \text{and} \quad h(x) = c_l x^l + \cdots + c_0$$

such that $f(x) = g(x)h(x)$. Because $a_0 = b_0 c_0$ is divisible by p, but not by p^2, so exactly one of b_0 and c_0 is divisible by p, let's suppose $p \mid c_0$. And $a_n = b_k c_l$ is not divisible by p, so $p \nmid c_l$. Now we can take the minimum subscript r to make $p \nmid c_r$. Obviously $r \neq 0$, and

$$a_r = b_0 c_r + b_1 c_{r-1} + \cdots + b_r c_0.$$

Because $p \nmid b_0 c_r$ and the rest terms in the sum are all divisible by p, so $p \nmid a_r$, which contradicts the condition in the theorem. $\qquad \square$

In particular, it can be seen from (4) that for any positive integer n, the polynomial $x^n + 2$ is irreducible in $\mathbb{Z}[X]$, so there are irreducible polynomials of any degree in $\mathbb{Z}[x]$ (and $\mathbb{Q}[x]$). In comparison, irreducible polynomials in $\mathbb{R}[x]$ are only of degree one or degree two, while irreducible polynomials in $\mathbb{C}[x]$ are all of degree one.

The following (5) is a typical application of Eisenstein's criterion, and the result itself is also very important.

(5) Let p be prime. Then $\Phi_p(x) = x^{p-1} + x^{p-2} + \cdots + x + 1$ is irreducible in $\mathbb{Q}[x]$.

We can't directly apply (4), but the polynomial

$$f(x) = \Phi_p(x + 1) = \frac{(x+1)^p - 1}{(x+1) - 1}$$

$$= x^{p-1} + \binom{p}{1} x^{p-2} + \cdots + \binom{p}{p-1}$$

meets the condition of (4) (refer to Example 7 in Chapter 7), therefore, $f(x)$ is irreducible in $\mathbb{Z}[x]$. On the other hand, the reducibility of $\Phi_p(x)$ in $\mathbb{Z}[x]$ is obviously equivalent to that of $f(x)$ in $\mathbb{Z}[x]$, so $\Phi_p(x)$ is irreducible in $\mathbb{Z}[x]$, and thus is irreducible in $\mathbb{Q}[x]$.

Complex Roots of Polynomials and their Irreducibility in $\mathbb{Z}[x]$

According to the Fundamental Theorem of Algebra, an nth degree polynomial in $\mathbb{Z}[x]$ has n roots in $\mathbb{C}$. Hence the factorization of a polynomial with integer coefficients in $\mathbb{C}[x]$ also helps to decide its irreducibility. Here is an illustrative example.

(6) Let $f(x)$ be a polynomial with integer coefficients, the leading coefficient is 1, and $f(0) \neq$. If $f(x)$ has only one single root α such that $|\alpha| \geq 1$, then $f(x)$ is irreducible in $\mathbb{Z}[x]$.

To prove it, let $f(x) = g(x)h(x)$, which is a polynomial with integer coefficients, and where $g(x)$ and $h(x)$ are both non constant. Since the leading coefficient of $f(x)$ is 1, it can be assumed that the leading coefficients of $g(x)$ and $h(x)$ are both 1, and $h(\alpha) = 0$. Therefore, it is known from the condition that the integer $g(0) \neq 0$, and the moduli of roots $\beta_1, \ldots, \beta_r$ of $g(x)$ (are all the roots of $f(x)$) are all less than 1. Because $g(x) = (x - \beta) \cdots (x - \beta_r)$,

$$1 \leq |g(0)| = |\beta_1| \cdots |\beta_r| < 1,$$

a contradiction. Therefore, $f(x)$ is irreducible in $\mathbb{Z}[x]$.

Remark 2. Let's briefly talk about the basic spirit of proving the irreducibility of a polynomial by using the information (distribution) of its roots.

In order to prove that $f(x)$ is irreducible in $\mathbb{Z}[x]$, we often use the proof by contradiction. Assume that $f(x)$ has a nontrivial decomposition: $f(x) = g(x)h(x)$. Sometimes, considering the value of $f(x)$ at a certain integer a from two aspects can produce contradictions. On one hand, $f(a)$ is the product of integers $g(a)$ and $h(a)$; On the other hand, the roots of $g(x)$ and $h(x)$ are all the roots of $f(x)$. Therefore, with the knowledge of the distribution of the roots of $f(x)$, we can get applicable estimates of $|g(a)|$ and $|h(a)|$ from the decompositions of $g(x)$ and $h(x)$ in $\mathbb{C}[x]$. Combine the results of the two aspects to derive a contradiction. (In (6), we actually studied the value of the polynomial at $x = 0$ from these two aspects.)

In elementary mathematics, there is no powerful tool to study the distribution of roots of a polynomial. The main method is the simple inequality estimation. Therefore, in the elementary field, it is not easy to give interesting results of irreducibility only from the information of roots. (For an example of using elementary methods to study the roots of polynomials, refer to Exercises 10–12 in Chapter 12.)

A polynomial (with integer coefficients) taking prime values is also related to its irreducibility. The following result provides a simple principle.

(7) Let $f(x)$ be a polynomial with integer coefficients of degree n ($n \geq 1$), and $\alpha_1, \ldots, \alpha_n$ be all its complex roots. If there is an integer $k > 1 + \text{Re}(\alpha_j)(j = 1, \ldots, n)$ such that $|f(k)|$ is prime, then $f(x)$ is irreducible in $\mathbb{Z}[x]$.

In fact, if $f(x)$ has a nontrivial decomposition $f(x) = g(x)h(x)$, make the standard factorization of $g(x)$ in $\mathbb{R}[x]$ (see Remark 7 in Chapter 12):

$$g(x) = b(x - a_1) \cdots (x - a_r)((x - b_1)^2 + c_1^2) \cdots ((x - b_s)^2 + c_s^2),$$

where b is a nonzero integer, and a_i, b_j, and c_j are all real numbers (obviously, a_i is the real root of $g(x)$, and b_j is the real part of an imaginary root of $g(x)$). Because the roots of $g(x)$ are all the roots of $f(x)$, we have $k > 1 + a_i$ and $k > 1 + b_j$ ($1 \leq i \leq r$, and $1 \leq j \leq s$). Hence $|g(k)| > 1$; similarly, $|h(k)| > 1$. And $g(x)$ and $h(x)$ are polynomials with integer coefficients, so $|g(k)|$ and $|h(k)|$ are both integers greater than 1, but $|g(k)| \cdot |h(k)| = |f(k)|$ is a prime number, a contradiction. (Refer to Remark 2.)

Illustrative Examples

Example 1. Determine whether the polynomials $x^4 + 1$ and $x^4 + 4$ are reducible in $\mathbb{Z}[x]$.

Solution Consider $x^4 + 1$ first. We can't use Eisenstein's criterion directly, but

$$(x + 1)^4 + 1 = x^4 + 4x^3 + 6x^2 + 4x + 2$$

meets the condition in (4) (taking prime $p = 2$), so it is irreducible in $\mathbb{Z}[x]$, thus $x^4 + 1$ is irreducible in $\mathbb{Z}[x]$.

A more elementary solution is the method of undetermined coefficients. Because $x^4 + 1$ obviously does not have a linear factor, if it is reducible in $\mathbb{Z}[x]$, it must be the product of two quadratic factors. We can suppose

$$x^4 + 1 = (x^2 + ax + b)(x^2 + cx + d),$$

where a, b, c, and d are all integers. By comparing the coefficients, we obtain that

$$a + c = 0, \quad ac + b + d = 0, \quad ad + bc = 0, \quad bd = 1.$$

Hence $b = d = \pm 1$, thus $a^2 = \pm 2$, which is impossible. Therefore, $x^4 + 1$ is irreducible in $\mathbb{Z}[x]$. For $x^4 + 4$, of course, (4) cannot be applied, because it is reducible in $\mathbb{Z}[x]$.

$$x^4 + 4 = (x^2 - 2x + 2)(x^2 + 2x + 2).$$

(If it can't be seen that $x^4 + 4 = x^4 + 4x^2 + 4 - 4x^2$, then the above undetermined coefficient method can be used to find the decomposition.)

Example 2. Determine whether the polynomial $x^5 - 3x^4 + 9x^3 - 6x^2 + 9x - 6$ is reducible in $\mathbb{Z}[x]$.

Solution Using Eisenstein's criterion (taking $p = 3$), we know that the polynomial is irreducible in $\mathbb{Z}[x]$.

This problem can also be solved by the method of undetermined coefficients: Because the leading coefficient of the polynomial in the problem is 1, its rational roots must be integers, which exactly divide the constant term -6. After checking through them one by one, we know that the polynomial has no integer roots, so there is no linear factor. Therefore, if the polynomial is reducible in $\mathbb{Z}[x]$, it must be the product of a quadratic polynomial and a cubic polynomial; As in Example 1, we can derive a contradiction (the details are omitted here).

Remark 3. The condition in Eisenstein's criterion is only a sufficient one for a polynomial (with integer coefficients) to be irreducible in $\mathbb{Z}[x]$, but not a necessary one; many irreducible polynomials do not satisfy this condition.

In elementary mathematics, the most basic method to determine whether a polynomial is reducible is of course the method of undetermined coefficients. By this method, we either derive a contradiction (i.e., irreducible polynomials) or give a required decomposition. This is especially effective for polynomials of lower degree. For polynomials of higher degree, the method of undetermined coefficients cannot be ignored, which is often combined with other tools to derive the required results (see Example 6 below).

Example 3. Let $a_1, \ldots, a_n$ be integers that are different from each other. Prove that the polynomial

$$(x - a_1) \cdots (x - a_n) - 1$$

is irreducible in $\mathbb{Z}[x]$.

Proof. Let $f(x)$ denote the polynomial in the problem and suppose $f(x)$ has a nontrivial decomposition in $\mathbb{Z}[x]$: $f(x) = g(x)h(x)$. Then, from $f(a_i) = -1$, we can know that one of $g(a_i)$ and $h(a_i)$ is $+1$ and the other is -1, so $g(a_i) + h(a_i) = 0$ $(i = 1, \ldots, n)$. Therefore, the polynomial $g(x) + h(x)$ has at least n different zeros, but the degrees of $g(x)$ and $h(x)$ are both not greater than $n - 1$. Therefore, from (4) in Chapter 12, we deduce that $g(x) + h(x) = 0$, that is, $f(x) = -g^2(x)$. Since the leading coefficient of $f(x)$ is positive, it is obviously impossible.   □

Example 4. Let $f(x)$ be a polynomial (with integer coefficients) of degree n. Prove: If there exist at least $2n + 1$ distinct integers m such that $|f(m)|$ is prime, then $f(x)$ is irreducible in $\mathbb{Z}[x]$.

Proof. Let $f(x) = g(x)h(x)$, where $g(x)$ and $h(x)$ are both non-constant polynomials with integer coefficients. Let $r = \deg g$ and $s = \deg h$. Since $g(x)$ is not a constant, both $g(x) + 1$ and $g(x) - 1$ have at most r roots, so at most r integers m make $g(m) = 1$, and also at most r integers m make $g(m) = -1$. Therefore, at most $2r$ integers m make $g(m) = \pm 1$. Similarly, at most $2s$ integers m make $h(m) = \pm 1$.

However, if an integer m makes $|f(m)| = |g(m)| \cdot |h(m)|$ be prime, then $g(m) = \pm 1$ or $h(m) = \pm 1$. Therefore, from the above proven results, it can be inferred that there are at most $2r + 2s = 2n$ different integers m, which make $|f(m)|$ be prime. This is contradictory to the hypothesis of the problem, so $f(x)$ is irreducible in $\mathbb{Z}[x]$.

The following Example 5 is an interesting result, called Cohen's theorem, which shows that given a prime number (in the decimal system) can naturally generate an irreducible polynomial (with integer coefficients) in $\mathbb{Z}[x]$.

Example 5. Let $p = \overline{a_n \cdots a_1 a_0} = a_n \cdot 10^n + \cdots + a_1 \cdot 10 + a_0$ be a prime number expressed in the decimal system, with $0 \le a_i \le 9$ $(i = 0, 1, \ldots, n)$, and $a_n \ne 0$. Then the polynomial

$$f(x) = a_n x^n + \cdots + a_1 x + a_0$$

is irreducible in $\mathbb{Z}[x]$.

Proof. In fact, the condition implies that $f(10)$ is prime, but Exercise 11 in Chapter 12 shows that the real part of any root of $f(x)$ is less than 9, so the conclusion can be deduced from (7).

Example 6. Let a polynomial $f(x) = a_0 x^n + a_1 x^{n-1} + \cdots + a_{n-1} x + a_n \in \mathbb{Z}[x]$ satisfiy the following conditions:

(i) $0 < a_0 < a_1 < \cdots < a_{n-1} < a_n$.
(ii) $a_n = p^m$, where p is a prime number (m is any positive integer) and $p \nmid a_{n-1}$.

Prove: $f(x)$ is irreducible in $\mathbb{Z}[x]$.

Proof. Prove by contradiction: Suppose there are non-constant polynomials with integer coefficients

$$g(x) = b_0 x^r + \cdots + b_r,$$
$$h(x) = c_0 x^s + \cdots + c_s$$

such that $f(x) = g(x)h(x)$. Comparing the coefficients, we get

$$a_{n-1} = b_{r-1} c_s + b_r c_{s-1}, \tag{1}$$

and

$$p^m = |a_n| = |b_r| \cdot |c_s|. \tag{2}$$

Since $p \nmid a_{n-1}$, it is known from 1 that b_r and c_s cannot be both divisible by p, so we may assume that $p \nmid b_r$. Then from 2 we can deduce $|c_s| = p^m$ and $|b_r| = 1$.

It can be seen from Exercise 12 in Chapter 12 that the moduli of the roots of $f(x)$ are all greater than 1, so the moduli of the roots $\alpha_1, \ldots, \alpha_r$ of $g(x)$ are all greater than 1. Combining them with Vieta's theorem, we can get

$$|b_r| = |b_0| |\alpha_1| \cdots |\alpha_r| > |b_0| \geq 1,$$

which contradicts the previous result $|b_r| = 1$. □

Remark 4. It is usually very difficult to determine whether a given polynomial with integer coefficients is irreducible in $\mathbb{Z}[x]$. However, we can construct an irreducible polynomial with integer coefficients in $\mathbb{Z}[x]$ effectively. Eisenstein's criterion and Example 6 above are two basic approaches (can also see Exercise 7 in this chapter).

Example 7. Let p be a prime number and ζ be a complex number such that $\zeta^p = 1$ and $\zeta \neq 1$. If $a_0, \ldots, a_{p-1}$ are all rational numbers satisfying

$$a_{p-1}\zeta^{p-1} + \cdots + a_1\zeta + a_0 = 0.$$

prove: $a_0 = a_1 = \cdots = a_{p-1}$.

Proof. From $\zeta^p - 1 = 0$, we get $(\zeta - 1)(\zeta^{p-1} + \cdots + \zeta + 1) = 0$. And $\zeta \neq 1$, so $\zeta^{p-1} + \cdots + \zeta + 1 = 0$. Thus, the equation in the problem is reduced to

$$(a_{p-1} - a_{p-2})\zeta^{p-2} + \cdots + (a_{p-1} - a_1)\zeta + (a_{p-1} - a_0) = 0. \qquad (3)$$

Let $b_k = a_{p-1} - a_k$ $(k = 0, 1, \ldots, p-2)$. Suppose

$$f(x) = b_{p-2}x^{p-2} + \cdots + b_1 x + b_0,$$

which is a polynomial with rational coefficients. It is known from 3 that $f(x)$ and $\Phi_p(x) = x^{p-1} + \cdots + x + 1$ have a common complex root ζ. But $\Phi_p(x)$ is irreducible in $\mathbb{Q}[x]$ (see (5)), so $\Phi_p(x) \mid f(x)$ in $\mathbb{Q}[x]$ (see (14) in Chapter 12). But $f(x)$ is either the zero polynomial or of degree $\leq p-2$, and the degree of $\Phi_p(x)$ is $p-1$, so $f(x)$ must be the zero polynomial, hence all $b_k = 0$, that is, $a_0 = a_1 = \cdots = a_{p-1}$. $\qquad\square$

Exercises

Group A

1. Prove that the following polynomials are irreducible in $\mathbb{Q}[x]$.
    (i) $5x^4 - 3x^2 + 24x + 3$;
    (ii) $x^7 - 2x^5 + 4x^2 + 16x + 2$;
    (iii) $x^7 - 14x^2 + 7$.

2. Determine whether $x^7 + 7x + 1$ is reducible in $\mathbb{Q}[x]$.

3. Determine whether $x^5 + 4x^4 + 2x^3 + 3x^2 - x + 5$ is reducible in $\mathbb{Q}[x]$

4. Prove: If $x^{n-1} + x^{n-2} + \cdots + x + 1$ is irreducible in $\mathbb{Q}[x]$, then n is prime $(n > 1)$.

5. Let p be prime. Prove: $x^{p-1} - x^{p-2} + \cdots - x + 1$ is reducible in $\mathbb{Q}[x]$.

Group B

6. Suppose an integer k is not divisible by 5. Prove that $x^5 - x + k$ is irreducible in $\mathbb{Z}[x]$.

7. Let $f(x) = a_n x^n + \cdots + a_1 x + p \in \mathbb{Z}[x]$, $a_n \neq 0$, p be prime, and

$$|a_1| + \cdots + |a_n| < p.$$

Prove that $f(x)$ is irreducible in $\mathbb{Z}[x]$.

8. Let $f(x) = a_n x^n + a_{n-1} x^{n-1} + \cdots + a_1 x + a_0$ be a polynomial with integer coefficients and $a_n \neq 0$. Denote

$$M = \max_{0 \leq i \leq n-1} \left| \frac{a_i}{a_n} \right|.$$

If there is an integer $m \geq M + 2$ such that $|f(m)|$ is prime, show that $f(x)$ is irreducible in $\mathbb{Z}[x]$.

Chapter 14

Interpolation and Difference of Polynomials

This chapter introduces Lagrange's interpolation formula of polynomials and the difference of polynomials.

Lagrange's Interpolation Formula

Given n different x-coordinates $x_1, \ldots, x_n$, and corresponding y-coordinates $y_1, \ldots, y_n$, this gives n different points in the coordinate plane:

$$(x_1, y_1), \ldots, (x_n, y_n). \tag{1}$$

The interpolation problem is to find a curve so that it passes through these n points in (1). In the analytic language, it is to find a function $f(x)$ such that

$$f(x_1) = y_1, \ldots, f(x_n) = y_n. \tag{2}$$

But obviously there is an infinite number of such functions.

Among all functions, there are many reasons to think that polynomial functions are the simplest ones (it only involves the addition, subtraction, and multiplication of constants and the variable x). Therefore, we hope to find a polynomial $f(x)$ with the lowest degree satisfying (2). The following Lagrange's interpolation formula gives a positive answer to this question.

(1) (**Lagrange's interpolation formula**) There is a unique polynomial $f(x)$ whose degree is not greater than $n-1$ that satisfies (2); and $f(x)$

177

can be expressed in the following form:

$$f(x) = y_1 \frac{(x - x_2)(x - x_3) \cdots (x - x_n)}{(x_1 - x_2)(x_1 - x_3) \cdots (x_1 - x_n)}$$

$$+ y_2 \frac{(x - x_1)(x - x_3) \cdots (x - x_n)}{(x_2 - x_1)(x_2 - x_3) \cdots (x_2 - x_n)} + \cdots$$

$$+ y_n \frac{(x - x_1)(x - x_2) \cdots (x - x_{n-1})}{(x_n - x_1)(x_n - x_2) \cdots (x_n - x_{n-1})}. \tag{3}$$

To prove it, we first note that if such a polynomial exists, then it must be unique. Because if there exist two such polynomials, then the difference of them will have at least n different zeros. Therefore, from (4) in Chapter 12, these two polynomials must be equal.

Lagrange gave an explicit construction for the existence of a solution (it is not difficult for the readers to verify that (3) meets the requirement of the problem), and his derivation method is as follows:

First, in the special case of $y_2 = \cdots = y_n = 0$, find the polynomial $f_1(x)$ satisfying the requirement. In this case, $f_1(x)$ has $n - 1$ different zeros $x_2, \ldots, x_n$, so it is divisible by $(x - x_2) \cdots (x - x_n)$ ((2) in Chapter 12), and we want the degree of $f_1(x)$ to be as low as possible, so suppose

$$f_1(x) = c(x - x_2) \cdots (x - x_n),$$

where c is a constant number determined by $f_1(x_1) = y_1$. It's not hard to find that

$$f_1(x) = y_1 \frac{(x - x_2) \cdots (x - x_n)}{(x_1 - x_2) \cdots (x_1 - x_n)}.$$

Similarly, when $y_1 = 0, \ldots, y_{i-1} = 0, y_{i+1} = 0, \ldots, y_n = 0$, it can be found that the $(n - 1)$st degree if $y_i \neq 0$ polynomial $f_i(x)$ is

$$f_i(x) = y_i \frac{(x - x_1) \cdots (x - x_{i-1})(x - x_{i+1}) \cdots (x - x_n)}{(x_i - x_1) \cdots (x_i - x_{i-1})(x_i - x_{i+1}) \cdots (x_i - x_n)}.$$

It is easy to verify that $f(x) = f_1(x) + \cdots + f_n(x)$ is the solution to the problem (i.e., (3)).

Remark 1. In the above method, the general problem is treated as a "superposition" of several special problems, which is worth pondering carefully. The readers who understand (n-dimensional) vector coordinates can see that its essence is

$$(y_1, \ldots, y_n) = (y_1, 0, \ldots, 0) + (0, y_2, \ldots, 0) + \cdots + (0, \ldots, y_n),$$

For the purpose of this book, we will not discuss it. For other applications of this technique, please refer to (5) in Chapter 19.

Remark 2. The degree of the polynomial (3) can be less than $n-1$, and it can also be identically zero. Therefore, the strict expression of the theorem should be "there exists one and only one polynomial, which either is zero or its degree does not exceed $n-1$". Traditionally, we call indiscriminately the interpolation polynomial (3) the polynomial with degree not greater than $n-1$.

Remark 3. We know that if $f(x)$ is a polynomial of degree n, then $f(x)$ is uniquely determined by its values at $n+1$ different points $x_1, \ldots, x_{n+1}$; Lagrange's interpolation formula (3) (n is replaced by $n+1$) gives an explicit expression of $f(x)$ by those x_i and $f(x_i)$, that is, all the $n+1$ coefficients of $f(x)$ can be expressed by x_i and $f(x_i)$. We note that the expressions of the leading coefficient and constant term of $f(x)$ are quite neat and symmetrical. These facts are often a basic starting point of some problems involving polynomials and their values (see Example 4 below).

Difference of Polynomials

Difference is an important mathematical concept, which is widely used. It is also closely related to interpolation. It is not discussed in this book. We just briefly talk about the difference of polynomials.

We first introduce the concept of difference of general functions. For a function $f(x)$ and a fixed $h \neq 0$,

$$f(x+h) - f(x)$$

is called **the first-order difference of $f(x)$ with step size h,** denoted as $\Delta_h^1 f(x)$ (or $\Delta h f(x)$). The difference of $\Delta_h^1 f(x)$ is

$$\Delta_h^1 f(x+h) - \Delta_h^1 f(x)$$
$$= (f(x+2h) - f(x+h)) - (f(x+h) - f(x))$$
$$= f(x+2h) - 2f(x+h) + f(x),$$

and is called the second-order difference of $f(x)$ (with step size h), denoted as $\Delta_h^2 f(x)$. In general, **the nth-order difference of $f(x)$ with step size h** is defined as $\Delta_h^n f(x) = \Delta_h^1 (\Delta_h^{n-1} f(x))$.

(2) We have the following formula

$$\Delta_h^n f(x) = \sum_{i=0}^{n} (-1)^{n-i} \binom{n}{i} f(x+ih). \tag{4}$$

In fact, when $n = 1$, the conclusion is obvious. Assume that (4) is true. Then

$$\Delta_h^{n+1} f(x) = \sum_{i=0}^{n} (-1)^{n-i} \binom{n}{i} f(x+(i+1)h) - \sum_{i=0}^{n} (-1)^{n-i} \binom{n}{i} f(x+ih)$$

$$= \sum_{i=1}^{n+1} (-1)^{n-i+1} \binom{n}{i-1} f(x+ih) - \sum_{i=0}^{n} (-1)^{n-i} \binom{n}{i} f(x+ih)$$

$$= \sum_{i=0}^{n+1} (-1)^{n-i+1} \left(\binom{n}{i-1} + \binom{n}{i} \right) f(x+ih)$$

$$= \sum_{i=0}^{n+1} (-1)^{n+1-i} \binom{n+1}{i} f(x+ih).$$

Therefore, (4) is true for all positive integers n.

When $f(x)$ is a polynomial, its difference has quite a good behavior (we have another way to describe it).

(3) Let $f(x)$ be a polynomial of degree m:

$$a_m x^m + a_{m-1} x^{m-1} + \cdots + a_1 x + a_0.$$

Then, $\Delta_h^n f(x)$ is a polynomial of degree $m - n$ for $n \le m$; but for $n > m$, $\Delta_h^n f(x)$ is always zero.

In fact, by definition we know that

$$\Delta_h f(x) = f(x+h) - f(x)$$

$$= a_m (x+h)^m + \cdots + a_0 - a_m x^m - \cdots - a_0$$

$$= mh a_m x^{m-1} + \text{lower powers},$$

which is a polynomial of degree $m - 1$, and the first term is $mh a_m x^{m-1}$. Similarly, $\Delta_h^{m-1} f(x)$ is a linear polynomial, the first term is $m! h^{m-1} a_m x$, and $\Delta_h^m f(x)$ is the constant $m! h^m a_m$. Hence the $(m+1)$st and higher order differences of $f(x)$ are all 0.

Combining (2) and (3) (taking step size $h = 1$), we get:

(4) Let $f(x)$ be a polynomial of degree m, and its leading coefficient is a_m. Then

$$\sum_{i=0}^{n} (-1)^{n-i} \binom{n}{i} f(x+i) = \begin{cases} 0, & \text{if } m < n; \\ m!a_m, & \text{if } m = n. \end{cases} \quad (5)$$

In particular, take $f(x) = x^m$, and take $x = 0$ in the above equality. Then we obtain Euler's identity ((6) in Chapter 4)

$$\sum_{i=0}^{n} (-1)^i \binom{n}{i} i^m = \begin{cases} 0, & \text{if } m < n; \\ (-1)^n n!, & \text{if } m = n. \end{cases}$$

Remark 4. When the step size $h = 1$ (it is the most common case), we usually abbreviate $\Delta_1^n f(x)$ to $\Delta^n f(x)$.

Remark 5. The simple fact that the first-order difference of a polynomial of degree m is a polynomial of degree $m - 1$ has many uses. For example, based on this, we can prove some propositions about polynomials by induction (on degrees of polynomials).

(5) Let $f(x)$ be a polynomial of degree n, and its leading coefficient is a_n. Then we have the following identities

$$\sum_{\substack{\varepsilon_i=0,1 \\ 1 \le i \le n}} (-1)^{\varepsilon_1 + \cdots + \varepsilon_n} f(\varepsilon_1 x_1 + \cdots + \varepsilon_n x_n) = (-1)^n n! a_n x_1 \cdots x_n. \quad (6)$$

$$\sum_{\substack{\varepsilon_i=0,1 \\ 1 \le i \le n}} (-1)^{\varepsilon_1 + \cdots + \varepsilon_n} f((-1)^{\varepsilon_1} x_1 + \cdots + (-1)^{\varepsilon_n} x_n) = n! 2^n a_n x_1 \cdots x_n.$$

$$(7)$$

First we prove (6). When $n = 1$, the conclusion is obvious. Assume that the conclusion is true for polynomials of degree $n - 1$. For a polynomial $f(x)$ of degree n, let $g(x) = f(x) - f(x + x_n)$ with $x_n \neq 0$. Then $g(x)$ is a polynomial of degree $n - 1$, and the coefficient of the first term is $-a_n n x_n$.

From the inductive hypothesis,

$$\sum_{\substack{\varepsilon_i=0,1 \\ 1\le i\le n-1}} (-1)^{\varepsilon_1+\cdots+\varepsilon_{n-1}} g(\varepsilon_1 x_1 + \cdots + \varepsilon_{n-1} x_{n-1})$$

$$= (-1)^{n-1} (n-1)! x_1 \cdots x_{n-1} \cdot (-a_n n x_n)$$

$$= (-1)^n n! a_n x_1 \cdots x_n.$$

The left side of the above formula is

$$\sum_{\substack{\varepsilon_i=0,1 \\ 1\le i\le n-1}} (-1)^{\varepsilon_1+\cdots+\varepsilon_{n-1}+0} f(\varepsilon_1 x_1 + \cdots + \varepsilon_{n-1} x_{n-1} + 0\cdot x_n)$$

$$- \sum_{\substack{\varepsilon_i=0,1 \\ 1\le i\le n-1}} (-1)^{\varepsilon_1+\cdots+\varepsilon_{n-1}} f(\varepsilon_1 x_1 + \cdots + \varepsilon_{n-1} x_{n-1} + x_n)$$

$$= \sum_{\substack{\varepsilon_i=0,1 \\ 1\le i\le n}} (-1)^{\varepsilon_1+\cdots+\varepsilon_n} f(\varepsilon_1 x_1 + \cdots + \varepsilon_n x_n).$$

The case of $x_n = 0$ is obvious. This proves that the conclusion is also true for polynomials of degree n.

Similarly, (7) can be proved (inductively), as long as we note that $g(x) = f(x + x_n) - f(x - x_n)$ is a polynomial of degree $n - 1$ (if $x_n \ne 0$) and the coefficient of the first term is $2na_n x_n$. We leave the details to the readers.

Integer-Valued Polynomials

If the value of a complex coefficient polynomial $f(x)$ is an integer when x is an integer, then $f(x)$ is called an **integer-valued polynomial**. All polynomials with integer coefficients are of course integer-valued polynomials. However, resembling the formulae of combination numbers, we also have integer-valued polynomials of any positive degree with non-integer coefficients.

(6) Let $\binom{x}{k} = \frac{x(x-1)\cdots(x-k+1)}{k!}$ (k is a positive integer). Then $\binom{x}{k}$ is an integer-valued polynomial of degree k.

In fact, the polynomial is obviously of degree k, and since the product of k consecutive integers is divisible by $k!$ (Remark 4 in Chapter 1), the value of $\binom{x}{k}$ is an integer when x is an integer.

(7) The polynomial $\binom{x}{k}$ satisfies $\binom{x+1}{k} = \binom{x}{k} + \binom{x}{k-1}$, that is $\Delta \binom{x}{k} = \binom{x}{k-1}$ (see Remark 4).

This can of course be checked directly as defined in (6). A more general method is: Since the equality is true when x values are positive integers, it is a polynomial identity from the identity theorem ((5) in Chapter 12).

The following results give a characterization of integer-valued polynomials.

(8) A sufficient and necessary condition for an nth degree complex coefficient polynomial $f(x)$ to be an integer-valued polynomial is that it can be expressed in the form

$$a_n \binom{x}{n} + a_{n-1} \binom{x}{n-1} + \cdots + a_1 \binom{x}{1} + a_0, \tag{8}$$

where each $a_i (i = 0, 1, \ldots, n)$ is an integer and $a_n \neq 0$.

Therefore, for integer-valued polynomials, $\binom{x}{k}$ are the most basic polynomials.

The condition is obviously sufficient. In order to prove the necessity, we first prove that $f(x)$ can be expressed in the form of (8). Divide $f(x)$ by the nth degree polynomial $\binom{x}{n}$ ((5) in Chapter 11). The quotient must be a constant, that is

$$f(x) - a_n \binom{x}{n} + f_1(x),$$

where a_n is a non-zero complex number, and the remainder $f_1(x)$ is either zero or a polynomial with degree less than n; then divide $f_1(x)$ by the $(n-1)$st degree polynomial $\binom{x}{n-1}$ to obtain

$$f_1(x) = a_{n-1} \binom{x}{n-1} + f_2(x),$$

where $f_2(x) = 0$ or $\deg f_2 < n - 1$, and a_{n-1} is a complex number. By repeating the process, we see that $f(x)$ is expressed in the form of (8) (please note that this representation is obviously unique).

Because $f(x)$ is an integer-valued polynomial, it is known from (2) (taking $h = 1$) that $\Delta^k f(x)$ is an integer-valued polynomial ($k \geq 1$). But clearly we know from (7) and (8) that $f(0) = a_0$ and

$$\Delta f(x) = a_n \binom{x}{n-1} + a_{n-1} \binom{x}{n-2} + \cdots + a_2 \binom{x}{1} + a_1.$$

Therefore, the value of $\Delta f(x)$ at $x = 0$ is a_1. Similarly, the value of $\Delta^k f(x)$ at $x = 0$ is $a_k (1 \le k \le n)$. Thus, $a_0, \ldots, a_n$ are all integers. This argument by the way gives an explanation of the coefficients $a_0, \ldots, a_n$ in (8).

A more direct method can also be used to prove that a_i is an integer. Since $f(0), f(1), \ldots, f(n)$ are integers, from (8), we can successively deduce that a_0 is an integer, a_1 is an integer, $\ldots$, and a_n is an integer.

Illustrative Examples

Example 1. Suppose an nth degree polynomial satisfies

$$f(k) = \frac{1}{\dbinom{n+1}{k}} (k = 0, 1, \ldots, n).$$

Find $f(n+1)$.

Solution   Since the values of $f(x)$ at $n+1$ different points are known, it can be uniquely obtained by Lagrange's interpolation formula that

$$f(x) = \sum_{k=0}^{n} f(k) \prod_{\substack{i \ne k \\ 0 \le i \le n}} \frac{x-i}{k-i}. \tag{9}$$

Thus for any integer $m > n$,

$$f(m) = \sum_{k=0}^{n} f(k) (-1)^{n-k} \frac{m(m-1)\cdots(m-n)}{(m-k)k!(n-k)!}$$

$$= \sum_{k=0}^{n} (-1)^{n-k} f(k) \binom{m}{k} \binom{m-k-1}{n-k}.$$

Therefore, under the condition of the problem,

$$f(n+1) = \sum_{k=0}^{n} (-1)^{n-k} = \begin{cases} 0, & \text{if } n \text{ is odd}; \\ 1, & \text{if } n \text{ is even}. \end{cases}$$

The following method can also be used: Since the $(n+1)$st-order difference of $f(x)$ is zero, we can get the result from formula (5) (replacing m and n

by n and $n+1$ respectively),

$$\sum_{i=0}^{n+1} (-1)^{n+1-i} \binom{n+1}{i} f(x+i) = 0. \tag{10}$$

Take $x = 0$ in (10) and substitute the value of $f(i)$ $(i = 0, 1, \ldots, n)$ to get the above result.

Remark 6. The significance of Lagrange's interpolation formula is mainly in theory. It is difficult to solve the interpolation problem in practice, and the result may not be expressed in a concise form.

It is more convenient to use the difference with step size h for the polynomial interpolation problem at points that form an arithmetic sequence (with common difference h) (refer to Exercise 4 in this chapter). Please compare the above two solutions of Example 1.

By the way, for this problem, we can find the value of $f(x)$ at any point, which can be obtained by the interpolation formula (9) (in principle), while by difference formula (10) (taking $x = \pm 1, \pm 2, \ldots$ in turn), we can only (recursively) obtain the values of $f(x)$ at integer points

Example 2. Let $f(x)$ be a polynomial of degree n, satisfying $f(k) = \frac{1}{k}$ $(k = 1, 2, \ldots, n+1)$. Find $f(n+2)$.

Solution This problem can be solved by two methods used in Example 1, but a simpler and more direct way is to consider (the polynomial of degree $n+1$)$g(x) = xf(x) - 1$. According to the known condition, $g(x)$ has $n+1$ distinct zeros $x = 1, 2, \ldots, n+1$. Because $g(0) = -1$, we get (refer to the solution of (1))

$$g(x) = \frac{(x-1)(x-2)\cdots(x-n-1)}{(-1)^n(n+1)!}.$$

Therefore $g(n+2) = (-1)^n$, and $f(n+2) = \begin{cases} 0, & n \text{ is odd}; \\ \frac{2}{n+2}, & n \text{ is even}. \end{cases}$

Example 3. Let $f(x)$ be an nth degree polynomial, satisfying $f(k) = 2^k$ $(k = 1, 2, \ldots, n+1)$. Find the value of $f(n+3)$. (Express the answer as an explicit function of n and in the simplest form.)

Solution Since $f(x)$ is a polynomial of degree n, its $(n+1)$st order-difference is zero. From (10) (taking $x = 1$) and substituting the values of

$f(k)$ $(k = 1, 2, \ldots, n+1)$, we obtain

$$f(n+2) = -\sum_{i=0}^{n} -(-1)^{n+1-i} \binom{n+1}{i} 2^{i+1}$$

$$= 2^{n+2} - 2 \left(\sum_{i=0}^{n+1} (-1)^{n+1-i} 2^i \binom{n+1}{i} \right)$$

$$= 2^{n+2} - 2(2-1)^{n+1} = 2^{n+2} - 2.$$

Then using (10) (taking $x = 2$), we can similarly get

$$f(n+3) = -\sum_{i=0}^{n-1} (-1)^{n+1-i} \binom{n+1}{i} 2^{i+2} + \binom{n+1}{n} (2^{n+2} - 2)$$

$$= 2^{n+3} - 4 \sum_{i=0}^{n+1} (-1)^{n+1-i} 2^i \binom{n+1}{i} - 2 \binom{n+1}{n}$$

$$= 2^{n+3} - 2n - 6.$$

For the particularity of this question, the following method can also be used (refer to (6)): The function

$$g(x) = 2 \left(\binom{x-1}{0} + \binom{x-1}{1} + \cdots + \binom{x-1}{n} \right)$$

is a polynomial of degree n, and obviously $g(k) = 2^k (1 \le k \le n+1)$. Since the values of two nth degree polynomials $f(x)$ and $g(x)$ are equal at $n+1$ different points, the identity theorem of polynomials shows that $f(x) = g(x)$. Now it's easy to find

$$f(n+3) = g(n+3)$$

$$= 2 \left(2^{n+2} - \binom{n+2}{n+1} - \binom{n+2}{n+2} \right)$$

$$= 2^{n+3} - 2n - 6.$$

Example 4. Let $x_1, x_2, \ldots, x_{n+1}$ be any $n+1$ distinct integers. Prove that among the $n+1$ values of any nth degree polynomial $x^n + a_1 x^{n-1} + \cdots + a_n$ taken at points $x_1, x_2, \ldots, x_{n+1}$, at least one of them has absolute value $\ge \frac{n!}{2^n}$.

Proof. Let the polynomial be $f(x)$. By Lagrange's interpolation formula,

$$f(x) = \sum_{i=1}^{n+1} f(x_i) \frac{\prod_{k \neq i}(x - x_k)}{\prod_{k \neq i}(x_i - x_k)}.$$

Since the leading coefficient of $f(x)$ is 1, from the above formula (refer to Remark 3),

$$\sum_{i=1}^{n+1} f(x_i) \frac{1}{\prod_{k \neq i}(x_i - x_k)} = 1.$$

Let M be the maximum value of $|f(x_i)|$ $(i = 1, \ldots, n+1)$. Then

$$M \cdot \sum_{i=1}^{n+1} \frac{1}{|\prod_{k \neq i}(x_i - x_k)|} \geq 1.$$

But $x_1, \ldots, x_{n+1}$ are different integers (suppose $x_1 < \cdots < x_{n+1}$), so

$$\left| \prod_{k \neq i}(x_i - x_k) \right|$$

$$= |(x_i - x_1) \cdots (x_i - x_{i-1})(x_i - x_{i+1}) \cdots (x_i - x_{n+1})|$$

$$\geq (i-1) \cdots 1 \cdots (n+1-i)$$

$$= (i-1)!(n+1-i)!.$$

Hence,

$$1 \leq M \sum_{i=1}^{n+1} \frac{1}{(i-1)!(n+1-i)!}$$

$$= M \sum_{i=0}^{n} \frac{1}{i!(n-i)!}$$

$$= \frac{M}{n!} \sum_{i=0}^{n} \binom{n}{i} = \frac{M}{n!} 2^n.$$

Thus $M \geq \frac{n!}{2^n}$.

Example 5. Let p be a prime number. Find the minimum positive integer d such that there exists an integer coefficient polynomial $f(x)$ of degree d with the leading coefficient 1, which satisfies that $p^{p+1} \mid f(n)$ for all integers n.

Solution   Let $\Delta f(x) = f(x+1) - f(x)$. It can be seen from the two aspects of $\Delta^d f(x)$ (see (5)) that

$$\sum_{i=0}^{d} (-1)^{d-i} \binom{d}{i} f(i) = d!.$$

If p^{p+1} exactly divides all $f(n)$ (n is an integer), then the left side of the above formula is divisible by p^{p+1}, so p^{p+1} exactly divides $d!$, which implies that $d \ge p^2$.

This is because if $d < p^2$, then the power of p in $d!$ is

$$\sum_{i=1}^{\infty} \left[\frac{d}{p^i} \right] < \sum_{i=1}^{\infty} \frac{d}{p^i} = \frac{d}{p-1} \le \frac{p^2-1}{p-1} = p+1,$$

which is contradictory to $p^{p+1} \mid d!$, so $d \ge p^2$.

In addition, $f(x) = (x-1) \cdots (x-p^2)$ is an integer coefficient polynomial of degree p^2 with the leading coefficient 1. For any integer n, the value $f(n)$ is the product of p^2 consecutive integers, so it is divisible by $p^2!$. And since the power of p in $p^2!$ is $p+1$, we have $p^{p+1} \mid f(n)$ (for all integers n).

According to the above results, the minimum positive integer $d = p^2$.

The following method can also be used to prove $p^{p+1} \mid d!$:

The known condition shows that $\frac{f(x)}{p^{p+1}}$ is an integer-valued polynomial, so it can be expressed as (see (8))

$$\frac{f(x)}{p^{p+1}} = a_d \binom{x}{d} + \cdots + a_1 \binom{x}{1} + a_0,$$

where $a_0, a_1, \ldots, a_d$ are all integers.

By comparing the coefficients of x^d in the above formula, we see that (note that the leading coefficient of $f(x)$ is 1) $\frac{1}{p^{p+1}} = \frac{a_d}{d!}$, that is, $d! = p^{p+1} a_d$. Since a_d is an integer, $p^{p+1} \mid d!$.

Exercises

Group A

1. Find a polynomial $f(x)$ of degree less than 4, which satisfies $f(-1) = 2i + 2$, $f(0) = 1$, $f(1) = 0$ and $f(2) = 2i - 1$, where $i = \sqrt{-1}$.
2. Prove that the polynomial $\frac{1}{24}x^4 + \frac{1}{12}x^3 - \frac{25}{24}x^2 + \frac{11}{12}x + 1$ is an integer-valued polynomial.

3. Let $f(x)$ be a polynomial of degree m. Prove that for any h and integer $n > m$,

$$\sum_{i=0}^{n} (-1)^{n-i} \binom{n}{i} f(x + ih) = 0.$$

4. Let $f(x)$ be a polynomial of degree n, whose values at consecutive $n+1$ integers are all integers. Prove that $f(x)$ is an integer-valued polynomial.

Group B

5. Let $f(x)$ be a polynomial of degree $2n$, $f(0) = f(2) = \cdots = f(2n) = 0$, $f(1) = f(3) = \cdots = f(2n-1) = 2$, and $f(2n+1) = -30$. Find the value of n.

6. Let $f(x)$ be a polynomial of degree n, satisfying $f(k) = \frac{k}{k+1}(k = 0, 1, \ldots, n)$. Find the value of $f(n+1)$.

7. Suppose the degree of a real coefficient polynomial $f(x)$ does not exceed $2n$, and for each integer k $(-n \le k \le n)$, there is $|f(k)| \le 1$. Prove: For each real number $x(-n \le x \le n)$, there is $|f(x)| \le 2^{2n}$.

8. Let $a_1, \ldots, a_n$ be distinct positive integers. Prove that for any positive integer k,

$$\sum_{i=1}^{n} \frac{a_i^k}{\prod_{j \ne i}(a_i - a_j)}$$

is an integer.

Chapter 15

Roots of Unity and Their Applications

Let $\zeta = e^{\frac{2\pi i}{n}} = \cos\frac{2\pi}{n} + i\sin\frac{2\pi}{n}$. The polynomial $x^n - 1$ has n distinct roots $\zeta, \zeta^2, \ldots, \zeta^n(=1)$, called the **$n$th roots of unity**. Clearly we see that in the complex plane, the nth roots of unity (or their corresponding points) are exactly the vertices of a regular n-sided polygon inscribed in the unit circle.

The nth roots of unity have the following basic properties:

(1) Let k and l be integers. Then $\zeta^k = \zeta^l$ if and only if $k \equiv l(\mathrm{mod}\, n)$.
(2) The product of any two nth roots of unity is still an nth root of unity; the reciprocal (i.e., (-1)st power) of any nth root of unity is also an nth root of unity.
(3) Let k be an integer, and $(k, n) = 1$. Then $\left(\zeta^{kl}\right)^l (l = 1, 2, \ldots, n)$ exactly give all nth roots of unity.

In fact, (1) and (2) are obvious corollaries of the definition. To prove (3), just note that because $(k, n) = 1$, the numbers $k, 2k, \ldots, nk$ form a complete system modulo n (Chapter 8, (11)). Combining it with (1), we know that $\zeta^{kl}(l = 1, 2, \cdots, n)$ give all nth roots of unity.

Remark 1. The complex number $\zeta = e^{\frac{2\pi i}{n}}$ is the most basic nth root of unity, because its integer powers give all nth roots of unity, and such a root of unity is called a **primitive nth root of unity**. (3) shows that for $(k, n) = 1(1 \leq k \leq n)$, ζ^k is also a primitive nth root of unity. It can be proved that these are all the primitive nth roots of unity, so the number of primitive nth roots of unity is $\varphi(n)$.

Because $\zeta, \zeta^2, \ldots, \zeta^n$ are n different roots of $x^n - 1$, there is the decomposition

$$x^n - 1 = (x - 1)(x - \zeta) \cdots (x - \zeta^{n-1}).$$

And $x^n - 1 = (x - 1)(1 + x + \cdots + x^{n-1})$. Thus

(4) $1 + x + \cdots + x^{n-1} = (x - \zeta) \cdots (x - \zeta^{n-1}).$

In (4), we make a reciprocal substitution (i.e., replace x by $\frac{1}{x}$, and multiply x^{n-1} to both sides) to obtain (the equivalent).

(5) $1 + x + \cdots + x^{n-1} = (1 - x\zeta) \cdots (1 - x\zeta^{n-1}).$

The following (6) and (7) are basic conclusions about the sum of roots of unity.

(6) $1 + \zeta + \zeta^2 + \cdots + \zeta^{n-1} = 0.$
(7) Let k be an integer. Then

$$\frac{1}{n} \sum_{t=1}^{n} \zeta^{kt} = \begin{cases} 1, & \text{if } n \mid k; \\ 0, & \text{if } n \nmid k. \end{cases}$$

The equality (6) is a special case of (7) and can also be derived from (4) (take $x = \zeta$). As for (7), when $n \mid k$, the conclusion is obvious; when $n \nmid k$, we have $\zeta^k \neq 1$ (see (1)), so by the formula of the sum of a geometric series, we can draw the conclusion.

Remark 2. From $\zeta^n = 1$ and (6), we can see that for any polynomial $f(x)$, the value $f(\zeta)$ is either 0 or can be expressed as a polynomial of ζ whose degree does not exceed $n - 2$, that is

$$f(\zeta) = a_0 + a_1\zeta + \cdots + a_{n-2}\zeta^{n-2}, \tag{1}$$

where a_k $(k = 0, 1, \ldots, n - 2)$ are obtained by adding and subtracting the coefficients of $f(x)$ (it is not difficult to describe the relationship between the two groups of coefficients using (1)). In particular, if $f(x)$ is a polynomial with integer coefficients, then the coefficients in (1) are also integers.

Remark 3. Let n be a fixed positive integer. For any integer k, define

$$f(k) = \begin{cases} 1, & \text{if } n \mid k; \\ 0, & \text{if } n \nmid k. \end{cases} \tag{2}$$

This is called the characteristic function of the numbers divisible by n (also called the characteristic function of the set of numbers divisible by n. Please refer to Chapter 17). In problems involving divisibility, (2) only gives a formal representation, so it is of limited use.

The formula (7) gives an analytic representation of (2), so it can be used to separate the numbers that form an arithmetic sequence in a (finite) set of integers, which is widely used:

Let n and r be fixed integers, $0 \leq r < n$, and S be a (finite) set of integers. Then the number N_r of the integer in S satisfying $\equiv r \pmod n$ can be expressed as

$$N_r = \sum_{K \in S} f(k - r) = \frac{1}{n} \sum_{K \in S} \sum_{t=1}^{n} \zeta^{(k-r)t}$$

$$= \frac{1}{n} \sum_{t=1}^{n} \zeta^{-rt} \sum_{K \in S} \zeta^{kt}. \tag{3}$$

(In the last step, we exchanged the orders of summations of t and k. For this, please refer to Remark 4 below.)

Let S be a set of subscripts of a sequence $\{a_i\}$. Then we can separate the terms of subscripts $\equiv r \pmod n$ from the sequence. For example, if the sum of these terms is A_r, then

$$A_r = \sum_{K \in S} a_k f(k - r) = \frac{1}{n} \sum_{K \in S} a_k \sum_{t=1}^{n} \zeta^{(k-r)t}$$

$$= \frac{1}{n} \sum_{t=1}^{n} \zeta^{-rt} \sum_{K \in S} a_k \zeta^{kt}. \tag{4}$$

In elementary problems, the inner sums of (3) and (4) can usually be decomposed into the product of simpler polynomials (with regard to ζ), which is easy to calculate, so that N_r and A_r can be obtained successfully.

Remark 4. In (3) and (4), we both exchanged the orders of summations. This technique is widely used. We will give a brief introduction here by the way.

Let $a_{i,j}$ be a sequence of numbers with two (ordered) subscripts ($a_{i,j}$ can also be denoted as a_{ij}). The double sum

$$\sum_{i=1}^{m} \sum_{j=1}^{n} a_{i,j}$$

can be interpreted as: For a fixed i, sum over j in the inner first, and then sum over i, which is to sum the numbers in the following number array by rows.

$$a_{1,1}, a_{1,2}, \ldots, a_{1,n};$$

$$\cdots \cdots$$

$$a_{i,1}, a_{i,2}, \ldots, a_{i,n};$$

$$\cdots \cdots$$

$$a_{m,1}, a_{m,2}, \ldots, a_{m,n}.$$

On the other hand, the double sum

$$\sum_{j=1}^{n}\sum_{i=1}^{m} a_{i,j}$$

(first sum over i, and then sum over j) is to sum the numbers in the number array by columns. Because the results of the two summations are the same (that is, the orders of summations can be changed),

$$\sum_{i=1}^{m}\sum_{j=1}^{n} a_{i,j} = \sum_{j=1}^{n}\sum_{i=1}^{m} a_{i,j}.$$

Generally speaking, if the ranges of i and j in $a_{i,j}$ are (finite) integer sets I and J respectively, then (similarly) we have

$$\sum_{i\in I}\sum_{j\in J} a_{i,j} = \sum_{j\in J}\sum_{i\in I} a_{i,j}. \tag{5}$$

Changing the orders of summations provides an alternative way of looking at the original sum, which often makes it easier to deal with (calculation, estimation, etc.).

Illustrative Examples

Example 1. Let $P(x)$, $Q(x)$, $R(x)$, and $S(x)$ be polynomials, and

$$P(x^5) + xQ(x^5) + x^2 R(x^5) = (x^4 + x^3 + x^2 + x + 1)S(x). \tag{6}$$

Prove: $x - 1$ is a common factor of $P(x)$, $Q(x)$, $R(x)$, and $S(x)$.

Proof. Let ζ be a 5th root of unity ($\zeta \neq 1$), and take $x = \zeta, \zeta^2, \zeta^3$, and ζ^4 in (6). Then

$$\left(\zeta^k\right)^2 R(1) + \zeta^k Q(1) + P(1) = 0 \ (k = 1, 2, 3, 4),$$

which means that the polynomial $x^2 R(1) + x Q(1) + P(1)$ has four different zeros, so $R(1) = Q(1) = P(1) = 0$ ((4) in Chapter 12). Substituting $x = 1$ into (6), we get $S(1) = 0$. Thus $P(x)$, $Q(x)$, $R(x)$, and $S(x)$ all have the factor $x - 1$. $\qquad\square$

Example 2. Let n be a positive integer. Prove that

$$\binom{n}{0} + \binom{n}{3} + \binom{n}{6} + \cdots = \frac{1}{3}\left(2^n + 2\cos\frac{n\pi}{3}\right).$$

$$\binom{n}{1} + \binom{n}{4} + \binom{n}{7} + \cdots = \frac{1}{3}\left(2^n + 2\cos\frac{(n-2)\pi}{3}\right).$$

$$\binom{n}{2} + \binom{n}{5} + \binom{n}{8} + \cdots = \frac{1}{3}\left(2^n + 2\cos\frac{(n-4)\pi}{3}\right).$$

Proof. Let ω be a cubic root of unity ($\omega \neq 1$). From formula (4) (take $n = 3$), we know that for $r = 0, 1$, and 2,

$$S_r = \binom{n}{r} + \binom{n}{r+3} + \binom{n}{r+2\cdot3} + \cdots$$

$$= \frac{1}{3}\sum_{t=1}^{3}\omega^{-rt}\sum_{k=0}^{n}\binom{n}{k}\omega^{kt}$$

$$= \frac{1}{3}\sum_{t=1}^{3}\omega^{-rt}\left(1+\omega^t\right)^n.$$

Since $\bar{\omega} = \omega^2$, we see that $\omega^{-r}(1+\omega)^n = (-1)^n \omega^{2n-r}$ and $\omega^{-2r}(1+\omega^2)^n = (-1)^n \omega^{n-2r}$ are conjugate, and the real part of $(-1)^n \omega^{n-2r}$ is $\cos\frac{(n-2r)\pi}{3}$. Therefore

$$S_r = \frac{2^n}{3} + \frac{2}{3}\cos\frac{(n-2r)\pi}{3}.$$

The following method can also be used for this problem (the essence is the same as above): In the binomial expansion

$$(1+x)^n = \binom{n}{0} + \binom{n}{1}x + \cdots + \binom{n}{n}x^n,$$

taking $x = 1, \omega$, and ω^2 in turn, we obtain

$$S_0 + S_1 + S_2 = 2^n,$$

$$S_0 + \omega S_1 + \omega^2 S_2 = (1 + \omega)^n,$$

$$S_0 + \omega^2 S_1 + \omega S_2 = (1 + \omega^2)n.$$

By solving this system of equations, S_0, S_1, and S_2 can be obtained.   □

Example 3. Prove that there doesn't exist four integer coefficient polynomials $f_k(x)$ ($k = 1, 2, 3$, and 4) such that the identity

$$9x + 4 = f_1^3(x) + f_2^3(x) + f_3^3(x) + f_4^3(x) \tag{7}$$

is true.

Proof. The problem seems ordinary, but it's not easy to do it.

Let ω be a cubic root of unity ($\omega \neq 1$). Then for any integer coefficient polynomial $f(x)$, from Remark 2, we have $f(\omega) = a + b\omega$ (a and b are integers). Thus (note $1 + \omega + \omega^2 = 0$)

$$f^3(\omega) = (a + b\omega)^3 = a^3 + b^3 - 3ab^2 + 3ab(a - b)\omega.$$

Since $ab(a - b)$ is always even, if there is an identity in the form of (7), then substituting $x = \omega$, we get

$$9\omega + 4 = A + B\omega. \tag{8}$$

Here both A and B are integers, and B is an even number. Comparing the real part and the imaginary part of (8), we get $B = 9$, which is obviously impossible.   □

Remark 5. If the number of polynomials on the right side of formula (7) is less than three, then the problem can be solved simply by considering congruence: Take $x = n$ as any integer, and then $9n + 4 = f_1^3(n) + f_2^3(n) + f_3^3(n)$.

In particular, this implies that any integer ($\equiv 4 \pmod 9$) can be expressed as the cubic sum of three integers. However, it's easy to know from modulo 9 that this is impossible (see Exercise 5 in Chapter 9). But this consideration can't work on this problem.

The basic idea of the proof of Example 3 is to take x as a root of unity, so as to greatly simplify the right side of (7). The purpose of taking $x = \omega$

is to use $\omega^3 = 1$ and expect that the cube of (simplified) $f(\omega)$ can produce applicable information, deriving contradictions or as a basis for a further demonstration. (See Remark 2 in Chapter 9 to compare with the effect of the congruence method.)

Remark 6. It is not difficult to see from the solution of Example 3 that the number of polynomials on the right side of formula (7) and the constant 4 on the left side are both irrelevant. The key is that the coefficient of the linear term (on the left side) is an odd number. Note that for some (linear) polynomials with an even coefficient of the linear term, there exists such an identity, for example,

$$6x = (x+1)^3 + (x-1)^3 + (-x)^3 + (-x)^3. \tag{9}$$

Remark 7. Example 3 is related to a famous problem in number theory. We will make a brief introduction here by the way. We know that there is an infinite number of integers that cannot be expressed as the cubic sum of three integers. It is conjectured that every integer can be expressed as the cubic sum of four integers. This problem is extremely difficult and has not been solved yet, but for the integer n such that $n \not\equiv \pm 4 (\mathrm{mod}\, 9)$, it has been proved that the conjecture is correct.

On the other hand, it is not difficult to prove that every integer n is the cubic sum of five integers. In fact, note that $n^3 - n$ is always a multiple of 6. In (9), replace x by $\frac{n-n^3}{6}$. Then the conclusion is true. This solution is based on the identity (9). The conclusion of Example 3 shows that it is not possible to prove that an integer of the form $9k \pm 4$ can be expressed as the cubic sum of four integers by establishing the identity (7).

From the geometric meaning of the roots of unity, we may expect that they can be used to solve some problems involving regular polygons. Let's give a simple example.

Example 4. Let $A_1 \cdots A_n$ be a regular n-sided polygon inscribed in the unit circle and P be a point on the circumference. Prove:

$$\sum_{k=1}^{n} |PA_k|^2 = 2n.$$

Proof. Let $\zeta = \mathrm{e}^{\frac{2\pi i}{n}}$, and $A_1, \ldots, A_n$ correspond to the complex numbers $1, \zeta, \zeta^2, \cdots, \zeta^{n-1}$. Let (the complex number of) the point P be $z = \mathrm{e}^{i\theta}$.

Then

$$\sum_{k=1}^{n} |PA_k|^2 = \sum_{k=0}^{n-1} |z - \zeta^k|^2 = \sum_{k=0}^{n-1} (z - \zeta^k)(\bar{z} - \zeta^{-k})$$

$$= \sum_{k=0}^{n-1} (|z|^2 - \zeta^k \bar{z} - \zeta^{-k} z + 1)$$

$$= 2n - \bar{z} \sum_{k=0}^{n-1} \zeta^k - z \sum_{k=0}^{n-1} \zeta^{-k} = 2n.$$

(The last step applied (6).)   □

The following problem is very difficult.

Example 5. For any n points $z_1, \ldots, z_n$ on the unit circle, prove:

$$\max_{|z|=1} |z - z_1| \cdots |z - z_n| \geq 2, \qquad (10)$$

and prove that the equality holds if and only if $z_1, \ldots, z_n$ constitute the vertices of a regular n-sided polygon.

Proof. By a proper rotation, we can suppose $z_1 \cdots z_n = (-1)^n$. Let

$$P(z) = (z - z_1) \cdots (z - z_n)$$

$$= z^n + a_1 z^{n-1} + \cdots + a_{n-1} z + 1$$

$$= z^n + f(z) + 1,$$

where $f(z)$ is either zero or its degree does not exceed $n - 1$. Let $\zeta_1, \ldots, \zeta_n$ be all the nth roots of unity. Then from (7),

$$\sum_{j=1}^{n} \zeta_j^k = 0 \, (k = 1, \ldots, n - 1).$$

Hence,

$$f(\zeta_1) + \cdots + f(\zeta_n) = 0.$$

Therefore, if $f(z)$ is not the zero polynomial, then there exists j that makes $f(\zeta_j) \neq 0$ and $\mathrm{Re} f(\zeta_j) \geq$, so $|P(\zeta_j)| = |2 + f(\zeta_j)| > 2$. If $f(z) = 0$, then of course $|P(\zeta_j)| = 2$. This proves (10).

The above argument also shows that if (10) is an equality, then $f(\zeta_j) = 0 \, (j = 1, 2, \ldots, n - 1)$, which implies that $f(z) = 0$, that is, $P(z) = 1 + z^n$, so $z_1, \ldots, z_n$ are the vertices of a regular n-sided polygon.

Exercises

Group A

1. Prove: $x^{44} + x^{33} + x^{22} + x^{11} + 1$ is divisible by $x^4 + x^3 + x^2 + x + 1$.
2. Let $f(x) = x^4 + x^3 + x^2 + x + 1$. Find the remainder when $f(x^5)$ is divided by $f(x)$.
3. Let $f(x)$ be a complex coefficient polynomial and n be a positive integer. If $(x-1) \mid f(x^n)$, prove that

$$(x^n - 1) \mid f(x^n).$$

4. Prove: $\binom{n}{0} + \binom{n}{4} + \binom{n}{8} + \cdots = \frac{1}{2}\left(2^{n-1} + 2^{\frac{n}{2}}\cos\frac{n\pi}{4}\right).$
5. Let $A_1 \cdots A_n$ be a regular n-sided polygon inscribed in the unit circle and P be a point on the circumference. Prove:

   (i) $|A_1 A_2| \cdot |A_1 A_3| \cdot \cdots \cdot |A_1 A_n| = n.$
   (ii) $\sum_{j,k=1}^{n} |A_j A_k|^2 = 2n^2.$

Group B

6. Let $\left(1 + x + x^2\right)^n = a_0 + a_1 x + \cdots + a_{2n} x^{2n}$. Prove:

   (i)

$$a_0 + a_2 + a_4 + \cdots = \frac{1}{2}(3^n + 1),$$

$$a_1 + a_3 + a_5 + \cdots = \frac{1}{2}(3^n - 1).$$

   (ii)

$$a_0 + a_3 + a_6 + \cdots = a_1 + a_4 + a_7 + \cdots$$

$$= a_2 + a_5 + a_8 + \cdots = 3^{n-1}.$$

7. Try to find all positive integers m and n such that the polynomial $1 + x + x^{2n} + \cdots + x^{mn}$ is divisible by $1 + x + x^2 + \cdots + x^m$.
8. Let $P(x)$ be a complex coefficient polynomial of degree $3n$ that satisfies

$$P(0) = P(3) = \cdots = P(3n) = 2,$$

$$P(1) = P(4) = \cdots = P(3n - 2) = 1,$$

$$P(2) = P(5) = \cdots = P(3n - 1) = 0,$$

and $P(3n + 1) = 730$. Try to determine n.

Chapter 16

Generating Function Method

The generating function method is widely used. The idea of this method is very simple, that is, to correspond discrete sequences to a kind of functions, and to determine the properties of a discrete sequence by studying the corresponding function.

In elementary mathematics, there are two kinds of basic correspondences.

The first kind is to correspond a (finite) sequence

$$a_0, a_1, \ldots, a_n \tag{1}$$

to the polynomial

$$A_c(x) = a_0 + a_1 x + \cdots + a_n x^n. \tag{2}$$

We call (2) the generating function of sequence (1) (The subscript c of $A_c(x)$ indicates that the coefficients of this polynomial correspond to a sequence).

The second kind is to correspond a (finite) positive integer (or integer) sequence

$$a_1, a_2, \ldots, a_n \tag{3}$$

to the function

$$A_e(x) = x^{a_1} + x^{a_2} + \cdots + x^{a_n}. \tag{4}$$

(4) is also called the generating function of (3), but now it refers to the exponents of the generating function rather than the coefficients (the subscript e of $A_e(x)$ indicates this). When the numbers in (3) are all non-negative, $A_e(x)$ is a polynomial, and if there is a negative integer in $a_1, a_2, \ldots, a_n$,

then $A_e(x)$ is a rational function (can be expressed as the quotient of two polynomials).

In the first kind of generating functions, the most common or perhaps the most important one is

$$(1+x)^n = \binom{n}{0} + \binom{n}{1} x + \cdots + \binom{n}{n} x^n, \tag{5}$$

which is a basic starting point for the study of the binomial coefficients.

The second kind of generating functions is mainly used to deal with some problems involving the sum of integers. The following basic properties provide such a connection.

(1) If the generating functions of integer sequences $a_1, a_2, \ldots, a_m$ and $b_1, b_2, \ldots, b_n$ are $A_e(x)$ and $B_e(x)$ respectively, then the generating function of the integer sequence

$$a_i + b_j \, (1 \le i \le m, \, 1 \le j \le n)$$

is $A_e(x)B_e(x)$ (that is, the sum of integer sequences corresponds to the product of the generating functions).

In addition, for each integer k, the coefficient of x^k in $A_e(x)B_e(x)$ is the number of the solutions of $k = a_i + b_j$.

Proving combinatorial identities is the most typical application of the generating function method. Let's give a simple example to illustrate the general idea.

Example 1. Prove:

(i) $\binom{m}{0}\binom{n}{k} + \binom{m}{1}\binom{n}{k-1} + \cdots + \binom{m}{k}\binom{n}{0} = \binom{m+n}{k}$.

(ii) $\binom{n}{0}^2 - \binom{n}{1}^2 + \binom{n}{2}^2 - \cdots + (-1)^n \binom{n}{n}^2 = \begin{cases} 0, & \text{if } n \text{ is odd;} \\ (-1)^{\frac{n}{2}} \binom{n}{\frac{n}{2}}, & \text{if } n \text{ is even.} \end{cases}$

Proof. To prove (i), consider the coefficient of x^k in $(1+x)^m (1+x)^n$. On one hand, in the polynomial

$$(1+x)^m (1+x)^n$$

$$= (1+x)^{m+n}$$

$$= \binom{m+n}{0} + \cdots + \binom{m+n}{k} x^k + \cdots + \binom{m+n}{m+n} x^{m+n},$$

the coefficient of x^k is $\binom{m+n}{k}$.

On the other hand, multiplying the expansions

$$(1+x)^m = \binom{m}{0} + \binom{m}{1} x + \cdots + \binom{m}{i} x^i + \cdots + \binom{m}{m} x^m,$$

$$(1+x)^n = \binom{n}{0} + \binom{n}{1} x + \cdots + \binom{n}{j} x^j + \cdots + \binom{n}{n} x^n,$$

we have the coefficient of x^k to be

$$\binom{m}{0}\binom{n}{k} + \binom{m}{1}\binom{n}{k-1} + \cdots + \binom{m}{k}\binom{n}{0}.$$

Combining the two results, we obtain (i).

The proof of (ii) is similar to (i). On one hand, multiply the expansions of $(1+x)^n$ and $(1-x)^n$, and it can be seen that the coefficient of x^n in $(1+x)^n (1-x)^n$ is

$$\binom{n}{0}\binom{n}{n} - \binom{n}{1}\binom{n}{n-1} + \cdots + (-1)^n \binom{n}{n}\binom{n}{0}$$

$$= \binom{n}{0}^2 - \binom{n}{1}^2 + \cdots + (-1)^n \binom{n}{n}^2.$$

On the other hand, replace x with $-x^2$ in (5), and we get

$$(1+x)^n (1-x)^n$$

$$= \left(1 - x^2\right)^n$$

$$= \binom{n}{0} - \binom{n}{1} x^2 + \cdots + (-1)^k \binom{n}{k} x^{2k} + \cdots + (-1)^n \binom{n}{n} x^{2n}.$$

Hence, when n is even, the coefficient of x^n is $(-1)^{\frac{n}{2}} \binom{n}{\frac{n}{2}}$; when n is odd, the coefficient of x^n is 0. This proves (ii). □

Example 1 shows a basic idea of proving combinatorial identities by the generating function method: (According to the characteristics of the problem.) Consider an appropriate generating function and use two methods to calculate the coefficient of a certain term, and the results are obtained by combining them. Here are two more examples.

Using two methods to calculate the same quantity, sometimes called double calculations, has appeared before in this book (refer to Remark 2 in Chapter 2).

Example 2. Let $\left(1 + x + x^2\right)^n = a_0 + a_1 x + \cdots + a_{2n} x^{2n}$. Prove:

$$a_k - \binom{n}{1} a_{k-1} + \cdots + (-1)^k \binom{n}{k} a_0 = \begin{cases} 0, & \text{if } 3 \nmid k; \\ (-1)^l \binom{n}{l}, & \text{if } k = 3l. \end{cases}$$

Proof. From the given expansion and the expansion of $(1 - x)^n$, it is easy to know that the left side of the equality to be proved is the coefficient of x^k in

$$(1 - x)^n \left(1 + x + x^2\right)^n.$$

On the other hand, in

$$\begin{aligned} &(1 - x)^n \left(1 + x + x^2\right)^n \\ &= \left(1 - x^3\right)^n \\ &= \binom{n}{0} - \binom{n}{1} x^3 + \binom{n}{2} x^6 - \cdots + (-1)^n \binom{n}{n} x^{3n}, \end{aligned}$$

the coefficient of x^k is (if $3 \nmid k$) or $(-1)^l \binom{n}{l}$ (if $k = 3l$). The result is obtained by combining the two aspects. $\qquad\square$

Example 3. Prove:

$$\sum_{k=0}^{n} \binom{n}{k} 2^{n-k} \binom{k}{\left[\frac{k}{2}\right]} = \binom{2n+1}{n}.$$

Proof. There are some skills to solve this problem. We note that the general term in the expansion of $\left(x^{-1} + x\right)^k$ is $\binom{k}{i} x^{2i-k}$. When k is odd, $\binom{k}{\left[\frac{k}{2}\right]} = \binom{k}{\frac{k-1}{2}}$ is the coefficient of x^{-1}; when k is even, $\binom{k}{\left[\frac{k}{2}\right]} = \binom{k}{\frac{k}{2}}$ is the constant term. Hence the constant term in $(1 + x)\left(x^{-1} + x\right)^k$ is $\binom{k}{\left[\frac{k}{2}\right]}$, and the left side of the equality to be proved is the constant term in

$$\sum_{k=0}^{n} \binom{n}{k} 2^{n-k} (1 + x) \left(x^{-1} + x\right)^k.$$

On the other hand, the above expression equals

$$(1 + x) \left(2 + x^{-1} + x\right)^n = \frac{1}{x^n} (1 + x) \left(x^2 + 2x + 1\right)^n = \frac{(x+1)^{2n+1}}{x^n},$$

so its constant term is obviously $\binom{2n+1}{n}$, which proves the result. $\qquad\square$

Example 4. Let p be prime. Prove:

$$\sum_{j=0}^{p} \binom{p}{j}\binom{p+j}{j} \equiv 2^p + 1 \pmod{p^2}.$$

Proof This problem is Example 7 in Chapter 10. Here we use the generating function method to solve it. The idea of demonstration is to convert the sum into an expression that is easier to deal with (in the sense of modulo p^2). For this problem, it is not very difficult to use the generating function (5) of combination numbers:

The sum in the problem is $\sum_{j=0}^{p} \binom{p}{j}\binom{p+j}{p}$, which is the coefficient of x^p in the polynomial

$$\sum_{j=0}^{p} \binom{p}{j}(1+x)^{p+j} = \left(\binom{p}{j}(1+x)^j\right)(1+x)^p$$

$$= ((1+x)+1)^p (1+x)^p$$

$$= (2+x)^p (1+x)^p.$$

Thus it equals

$$\sum_{j=0}^{p} \binom{p}{j}\binom{p}{p-j}2^j = \sum_{j=0}^{p}\binom{p}{j}^2 2^j.$$

(So far, we don't need p to be prime.) When p is prime, $p\,\Big|\,\binom{p}{j}$ ($j \neq 0$, and p), so

$$\sum_{j=0}^{p}\binom{p}{j}\binom{p+j}{j} \equiv \binom{p}{0}^2 2^0 + \binom{p}{p}^2 2^p \equiv 1 + 2^p \pmod{p^2}.$$

Example 5 is a combination problem that allows a limited number of repetitions, which is more complex than the combination with unlimited repetitions in (6) of Chapter 1, and it is more convenient to deal with it by the generating function method.

Example 5. There are 2 red balls, 3 black balls, and 5 white balls (no difference among the balls in the same color). How many different ways are there to select 6 balls from them?

Solution One selection method corresponds to a group of nonnegative integer solutions of the equation $x + y + z = 6$ ($x \leq 2$, $y \leq 3$, and $z \leq 5$), and vice versa. It can be seen from (1) that the number of ways is the coefficient of x^6 in the expansion of the polynomial

$$(1 + x + x^2)(1 + x + x^2 + x^3)(1 + x + x^2 + x^3 + x^4 + x^5)$$

The above expression is equivalent to

$$(1 + 2x + 3x^2 + 3x^3 + 2x^4 + x^5)(1 + x + x^2 + x^3 + x^4 + x^5),$$

so we clearly see that the coefficient of x^6 is $2 + 3 + 3 + 2 + 1 = 11$, that is, there are 11 ways in total.

Remark 1. Generally speaking, suppose n elements are divided into k groups, such that the elements in the same group are considered the same as each other and the elements in different groups are different. Let the number of elements in the k groups be $n_1, \ldots, n_k$ $(n_1 + \cdots + n_k = n)$, and let $a_r (0 \leq r \leq n)$ be the total number of the different methods to take r from these n elements each time. Then the generating function of the sequence

$$a_0, a_1, \ldots, a_n$$

is

$$(1 + x + \cdots + x^{n_1})(1 + x + \cdots + x^{n_2}) \cdots (1 + x + \cdots + x^{n_k}), \qquad (6)$$

this is, a_r is the coefficient of x^r in the expansion of (6).

At the end of this chapter, we introduce some (combinatorial) problems that can be solved by generating functions. The readers will see that linking a problem to a (appropriate) generating function is only the first step in solving the problem. The key is to further study the relevant generating function to produce the required result. There are few ways to deal with generating functions in elementary mathematics. The basic way is to apply some properties of polynomials and the roots of unity.

Example 6. Let $n > 1$. Suppose two n-element sets of rational numbers (elements can be repeated) satisfy

$$\{a_1, \ldots, a_n\} \neq \{b_1, \ldots, b_n\}, \qquad (7)$$

and

$$\{a_i + a_j | 1 \leq i < j \leq n\} = \{b_i + b_j | 1 \leq i < j \leq n\}. \qquad (8)$$

(The equality here includes the multiplicity of elements, that is, the numbers of times that any number appears on the left and right sides of (8) are the same.) Prove: n is a power of 2.

Proof. First of all, for any $c \neq 0$, the sets $\{a_1 c, \ldots, a_n c\}$ and $\{b_1 c, \ldots, b_n c\}$ still have the properties in the problem. In particular, let c be a common denominator of rational numbers $a_1, \ldots, a_n, b_1, \ldots, b_n$. Then we can assume that $a_1, \ldots, a_n, b_1, \ldots, b_n$ are all integers. Furthermore, adding a large enough positive integer to all a_i and b_i will not change the problem, so we can assume that a_i and b_i ($i = 1, \ldots, n$) are all positive integers. The purpose of these procedures is to demonstrate the problem by the knowledge of polynomials.

Now consider the generating functions

$$f(x) = \sum_{i=1}^{n} x^{a_i}, \ g(x) = \sum_{i=1}^{n} x^{b_i}. \tag{9}$$

Then (refer to Remark 1)

$$f^2(x) = \sum_{i=1}^{n} x^{2a_i} + 2 \sum_{1 \le i < j \le n} x^{a_i + a_j} = f(x^2) + 2 \sum_{1 \le i < j \le n} x^{a_i + a_j}$$

and

$$g^2(x) = g(x^2) + 2 \sum_{1 \le i < j \le n} x^{b_i + b_j}.$$

Hence from (8),

$$f^2(x) - g^2(x) = f(x^2) - g(x^2).$$

From (7), $f(x) - g(x)$ is not always zero, so

$$f(x) + g(x) = \frac{f(x^2) - g(x^2)}{f(x) - g(x)}. \tag{10}$$

Furthermore, take $x = 1$ in (9), and then $f(1) = g(1) = n$, that is, $x = 1$ is a zero of $f(x) - g(x)$. Since $f(x) - g(x)$ is a polynomial, (the reason to convert a_i and b_i into positive integers is now clear)

$$f(x) - g(x) = (x - 1)^k p(x), \tag{11}$$

where k is a positive integer, $p(x)$ is a polynomial, and $p(1) \neq 0$. (Refer to the last paragraph of the first section of Chapter 12.)

From (10) and (11),

$$f(x) + g(x) = \frac{(x^2 - 1)^k p(x^2)}{(x - 1)^k p(x)} = (x + 1)^k \frac{p(x^2)}{p(x)}.$$

Take $x = 1$ in the above formula. Then

$$2n = f(1) + f(1) = 2^k,$$

i.e., $n = 2^{k-1}$, which is a power of 2.   □

Remark 2. When n is a power of 2, there exist positive integers $a_1, \ldots, a_n$ and $b_1, \ldots, b_n$ satisfying equations (7) and (8). This construction is a little troublesome, which will not be discussed here.

Example 7. Weights can be placed on both trays of a balance to measure objects of weight $1, 2, \ldots, n$.

  (i) Determine the minimum number $f(n)$ of weights needed.
  (ii) For which n, the weights of $f(n)$ numbers of weights in (i) are unique?

Solution Let the weights of weights be $a_1, \ldots, a_s$. An object of weight k can be measured if and only if k can be expressed in the form

$$\varepsilon_1 a_1 + \cdots + \varepsilon_s a_s, \tag{12}$$

where $\varepsilon_i = 0$ and ± 1 for $i = 1, 2, \ldots, s$. (When measuring an object, a_i with coefficient -1 means that the corresponding weight is placed in the same tray as the object, a_i with coefficient $+1$ means that the corresponding weight is placed in the other tray, and a_i with coefficient 0 means that the corresponding weight is not needed.)

Because there are three choices for each ε_i, there are 3^s numbers in (12). Apart from 0, the other numbers are all in positive-negative pairs. Therefore, there are $\frac{3^s-1}{2}$ positive numbers. Hence, s weights can weigh at most $\frac{3^s-1}{2}$ different weights of objects, that is, when $n > \frac{3^{s-1}-1}{2}$, the number of $s - 1$ weights is not enough.

We prove that if we take $a_i = 3^{i-1}$ $(i = 1, 2, \ldots, s)$, then all integers k satisfying $-\frac{3^s-1}{2} \le k \le \frac{3^s-1}{2}$ can be expressed in the form of (12). Thus, we have proved that $f(n) = s$ when n satisfies $\frac{3^{s-1}-1}{2} < n \le \frac{3^s-1}{2}$.

In fact, let $N = \frac{3^s-1}{2}$. Then from the equality

$$\left(x^{-1} + 1 + x\right)\left(x^{-3} + 1 + x^3\right) \cdots \left(x^{-3^{s-1}} + 1 + x^{3^{s-1}}\right)$$

$$= x^{-1}\frac{x^3-1}{x-1} \cdot x^{-3} \cdot \frac{x^9-1}{x^3-1} \cdot \cdots \cdot x^{-3^{s-1}}\frac{x^{3^s}-1}{x^{3^{s-1}}-1}$$

$$= x^{-N} \frac{x^{3^s} - 1}{x - 1}$$

$$= x^{-N} + x^{-N+1} + \cdots + 1 + \cdots + x^{N-1} + x^N,$$

we see that the above assertion is correct.

Now we solve the problem (ii). Firstly, it is not difficult to see that if $n \neq \frac{3^s - 1}{2}$, then the weights of the weights are not unique. Because when $\frac{3^{s-1} - 1}{2} < n < \frac{3^s - 1}{2}$, it has been proved that $a_i = 3^{i-1}$ $(1 \leq i \leq s)$ is a group of solutions. If we take $a_i' = 3^{i-1}$ $(i = 1, \ldots, s-1)$ and $a_s' = 3^{s-1} - 1$, then the integers with absolute value $\leq \frac{3^{s-1} - 1}{2}$ can be expressed by them and for the integers k satisfying $\frac{3^{s-1} - 1}{2} \leq k \leq n \left(\leq \frac{3^s - 1}{2} - 1 \right)$,

$$-\frac{3^{s-1} - 1}{2} \leq k - a_s' \leq \frac{3^{s-1} - 1}{2},$$

so $k - a_s'$ can be represented by $a_1', \ldots, a_{s-1}'$. Thus $1, 2, \ldots, n$ can also be represented by $a_1', \ldots, a_s'$.

Then we prove that when $n = \frac{3^s - 1}{2}$, the weights of $f(n) = s$ weights $a_1, a_2, \ldots, a_s$ must be (a permutation of) $1, 3, 3^2, \ldots, 3^{s-1}$.

Because there are 3^s numbers in (12), there is exactly one way to express the 3^s numbers $-n, \ldots, 0, 1, \ldots, n$, and in particular

$$a_1 + \cdots + a_s = n. \tag{13}$$

Hence

$$\left(x^{a_1} + 1 + x^{-a_1} \right) \cdots \left(x^{a_s} + 1 + x^{-a_s} \right) = x^{-n} + \cdots + x^{-1} + 1 + x + \cdots + x^n.$$

From (13),

$$\left(x^{2a_1} + x^{a_1} + 1 \right) \cdots \left(x^{2a_s} + x^{a_s} + 1 \right) = \frac{x^{2n+1} - 1}{x - 1} = \frac{x^{3^s} - 1}{x - 1}.$$

Take $x = e^{\frac{2\pi i}{3a_j}} = \cos \frac{2\pi}{3a_j} + i \sin \frac{2\pi}{3a_j}$ $(j = 1, \ldots, s)$ in the above formula. Obviously, the left side is 0, so the right side is also 0, that is $e^{\frac{2\pi i}{a_j} 3^{s-1}} = 1$. Consequently, it is easy to deduce that $a_j \mid 3^{s-1}$, that is, a_j is a power of 3 $(1 \leq j \leq s)$.

On the other hand, there are no two equal numbers in all a_j, so we can assume that $a_1 < a_2 < \cdots < a_s$, and since a_j is a power of 3, we have $a_j \geq 3^{j-1}$ $(1 \leq j \leq s)$. Thus, combining them with (13), we obtain $n = \sum_{j=1}^{s} a_j \geq \sum_{j=0}^{s-1} 3^j = n$, hence $a_j = 3^{j-1}$ $(j = 1, 2, \ldots, s)$.

Remark 3. Suppose there are s weights, whose weights are $a_1, \ldots, a_s$ (all positive integers).

(i) If we specify that the weights can only be placed on one end of the balance, and let b_n be the number of different ways to weigh the object of weight n, then the generating function of $\{b_n\}$ $(n \geq 1)$ is (refer to (1))

$$(1 + x^{a_1})(1 + x^{a_2}) \cdots (1 + x^{a_s}).$$

(ii) If the weights can be placed on both ends of the balance, and let c_n be the number of different ways to weigh the object of weight n, then the generating function of $\{c_n\}$ $(n \geq 1)$ is

$$\left(x^{-a_1} + 1 + x^{a_1}\right)\left(x^{-a_2} + 1 + x^{a_2}\right) \cdots \left(x^{-a_s} + 1 + x^{a_s}\right).$$

Example 8. Let p be an odd prime number. Find the number of p-element subsets of the set $\{1, 2, \ldots, 2p\}$ whose sum of elements is divisible by p.

Solution   The generating function of the sum of subsets of the set $\{1, 2, \ldots, p\}$ is

$$f(x) = (1 + x)\left(1 + x^2\right) \cdots (1 + x^p)$$

$$= 1 + a_1 x + \cdots + a_n x^n, \ n = \frac{1}{2}p(p - 1). \tag{14}$$

(The power k of each x^k on the right side uniquely corresponds to the sum of the elements of a subset; the coefficient of x^k is the number of subsets whose sum is k.)

Therefore, the problem is to find the sum of all a_k whose subscripts are divisible by p. From Remark 3 in Chapter 15 and the above (14),

$$\sum_{\substack{p \mid k \\ 1 \leq k \leq n}} a_k = \frac{1}{p} \sum_{t=1}^{p} \sum_{k=1}^{n} a_k \zeta^{kt}$$

$$= \frac{1}{p} \sum_{t=1}^{p} \left[(1 + \zeta^{1 \cdot t})(1 + \zeta^{2 \cdot t}) \cdots (1 + \zeta^{p \cdot t}) - 1 \right]$$

$$= \frac{2^p}{p} + \frac{1}{p} \sum_{t=1}^{p-1} \left(1 + \zeta^t\right)\left(1 + \zeta^{2t}\right) \cdots \left(1 + \zeta^{pt}\right) - 1$$

$$= \frac{2^p - p}{p} + \frac{2}{p} \sum_{t=1}^{p-1} \left(1 + \zeta^t\right)\left(1 + \zeta^{2t}\right) \cdots \left(1 + \zeta^{(p-1)t}\right)).$$

Since p is prime, $t, 2t, \ldots, (p-1)t$ form a reduced system modulo p ((11) in Chapter 8) for any t $(1 \le t \le p-1)$, thus the number is

$$\frac{2^p - p}{p} + \frac{2}{p} \sum_{t=1}^{p-1} (1 + \zeta)(1 + \zeta^2) \cdots (1 + \zeta^{p-1})$$

$$= \frac{2^p - p}{p} + \frac{2}{p} \cdot (p-1) = \frac{2^p - 2}{p} + 1.$$

(We applied $(1 + \zeta)(1 + \zeta^2) \cdots (1 + \zeta^{p-1}) = 1$, which can be obtained by taking $x = -1$ in the decomposition of (4) in Chapter 15.)

Example 9. Let p be an odd prime number. Find the number of (non-empty) p-element subsets of the set $\{1, 2, \ldots, 2p\}$ whose sum of elements is divisible by p.

Solution This problem is related to Example 8, but it is more difficult. Our method is actually to use the binary generating function.

Let $\zeta = e^{\frac{2\pi i}{p}}$. Then the number to be found is the coefficient of x^p in the polynomial

$$f(x) = \frac{1}{p} \sum_{k=1}^{p} (1 + \zeta^k x)(1 + \zeta^{2k} x) \cdots (1 + \zeta^{2pk} x) \tag{15}$$

(refer to (3) in Chapter 15).

The expression (15) can be converted into (notice that $\zeta^{(p+r)k} = \zeta^{rk}$, $r = 1, \ldots, p$)

$$f(x) = \frac{1}{p} \sum_{k=1}^{p} (1 + \zeta^k x)^2 \cdots (1 + \zeta^{pk} x)^2$$

$$= \frac{1}{p} \sum_{k=1}^{p} (1 + \zeta^k x)^2 \cdots \left(1 + \zeta^{(p-1)k} x\right)^2 (1 + x)^2$$

$$= \frac{1}{p} \sum_{k=1}^{p-1} (1 + \zeta^k x)^2 \cdots \left(1 + \zeta^{(p-1)k} x\right)^2 (1 + x)^2 + \frac{1}{p} (1 + x)^{2p}$$

$$= \frac{(p-1)}{p} (1 + \zeta x)^2 \cdots \left(1 + \zeta^{p-1} x\right)^2 (1 + x)^2 + \frac{1}{p} (1 + x)^{2p}$$

$$= \frac{(p-1)}{p} \left(x^{p-1} - x^{p-2} + \cdots - x + 1 \right)^2 (1+x)^2 + \frac{1}{p} (1+x)^{2p}$$

$$= \frac{p-1}{p} (x^p + 1)^2 + \frac{1}{p} (x+1)^{2p} .$$

Therefore, the coefficient of x^p is $\frac{p-1}{p} \times 2 + \frac{1}{p} \binom{2p}{p} = \frac{1}{p} \left(\binom{2p}{p} - 2 \right) + 2$.

In the above we applied the following equality:

$$x^{p-1} - x^{p-2} + \cdots - x + 1 = (1 + \zeta x) \left(1 + \zeta^2 x \right) \cdots (1 + \zeta^{p-1} x).$$

This equality can be obtained by replacing x with $-x$ in (5) of Chapter 15.

Exercises

Group A

1. Prove by the generating function method that $\binom{n}{k} = \binom{n-1}{k} + \binom{n-1}{k-1}$.
2. Prove: $\sum_{k=0}^{n} \binom{n}{k} 2^{n-k} (-1)^k \binom{k}{[\frac{k}{2}]} = \binom{2n}{n} - \binom{2n}{n-1}$.
3. There are 12 balls, including 3 red, 3 white, and 6 black. How many different ways are there to take 8 balls from them? If at least 2 black balls and 1 red ball need to be chosen, how many different ways are there? (Assume that there is no difference among the balls in the same color.)
4. What kind of weights of objects can be measured with weights of 1, 3, 5, and 7 under the following conditions?

   (i) Weights shall only be placed at one end of the balance.
   (ii) Weights can be placed on both ends of the balance.

Group B

5. Find the coefficient of x^k in the expansion of $\left(1 + x + x^2 + \cdots + x^{n-1} \right)^2$.
6. Try to make two dice that are different from ordinary dice. Mark positive integers on the six faces of each dice, such that they can be used (together) with the same effect as two ordinary dice (i.e., count once for sum 2, count twice for sum 3, etc.).
7. Prove the identity

$$\sum_{k=1}^{n} k \binom{n}{k}^2 = n \binom{2n-1}{n-1}$$

by the generating function method. (See Exercise 11 in Chapter 2.)

8. Color each number in the set $M = \{1, 2, \ldots, n\}$ with one of the three colors. Suppose
$A = \{(x, y, z) | x, y, z \in M, x, y, z$ are in the same color, $n \mid x + y + z\}$,
$B = \{(x, y, z) | x, y, z \in M, x, y, z$ are in pairwise different colors, $n \mid x + y + z\}$.

Prove: $2|A| \geq |B|$.

Chapter 17

Sets and Families of Subsets

Let X be a finite set. The set composed of some of its subsets is called a subset family of X. This chapter introduces some basic ideas and techniques in combinatorial mathematics by some problem involving subset family.

Example 1. Let a set X have $n(n \geq 1)$ elements $(n \geq 1)$, $A_1, A_2, \ldots, A_k$ form a family of different subsets of X, their intersections of each other are not empty, and any other subsets of X cannot intersect all of $A_1, A_2, \ldots, A_k$. Prove: $k = 2^{n-1}$.

Proof. In this problem, not only the k subsets $A_1, \ldots, A_k$ but also other subsets of X are also involved, so the problem is related to all subsets of X.

There are 2^n elements in the family of all subsets of X (each element is a subset of X). They can be matched into 2^{n-1} pairs with their complements in X. This "intermediate quantity" related to the problem is the starting point of the argument.

Firstly, there must be $k \leq 2^{n-1}$. This is because if $k > 2^{n-1}$, then there are two subsets in $A_1, \ldots, A_k$, which are complementary, so their intersection is the empty set, which is contradictory to the condition.

Then we prove $k \geq 2^{n-1}$. If $k < 2^{n-1}$, then apart from $A_1, \ldots, A_k$, there will be more than 2^{n-1} subsets of X, so there is also a pair of complementary subsets, denoted as C and D. According to the second condition of the problem, there exist A_i and A_j $(1 \leq i, j \leq k$, and $i \neq j)$, such that $A_i \cap C = \varnothing$ and $A_j \cap D = \varnothing$, but C and D are complementary, so $A_i \subseteq D$

and $A_j \subseteq C$, and hence

$$A_i \cap A_j \subseteq C \cap D = \emptyset,$$

i.e., $A_i \cap A_j = \emptyset$, which conflicts with the condition, so $k \geq 2^{n-1}$. Based on the proven results, $k = 2^{n-1}$.   □

Remark 1. There are many combinatorial problems where the relationship between the condition and the conclusion is quite hidden. The starting point of solving a problem is often "the intermediate quantity related to the problem". If such a quantity is selected properly, then condition in the problem will naturally come into use. Example 1 can be seen as an illustrative example.

Example 2. Let X be an n-element set. Find $\sum_{A,B \subseteq X} |A \cap B|$ and $\sum_{A,B \subseteq X} |A \cup B|$. Here A and B independently take all subsets of X.

Solution This problem is quite easy, and there are several solutions, two of which are introduced here.

Calculate $\sum_{A,B \subseteq X} |A \cap B|$ first. We rewrite it into a double sum

$$\sum_{A \subseteq X} \sum_{B \subseteq X} \left| A \bigcap B \right|. \tag{1}$$

(First fix A and sum over B. Then sum over A.) For each fixed A, consider in pairs of B and $\overline{B}$ (2^{n-1} pairs in total). Because

$$\left| A \bigcap B \right| + \left| A \bigcap \overline{B} \right| = |A|,$$

the inner sum of (1) is equal to $2^{n-1}|A|$, that is, (1) becomes

$$2^{n-1} \sum_{A \subseteq X} |A|.$$

(The effect of pairing is to "eliminate" the variable B.) Now, classify $A \subseteq X$ according to the number of elements. Since there are $\binom{n}{k}$ numbers of k-element subsets of X, the sum is (using Example 1 of Chapter 2)

$$2^{n-1} \sum_{k=0}^{n} k \binom{n}{k} = n \cdot 2^{n-1} \cdot 2^{n-1} = n \cdot 4^{n-1}.$$

To calculate $\sum_{A,B \subseteq X} |A \cup B|$, we apply

$$\overline{A \bigcup B} = \overline{A} \bigcap \overline{B}.$$

Then $|A \cup B| = n - |\overline{A} \cap \overline{B}|$, and because A and B take over all subsets of X, we see that $\overline{A}$ and $\overline{B}$ also take over all subsets of X. Thus

$$\sum_{A,B \subseteq X} \left| A \bigcup B \right| = \sum_{A,B \subseteq X} \left(n - \left| \overline{A} \bigcap \overline{B} \right| \right)$$

$$= n \cdot 2^n \cdot 2^n - n \cdot 4^{n-1} = 3n \cdot 4^{n-1}.$$

Remark 2. Before giving another solution of Example 2, we briefly introduce the characteristic function of a subset of a finite set.

Let X be a set. For a subset $A \subseteq X$, the characteristic function $f_A(x)$ of A is a function on X, defined as

$$f_A(x) = \begin{cases} 1, & \text{if } x \in A; \\ 0, & \text{if } x \notin A. \end{cases}$$

Therefore, the characteristic function of X is the constant 1, and the characteristic function of the empty set is the constant 0.

There are two intentions of choosing 0 and 1 in the definition of characteristic functions. One is to distinguish whether x belongs to A or not. The other is that it makes the characteristic function describe basic operations of the subsets conveniently, and the sum (in some forms) of characteristic functions has applicable combinatorial significance:

Let A and B be subsets of X.

(1) A is a subset of B if and only if their characteristic functions satisfy $f_A(x) \leq f_B(x)$ (for all $x \in X$).
(2) The characteristic function of the complement of A with respect to X is $1 - f_A(x)$.
(3) The characteristic function of the intersection of A and B is $f_A(x)f_B(x)$.
(4) The characteristic function of the union of A and B is $f_A(x) + f_B(x) - f_A(x)f_B(x)$.
(5) If A is a finite set, then $\sum_{x \in X} f_A(x) = |A|$.
(6) Let x be a fixed element in X and $\{A_1, \ldots, A_k\}$ be a family of subsets of X. Then $\sum_{i=1}^{k} f_{A_i}(x)$ is the number of subsets in $A_1, \ldots, A_k$ containing x. In particular, if X is an n-element set, then $\sum_{A \subseteq X} f_A(x) = 2^{n-1}$.

All proofs of the above properties are trivial. We leave them to the readers.

Now we give another summation method of the sum $\sum_{A,B \subseteq X} |A \cap B|$ in Example 2. It can be seen from (3), (5), and (6) in Remark 2 that the sum is equal to

$$\sum_{A,B \subseteq X} \sum_{x \in X} f_A(x) f_B(x) = \sum_{x \in X} \sum_{A,B \subseteq X} f_A(x) f_B(x)$$

$$= \sum_{x \in X} \left(\sum_{A \subseteq X} f_A(x) \right) \left(\sum_{B \subseteq X} f_B(x) \right)$$

$$= \sum_{x \in X} 2^{n-1} \cdot 2^{n-1} = n \cdot 4^{n-1}.$$

(In the first step, we changed the order of summations, which makes it easy to work out the inner sum. Please refer to Remark 4 in Chapter 15, especially (5).)

Example 3. Let $A_i (i = 1, \ldots, n)$ be a family of finite sets, and $|A_1| = \cdots = |A_n| = r$ with $\overset{n}{\underset{i=1}{\cup}} A_i = S$. Suppose for a fixed positive integer k, the union of any k sets in $A_1, \ldots, A_n$ is equal to S and the union of any $k - 1$ sets is a proper subset of S. Prove: $|S| \geq \binom{n}{k-1}$, and determine the value of r for which the equality holds.

Proof. Suppose $S = \{a_1, \ldots, a_l\}$ and $I = |S|$. Construct a numerical table as shown in Figure 17.1: If $a_i \in A_j$, fill in 1 in the grid where row i intersects column j; otherwise fill in 0.

Since the union of any $k - 1$ sets in $A_1, \ldots, A_n$ is not equal to S (i.e., at least one element in S does not belong to this union), if one takes any $k - 1$ columns in the table, then there must be a row such that the numbers in the cross grids of this row and the $k - 1$ columns are all zero. Thus there

Figure 17.1

are at least $k-1$ zeros in this row (of the table). For convenience, we call such a row "zero row".

Because the union of any $A_1, \ldots, A_n$ sets in $A_1, \ldots, A_n$ is equal to S, the "zero rows" generated by different ways of choosing the above different $k-1$ columns cannot be the same row. (And there are exactly $k-1$ zeros in each "zero rows".)

Therefore, we generate an injection from the $(k-1)$-combinations of n columns to l rows, so $\binom{n}{k-1} \leq l$, i.e., $|S| \geq \binom{n}{k-1}$ (refer to Remark 4 in Chapter 3).

When $|S| = \binom{n}{k-1}$, each row in the numerical table is a "zero row," and calculate the number of ones in the numerical table by rows and by columns respectively (note that there are exactly r ones in each column), so

$$nr = \binom{n}{k-1}(n-(k-1)),$$

i.e., $r = \binom{n-1}{k-1}$. $\square$

Example 4. Let X be a finite set, and $A_1, \ldots, A_m$ are its subsets such that $|A_i| = r(i = 1, \ldots, m)$ and $|A_i \cap A_j| \leq k(1 \leq i < j \leq k)$. Prove: $|X| \geq \frac{r^2 m}{r+(m-1)k}$.

Proof. Let $f_i(x)$ be the characteristic function of the set A_i, and let $d(x) = \sum_{i=1}^m f_i(x)$, which is the number of times that x appears in $A_1, \ldots, A_m$. We have (refer to Remark 2)

$$\sum_{x \in X} d^2(x) = \sum_{x \in X} \sum_{i=1}^m f_i^2(x) + 2 \sum_{x \in X} \sum_{1 \leq i < j \leq m} f_i(x) f_j(x)$$

(change the summation order)

$$= \sum_{i=1}^m \sum_{x \in X} f_i(x) + 2 \sum_{1 \leq i < j \leq m} \sum_{x \in X} f_i(x) f_j(x)$$

$$= \sum_{i=1}^m |A_i| + 2 \sum_{1 \leq i < j \leq m} \left| A_i \cap A_j \right|$$

$$\leq rm + 2 \binom{m}{2} k$$

$$= rm + m(m-1)k. \tag{2}$$

On the other hand, from Cauchy's inequality,

$$\sum_{x \in X} d^2(x) \geq \frac{1}{|X|} \left(\sum_{x \in X} d(x) \right)^2$$

$$= \frac{1}{|X|} \left(\sum_{i=1}^{m} |A_i| \right)^2 = \frac{r^2 m^2}{|X|}. \tag{3}$$

Combining (2) and (3), we get the result. In this solution, we started from considering two aspects of $\sum_{x \in X} d^2(x)$ (refer to Remark 1).

Please note that when deriving (3), we used

$$\sum_{x \in X} d(x) = \sum_{i=1}^{m} |A_i|. \tag{4}$$

This is equivalent to that in the table representing the subordinate relationship between the elements of X and $A_1, \ldots, A_m$ (see Example 3), the number of 1 calculated by rows is equal to the number of 1 calculated by columns. (4) can be easily derived by changing the summation order (refer to solution 2 of Example 2):

$$\sum_{x \in X} d(x) = \sum_{x \in X} \sum_{i \in 1}^{m} f_i(x) = \sum_{i \in 1}^{m} \sum_{x \in X} f_i(x) = \sum_{i=1}^{m} |A_i|.$$

This problem can also be proved by the following method: Let $X = \{x_1, \ldots, x_l\}$. Similar to Example 3, make an $l \times m$ numerical table representing the subordinate relationship between the elements of X and $A_1, \ldots, A_m$ (please draw the table by yourself). Assume that row i has a_i ones (a_i is $d(x_i)$ in the previous solution). We call two numbers of 1 (not necessarily adjacent) in any row the 1×2 sub table. Therefore, the number of 1×2 sub tables in the table is

$$\sum_{i=1}^{l} \binom{a_i}{2} = \frac{1}{2} \sum_{i=1}^{l} a_i^2 - \frac{1}{2} \sum_{i=1}^{l} a_i. \tag{5}$$

On the other hand, each 1×2 sub table corresponds to an element in the intersection of a pair of sets in $A_1, \ldots, A_m$; conversely, an element a_r in any intersection of A_i and $A_j (i \neq j)$ corresponds to a 1×2 sub table (in row r of the table). Therefore, the number of 1×2 sub tables is

$$\sum_{1 \leq i < j \leq m} \left| A_i \bigcap A_j \right| \leq k \binom{m}{2}. \tag{6}$$

Combining (5) and (6), we obtain

$$\frac{1}{2}\sum_{i=1}^{l} a_i^2 - \frac{1}{2}\sum_{i=1}^{l} a_i \le k \binom{m}{2},$$

which is actually (2). Now (using (4)) we can derive the conclusion in the same way as the previous solution. □

The starting point of the second proof is to consider the 1×2 sub tables in the numerical table, which is similar to the consideration of $\sum d^2(x)$ (in the former solution).

Example 5. Let X be an n-element set, and $A_1, \ldots, A_m$ be its subsets such that $|A_i| = r (i = 1, \ldots, m)$. Prove: X has a subset Y such that the number of sets in $A_1, \ldots, A_m$ which is contained in Y or in $\overline{Y}$ is less than or equal to $\frac{m}{2^{r-1}}$.

Proof. For any $Y \subseteq X$, let $f(Y)$ be the number of sets in $A_1, \ldots, A_m$ contained in Y or not intersecting Y (i.e., contained in $\overline{Y}$). Let

$$f_j(Y) = \begin{cases} 1, & \text{if } A_j \subseteq Y \text{ or } A_j \subseteq \overline{Y}; \\ 0, & \text{otherwise.} \end{cases}$$

Then

$$\sum_{Y \subseteq X} f(Y) = \sum_{Y \subseteq X} \sum_{j=1}^{m} f_j(Y) \text{ (change the summation order)}$$

$$= \sum_{j=1}^{m} \sum_{Y \subseteq X} f_j(Y). \tag{7}$$

For a fixed A_j, the set $X \backslash A_j$ has 2^{n-r} subsets, and the unions of each subset and A_j are actually all subsets Y of X containing A_j; and all subsets of $X \backslash A_j$ are actually the subsets of X which do not intersect A_j. Therefore, the inner sum of (7) is $2^{n-r} + 2^{n-r} = 2^{n-r-1}$, thus

$$\sum_{Y \subseteq X} f(Y) = m2^{n-r+1},$$

i.e.,

$$\frac{1}{2^n} \sum_{Y \subseteq X} f(Y) = \frac{m}{2^{r-1}}. \tag{8}$$

The above formula shows that the average value of 2^n numbers of $f(Y)$ is equal to $\frac{m}{2^{r-1}}$, hence $Y \subseteq X$, which makes $f(Y) \le \frac{m}{2^{r-1}}$, therefore, this Y meets the requirements (it can also be proved by contradiction from (8)). □

Remark 3. The method of Example 5 is often called the average argument, whose spirit is to give the properties of a single item from the average result (so-called "from average to single").

It may not be easy to prove directly that a certain item has a certain property, but the sum (arithmetic mean), or sum of squares (square mean), or product (geometric mean) of some items may have good properties. According to the characteristics of specific problems, it is a very basic technique to take some average the an "intermediate quantity" (refer to Remark 1). There are such examples later in this book.

Example 6. Let X be an n-element set and F be a family of some 3-element subsets of X, in which every two subsets have at most one common element. Prove: X has a subset, which does not contain any 3-element subset in F, and the number of elements is at least $[\sqrt{2n}]$.

Proof. If only one of the two conditions is required, the problem is extremely trivial.

In order to prove the result of this problem, we (naturally) take the largest subset of X that does not contain any 3-element subset in F, denoted as S. (Because X has a subset that does not contain any subset in F, such as the empty set or a 2-element set, and X has only a finite number of subsets, the largest subset must exist, but it is not necessarily unique.)

Let $|S| = r$. Then $|X \backslash S| = n - r$. Because of the maximality of S, for any element x in $X \backslash S$, the set $S \cup (x)$ must contain a certain 3-element subset in F, that is, corresponding to this x, the set S has a (different) family S_x^* of 2-element subsets, and the union of any 2-element subset A_x in it and x belongs to F.

In this way, we generate a mapping f from $X \backslash S$ to a set in 2-element subset family of S.

Now we prove that this mapping is an injection. Let $x, y \in X \backslash S$ with $x \neq y$. Then there are no common elements in the corresponding families S_x^* and S_y^*, because if there exist $A_x \in S_x^* \ A_y \in S_y^*$ such that $A_x = A_y$, then both $\{x\} \cup A_x$ and $\{y\} \cup A_y$ belong to the family F, and they have two common elements, which contradicts the assumption on the family F. Hence the conclusion is proved, from which we get

$$n - r = |X \backslash S| \leq \sum_{x \in S} |S_x^*|$$

$$\leq \text{the number of all 2 - element subsets of } S = \binom{r}{2},$$

i.e., $n - r \leq \frac{r(r-1)}{2}$. To prove $r \geq [\sqrt{2n}]$ by this inequality, we rearrange it to $2n \leq r(r+1)$, so

$$\sqrt{r(r+1)} \geq \sqrt{2n},$$

and $r = [\sqrt{r(r+1)}] \geq [\sqrt{2n}]$. □

At the end of this chapter, we introduce the famous **Sperner's theorem**, which is considered to be the most fundamental result in the theory of family of subsets.

Example 7. Let X be an n-element set, and $A_1, \ldots, A_m$ be subsets of X that do not contain each other. Prove:

$$m \leq \binom{n}{\left\lceil \frac{n}{2} \right\rceil}.$$

Proof. If subsets S_i $(i = 0, 1, \ldots, n)$ of X satisfy $|S_i| = i$ and

$$\varnothing = S_0 \subset S_1 \subset \cdots \subset S_n = X,$$

then $\{S_0, S_1, \ldots, S_n\}$ is said to be the maximal chain of subsets of X. If $S_k \subset S_{k+1} \subset \cdots \subset S_n = X$ has been determined, then there are k different choices of S_{k-1} such that $S_{k-1} \subset S_k$ $(1 \leq k \leq n)$. From the multiplication principle, the subsets of X have $n!$ maximal chains. In addition, for any fixed $(k-1)$-elements subset S, since S contains k subsets of $(k-1)$-elements and is included in $n - k$ subsets of $(k+1)$-elements of X, the number of maximal chains containing S is

$$k \times (k-1) \times \cdots \times 1 \times (n-k) \times (n-k-1) \times \cdots \times 1 = k!(n-k)!.$$

We suppose the $n!$ maximal chains are $C_i(i = 1, 2, \ldots, n!)$, and construct an $n! \times m$ numerical table as shown in Figure 17.2.

If the maximal chain C_i contains the subset A_j in the problem, fill in 1 in the cross grid of row i and column j; otherwise fill in 0. Therefore, the

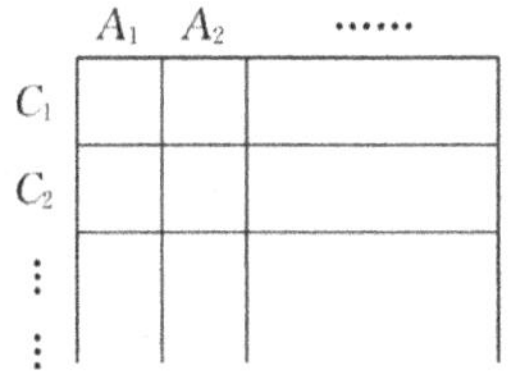

Figure 17.2

number of 1 in column j is the number of the maximal chains containing A_j, which is $|A_j|!(n - |A_j|)!$. However, there is at most one 1 in each row (because there is no subset in $A_1, \ldots, A_m$ containing another subset in $A_1, \ldots, A_m$). Therefore, calculating the numbers of 1 in the table by rows and by columns, we obtain

$$\sum_{j=1}^{m} |A_j|!(n - |A_j|)! \leq n!,$$

i.e.,

$$\sum_{j=1}^{m} \frac{1}{\binom{n}{|A_j|}} \leq 1.$$

But $\binom{n}{k}$ reaches its maximum value when $k = \left[\frac{n}{2}\right]$ (see (6) in Chapter 2), so $m \leq \binom{n}{\left[\frac{n}{2}\right]}$ is deduced from the above formula.   □

The above proof belongs to Lubell et al. It is extremely ingenious and worth pondering.

Exercises

Group A

1. In a campaign, each party makes n different promises ($n > 0$). It is known that any two parties make at least one same promise, but no two parties have exactly the same promises. Determine the maximum possible number of parties.

2. For each non-empty subset of the set $1, 2, \ldots, n$, define an "alternating sum" as follows: Arrange the elements of a subset in descending order and then subtract or add the succeeding number alternately from the largest number (e.g., the alternating sum of the subset $\{1, 2, 4, 7\}$ is $7 - 4 + 2 - 1 = 4$ and the alternating sum of $\{3\}$ is 3.) Find the sum S of all alternating sums.

3. Let the set $A = \{0, 1, 2, \ldots, 9\}$, and let $B_1, \ldots, B_k$ be non-empty subsets of A, and if $i \neq j$, $B_i \cap B_j$ has at most two elements. Find the maximum value of k.

4. Let $S = \{1, 2, \ldots, 10\}$ and $A_1, \ldots, A_k$ be k subsets of S, satisfying $|A_i| = 5(i = 1, 2, \cdots, k)$ and $|A_i \cap A_j| \leq 2$ ($1 \leq i, j \leq k$ and $i \neq j$). Find the maximum value of k.

Group B

5. In order to open a safe, a committee of 11 members is organized. Several locks are added to the safe, and the keys to these locks are assigned to the committee members (one lock may have multiple keys). How many locks are needed at least so that any six members can open the safe, but no five members can open the safe? In addition, in the case of the minimum number of locks, how should you allocate keys to members to meet the above requirements?

6. Let $A_1, A_2, \ldots, A_n$ be all 2-element subsets of $S = \{a_1, \ldots, a_n\}$, and when $A_i \cap A_j \neq \emptyset$, one of A_i and A_j is $\{a_i, a_j\}$ ($1 \leq i, j \leq n$ and $i \neq j$). Prove: Each element in S is exactly in two A_i.

7. Let $S = \{1, 2, 3, \ldots, 20\}$. For any 9-element subset T of S, a function $f(T)$ takes integer values between 1 and 20. Prove: No matter what a function f is, there always exists a 10-element subset S' of S, such that $f(S' \backslash \{k\}) \neq k$ for all $k \in S'$.

8. Let $A_1, \ldots, A_n$ be n distinct finite subsets ($n \geq 2$), with the property: For any A_i and A_j, there always exists A_k such that $A_k = A_i \cup A_j$. Let $m = \min_{1 \leq i \leq n} |A_i| \geq 2$.

Prove: There exists $x \in \bigcup_{i=1}^{n} A_i$ such that x belongs to at least $\frac{n}{m}$ sets in $A_1, \ldots, A_n$.

Chapter 18

Graph Theory Problems

Graph theory, an important branch of combinatorics, develops rapidly and is widely used.

Problems related to graph theory often appear in mathematics competitions. These problems do not require much knowledge, but are intelligent, flexible and have most competitive features.

This chapter only deals with the most basic concepts and terms in graph theory. We provide some problems solved by graphs. These solutions also show some basic ideas and techniques in combinatorics.

The said graph usually refers to a **simple graph** (unless otherwise stated), which is a graph composed of several different points and segments connecting some of them. Some points in the graph may not be connected with other points, but any point is not connected with itself, and at most one segment is connected between any two end points. In addition, the two points of a segment are usually considered to be unordered (that is, the direction of the line segment is not considered). Such a graph is called an undirected graph. If the direction of line segments of a graph is considered, it is called a directed graph. We only consider undirected graphs.

The above graph is usually denoted by the letter G. The points in graph G are called **vertices**. The set of all vertices is denoted as $V(G)$ or V. A segment in a graph G is called an **edge**, and the set of all edges is denoted as $E(G)$ or E.

Please note that the graph mentioned here does not involve usual geometric properties, so the position of a vertex and the length or the curvature of an edge in the graph are all irrelevant. We only care about the numbers of vertices and edges in a graph, the number of edges incident to a vertex, and which vertices have edges between them, etc.

Let V be the set of vertices and $x \in V$ be a vertex. The number of edges incident to x is very important. We use $d(x)$ to denote this number, which is called the **degree** of x.

Many problems can be restated in the language of graph theory: Use vertices to represent the objects under study and use edges that connect pairs of vertices to represent the relationship between the corresponding objects, thus making a graph.

"A graph is worth a thousand words." Graphs often make a problem intuitive, clear, and help think and express arguments.

Example 1. Ten people attend a meeting. After the meeting, we obtain the statistics that the numbers of each person's friends are

(i) $3, 3, 3, 3, 5, 6, 6, 6, 6, 6$;
(ii) $1, 1, 3, 3, 3, 3, 5, 6, 8, 9$.

Prove: Both sets of statistics are wrong.

Proof. We represent people as vertices. If two people are friends, then form an edge between the corresponding two vertices.

The sum of the degrees of each vertex should be twice the number of edges (because each edge is calculated twice). In particular, the sum of degrees is an even number, but the sum of numbers in (i) is an odd number, so the statistics of (i) are wrong.

The sum of numbers in (ii) is even, but they can't be used as the degrees of vertices of a graph. Because the vertex of degree 9 is connected with all other vertices, it is in particular connected with two vertices of degree 1. Therefore, the vertex of degree 8 cannot be connected with any vertex of degree 1. In this way, there are only 7 vertices left for connection, which conflicts the degree of 8.   □

Remark 1. As for the solution in (i), it can be proved: If the set of vertices and the set of edges of a graph G are V and E respectively, then

$$\sum_{x \in V} d(x) = 2|E|.$$

This is the most basic conclusion in graph theory and has many uses.

Remark 2. Generally speaking, it is not easy to determine whether a sequence of non-negative integers equals the degrees of the vertices of a graph. We will not discuss it here. For some specific problems, if the answer is yes, then is usually necessary to construct a graph that meets the requirement; if the answer is no, use the necessary condition: The sum of the degrees is even, as shown in (i) in Example 1; we can also use the enumeration method as shown in (ii).

Example 2. Let $S = \{x_1, x_2, \ldots, x_n\}$ be a set of points in a plane, where the distance between any two points is at least 1. Prove: There are at most $3n$ pairs of points in S, such that the distance between each pair of points is exactly 1.

Proof. Take n points in S as vertices, if the distance between two vertices is 1, then an edge is formed between the corresponding vertices. Then we obtain a graph G. Our problem is equivalent to proving that the number of edges in G satisfies $|E| \leq 3n$.

In the graph G, the vertices connected with the vertex x_i are all on the circumference with x_i as the center and 1 as the radius. Because the distance between any two points in the set S is ≥ 1, there are at most 6 points in S on this circumference, so $d(x_i) \leq 6$.

From the theorem in Remark 1,

$$d(x_1) + \cdots + d(x_n) = 2|E|,$$

so $6n \geq 2|E|$, i.e., $|E| \leq 3n$. $\qquad\square$

The following Example 3 is a little difficult.

Example 3. The 20 members of a tennis club play 14 single games, and each member takes part in at least one game. Prove that there must be 6 games, such that the 12 players are all different.

Proof. Represent these 20 members by vertices $v_1, \ldots, v_{20}$. If there is a game between two members, form an edge between the corresponding vertices, and we get a graph G.

It is known from the condition that G has 14 edges. Let d_i denote the degree of v_i. Then $d_i \geq 1$ $(i = 1, 2, \ldots, 20)$. From the conclusion in Remark 1,

$$d_1 + d_2 + \cdots + d_{20} = 2 \times 14 = 28.$$

Erase $d_i - 1$ edges at each vertex v_i. Since an edge may be erased at both ends, the number of erased edges shall not exceed

$$(d_1 - 1) + (d_2 - 1) + \cdots + (d_{20} - 1) = 28 - 20 = 8,$$

after erasing these edges, we have at least $14 - 8 = 6$ edges in the graph, and now the degree of each vertex is at most 1, thus, the 12 vertices of these six edges are different from each other, which indicates that there exist 6 edges (games) that meet the requirements.   □

Remark 3. A graph obtained by erasing some edges or vertices (along with the edges derived from these vertices) in a graph G is called a **subgraph** of G. Erasing vertices and edges is a very basic technique in graph theory. Its main purpose is to produce a subgraph with some properties from G (see Example 3); or, we often infer the properties of G itself from the properties of some subgraphs of G. We often do this when we use an inductive argument.

Example 4. There are 18 teams attending a tournament. In each round, each team plays one game against another, and in subsequent rounds, these two teams won't compete again. Now it's the ninth round, and prove that in the previous eight rounds, there must be three teams that have not yet played against each other.

Proof. We use 18 vertices $v_1, \ldots, v_{18}$ to denote the 18 teams. If two teams have played in the previous eight rounds, then form an edge between the corresponding two vertices. It should be proved that there must be three vertices that are not connected with each other in the graph.

Since each team only plays one game in each round, the degree of each vertex in the graph is 8, so there must be two vertices that are not connected. Suppose that v_1 is not connected with v_{10}, and v_1 is connected with $v_2, \ldots, v_9$.

If there is a vertex in $v_2, \ldots, v_9$ that is connected to v_{10}, then there must be a vertex in the 8 vertices $v_{11}, v_{12}, \ldots, v_{18}$ that is not connected to v_{10} (because $d(v_{10}) = 8$). It may be assumed that v_{11} is not connected with v_{10}. Then v_1, v_{10}, and v_{11} are the three vertices that meet the requirement.

If none of $v_2, \ldots, v_9$ is connected with v_{10}, consider the first round of competition, and suppose v_1 and v_2 have played in the first round. In this round, there are 7 teams in $v_3, \ldots, v_9$, so at least one of them needs to play against one of the teams $v_{11}, v_{12}, \ldots, v_{18}$. Without loss of generality, suppose v_3 is connected with v_{11}. Since $d(v_3) = 8$ and v_3 is already connected with v_1 and v_{11}, we know that v_3 is not connected with some vertex in $v_2, v_4, v_5, v_6, v_7, v_8$, and v_9. Suppose v_3 is not connected with v_4. Then v_3, v_4, and v_{10} are three vertices that are not connected with each other.   □

In the solution of Example 4, the enumeration method was used to divide the problem into various situations, discuss and analyze them one by one, and finally draw the conclusion.

The enumeration method usually does not look impressive, but it is the most basic method in mathematics, especially in combinatorics.

The key of the enumeration method is to find the right object to discuss. If we choose the right starting point for the discussion, the argument is often well organized. Otherwise, it may be confusing and hard to proceed.

The following Example 5 is a famous problem.

Example 5. Six points are given, any two of which are connected by an edge. Now color the edges red or blue. Prove that there must be a homochromatic triangle (i.e., three edges have the same color) in the graph.

Proof. We use the enumeration method (but simpler than the previous one). Let $A_1, \ldots, A_6$ denote the 6 points.

Consider the five edges $A_1A_2, A_1A_3, A_1A_4, A_1A_5$, and A_1A_6 derived from A_1. Since the five edges only have two colors, at least three edges have the same color. We suppose that A_1A_2, A_1A_3, and A_1A_4 are all red. If the three edges of the triangle $A_2A_3A_4$ are all blue (as shown in Figure 18.1, the solid line represents red and the dotted line represents blue), then the conclusion is true. If at least one of the three edges of the triangle $A_2A_3A_4$ is red, assume that the edge A_2A_3 is red. Then the triangle $A_1A_2A_3$ is a homochromatic triangle.

This problem can also be solved by the following method: If there are x homochromatic triangles (note that these triangles may have common edges), then there are $\binom{6}{3} - x$ heterochromatic triangles.

Now consider the number S of homochromatic angles (that is, the angle composed of two edges in the same color).

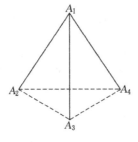

Figure 18.1

On one hand, each homochromatic triangle has three homochromatic angles, and a heterochromatic triangle has one homochromatic angle, so

$$S = 3x + \left(\binom{6}{3} - x \right) = 2x + 20. \tag{1}$$

On the other hand, if a vertex is contained in k red edges, then it is contained in $5 - k$ blue edges, so the number of homochromatic angles with this point as its vertex is

$$\binom{k}{2} + \binom{5-k}{2}.$$

It is easy to know that the minimum value of the above expression is $\binom{3}{2} + \binom{2}{2} = 4$. Thus $S \geq 6 \times 4 = 24$. Combining it with (1), we know that $x \geq 2$, that is, there are at least two homochromatic triangles in the graph. $\qquad \square$

In the above argument, the same quantity is calculated (considered) in two ways, which has appeared many times before.

Remark 4. There are n vertices in a graph. If any two of them are connected by one edge, then the graph is called a **complete graph**, denoted as K_n, which is often concerned in graph theory.

The conclusion of Example 5 can be put in another way: "after each edge of K_6 is arbitrarily colored red or blue, there must be a homochromatic K_3 in K_6." Obviously, if K_6 is replaced with $K_n (n > 6)$, then the above conclusion still holds, but if K_6 is replaced with K_5, the conclusion may not be correct (please give a counter example).

More generally, let $q_1 \cdots q_k$ be given positive integers such that $q_i \geq 2 (i = 1, \ldots k)$ and $k \geq 2$. Let $c_1 \cdots c_k$ be k different colors, and color each edge of K_n with one of these k colors. A deep theorem of **Ramsey** asserts that there exists a positive integer n_0 (depending on $q_1 \cdots q_k$), such that when $n \geq n_0$, at least one of the following is contained in the above K_n: a K_{q_1} of color c_1, a K_{q_2} of color $c_2, \ldots$, or a K_{q_k} of color c_k.

The minimum n_0 with the above properties, denoted as $R(q_1, \ldots, q_k)$, is called the **Ramsey number**, which is very important but very difficult to be determined. At present, only a few cases have been solved. Our previous discussion shows that $R(3, 3) = 6$.

There are some problems about the Ramsey number in mathematics competitions. Please refer to Exercise 7 in this chapter.

Whether a graph contains complete subgraphs is an interesting topic. The following Example 6 is a classical result, which shows a restriction of the degree of vertices in a graph not containing K_r.

Example 6. If a graph G has n vertices and does not contain K_r (as a subgraph), where r is a fixed positive integer, then the degrees of vertices in G satisfy the inequality

$$\min_{x \in V} d(x) \leq \left[\frac{(r-2)n}{r-1} \right].$$

Proof. Prove by contradiction. Let $(r-2)n = (r-1)q + t$ with $0 \leq t \leq r-2$. Assume that the conclusion is wrong. Then for each $x \in V$, degree $d(x) \geq q+1$.

Take any vertex $x_{i_1} \in V$, and let S_{i_1} denote the set of vertices in G connected with x_{i_1} by edges. Then take another vertex $x_{i_2} \in S_{i_1}$ and the corresponding set S_{i_2}. From the inclusion and exclusion principle, we obtain that

$$\left| S_{i_1} \bigcap S_{i_2} \right| = |S_{i_1}| + |S_{i_2}| - \left| S_{i_1} \bigcup S_{i_2} \right| \geq 2q + 2 - n,$$

which follows since both S_{i_1} and S_{i_2} contain at least $q+1$ elements. Then take another element $x_{i_3} \in S_{i_1} \cap S_{i_3}$ and the corresponding set S_{i_3}, and we obtain that

$$\left| S_{i_1} \bigcap S_{i_2} \bigcap S_{i_3} \right| \geq 3(q+1) - 2n,$$

and proceed (inductively) in this way. Finally, for $x_{i_{r-1}} \in \cap_{j=1}^{r-2} S_{i_j}$ and the corresponding set $S_{i_{r-1}}$, we have

$$\left| \bigcap_{j=1}^{r-1} S_{i_j} \right| = |S_{i_{r-1}}| + \left| \bigcap_{j=1}^{r-2} S_{i_j} \right| - \left| S_{i_{r-1}} \bigcup \left(\bigcap_{j=1}^{r-2} S_{i_j} \right) \right|$$

$$\geq q + 1 + (r-2)(q+1) - (r-3)n - n$$

$$= (r-1)(q+1) - (r-2)n$$

$$= r - 1 - t > 0,$$

so there is at least one vertex $x_{i_r} \in \cap_{j=1}^{r-1} S_{i_j}$. As can be seen from the above construction of S_{i_j}, we obtain a complete graph K_r whose set of vertices is $\{x_{i_1}, x_{i_2}, \ldots, x_{i_r}\}$, which contradicts with the known. $\square$

If a graph does not contain K_r, then there is an important result about the restriction of the number of its edges, which is called Turan's theorem. This book will not discuss it. The following Example 7 is its special case.

Example 7. If a graph G has n vertices and does not contain K_3 as a subgraph, then the number of edges in the graph G satisfies $|E| \le \frac{n^2}{4}$.

Proof. We introduce several different proofs.

Proof 1: Prove by induction on the number n of vertices. We only consider the case when n is an even number (when n is odd it can be carried out similarly). When $n = 2$, the conclusion is obvious. Assume that the conclusion is true for all even numbers n not exceeding $2k$. Now we prove that for $n = 2k + 2$, the conclusion is also true.

Let G be a graph with $n = 2k + 2$ vertices and no triangles. If there is no edge in G, there is no need to prove it. Otherwise let G have an edge connecting vertices u and v. Erase u and v and the edge connected with u and v in G, and get a subgraph G'. Then G' has $2k$ vertices and no triangles. By the inductive hypothesis, G' has at most $\frac{4k^2}{4} = k^2$ edges. Now let's see how many edges G may have beyond that. Note that any other vertex w cannot be connected with both u and v, since otherwise u, v and w are the vertices of a triangle in G. Therefore, if u is connected with l vertices in G', then v can only be connected with $2k - l$ vertices in G' at most. Therefore, the number of edges in G is at most

$$k^2 + l + (2k - l) + 1 = \frac{n^2}{4}.$$

Proof 2: Let the vertices of G be $x_1, \ldots, x_n$. Let's note that if two vertices x_i and x_j are connected, since there is no triangle in the graph, then

$$d(x_i) + d(x_j) \le n \tag{2}$$

(refer to the last paragraph of the previous solution). Now we will sum over all the connected vertices x_i and x_j of (2). Because each x_i is connected with $d(x_i)$ vertices x_j, each $d(x_i)$ appears $d(x_i)$ times in the sum, and (on the right side of the inequality) n contributes $|E|$ times (if there is one edge, then it contributes one n), so

$$\sum_{i=1}^{n} d^2(x_i) \le n|E|. \tag{3}$$

Then from Cauchy's inequality and the theorem in Remark 1,

$$\sum_{i=1}^{n} d^2(x_i) \geq \frac{1}{n} \left(\sum_{i=1}^{n} d(x_i) \right)^2 = \frac{1}{n}(2|E|)^2. \tag{4}$$

Combining (3) and (4), we can get the result.

Proof 3: Let x_1 be (one of) the vertex (vertices) with the largest degree in G, let $d(x_1) = d$, and let d vertices connected with x_1 be $x_n, x_{n-1}, \ldots, x_{n-d+1}$.

Since G does not contain triangles, any two vertices in $x_n, x_{n-1}, \ldots, x_{n-d+1}$ are not connected, so it is easy to know that the number of edges in G satisfies

$$|E| \leq d(x_1) + d(x_2) + \cdots + d(x_{n-d}). \tag{5}$$

Now, from the selection of $d = d(x_1)$, there is $d(x_i) \leq d(i = 1, \ldots, n - d)$, so from (5) we can get the conclusion that

$$|E| \leq (n - d)d \leq \left(\frac{n - d + d}{2} \right)^2 = \frac{n^2}{4}.$$

$\square$

Remark 5. For the first proof of Example 7, please refer to Remark 3 and Example 8 below.

Remark 6. The average inequality (3) in Proof 2 is generated from the "single" inequality (2) by summation (this method and Remark 3 of the previous chapter are two sides of the same coin). This technique is also commonly used in combinatorial mathematics and is also a "trick" in algebra to prove equalities and inequalities.

Remark 7. It is not difficult to see that the inequality (5) holds for any vertex x_1. Therefore, at the beginning of Proof 3, x_1 can be treated as a (undetermined) parameter. After deriving (5), if $d(x_1)$ is selected as the maximum, then the upper bound of the right side of (5) can be estimated effectively.

Example 8. Suppose a graph G has n vertices and q edges. Mark the edges with the numbers $1, 2, \ldots, q$. Prove that there must be a vertex, such that starting from this vertex, one can go through $\left[\frac{2q}{n} \right]$ edges whose labels form an ascending sequence.

Proof. Let $L(x)$ denote the maximum length of a valid path (meaning that the edges have ascending labels) if we start at vertex x. Next we prove that there exists x such that $L(x) \geq \left[\frac{2q}{n}\right]$. Since the condition contains little information in this regard, we consider the average of all $L(x)$. If we can prove that

$$\sum_{x \in V(G)} L(x) \geq 2q, \tag{6}$$

then we can immediately draw the conclusion (refer to Remark 3 in the previous chapter).

Now we prove (6) by induction on the number q of edges. If $q = 1$, then there is only one edge. Let the vertices connected by it be x_1 and x_2. Then the left side of (6) is $L(x_1) + L(x_2) = 2 \geq 2 \times 1$. Assume that (6) is true for $q - 1$. Then consider the case of q edges.

Let x_1 and x_2 be connected by the edge labeled 1 and erase the edge (but x_1 and x_2 are preserved). The resulting graph G' has $q - 1$ edges. If $L'(x)$ is defined in G' similarly as $L(x)$ in G, then from the inductive hypothesis,

$$\sum_{x \in V(G')} L'(x) \geq 2(q - 1). \tag{7}$$

On the other hand,

$$L(x_1) \geq L'(x_2) + 1, \tag{8}$$

(since starting from x_1, one can go to x_2 first, and then go through $L'(x_2)$ edges starting from x_2) and

$$L(x_2) \geq L'(x_1) + 1, \tag{9}$$

and for the vertex x different from x_1 and x_2,

$$L(x) \geq L'(x). \tag{10}$$

Add (7), (8), (9), and (10), and then we obtain (6).   □

Exercises

Group A

1. Determine whether the sequence $6, 6, 5, 3, 3, 2, 1$ can be the degrees of vertices of a graph.
2. Prove that among $n(>2)$ people, at least two of them have the same number of friends (within these people).

3. Suppose $2n$ people attend a party, where each person knows at least n others. Prove: There exist 4 people who can sit around the round table, so that everyone knows the people next to him.

4. Given any 6 points in a plane, no three points are collinear. Prove: Among triangles with these points as vertices, there must be a triangle whose longest side is the shortest side of another triangle.

5. There are 500 participants in a meeting. (i) If everyone knows 400 people, can we select 6 people, such that any two of them know each other? (ii) If everyone knows more than 400 people, prove: We can always find 6 people, such that any two of them knows each other.

6. Prove that it is possible to color the edges of K_5 red or blue, so that it does not contain a homochromatic K_3.

Group B

7. There are 17 scientists, each of whom communicates with the others. They only discuss three problems in their communication; and only one problem is discussed between each two scientists. Prove: There are at least three scientists discussing the same problem with each other.

8. Prove: If a graph G has n vertices and m edges, then the graph contains at least $\frac{4m}{3n}\left(m - \frac{n^2}{4}\right)$ triangles.

Chapter 19

Congruence (2)

This chapter introduces two important topics in congruence: The order of an integer modulo m and the famous Chinese remainder theorem.

The Order of a Modulo m

Let $m > 1$ be a fixed integer, and a be an integer that is coprime to m. From Example 3 of Chapter 8, we see that there exists an integer k with $1 \leq k < m$ such that $a^k \equiv 1 \pmod{m}$. We call the minimum positive integer (still denoted as k) with this property the **order of a modulo m**.

The following (1) shows that the order of a modulo m has a very sharp property.

(1) Let $(a, m) = 1$, k be the order of a modulo m, and u and v be any integers. Then $a^u \equiv a^v \pmod{m}$ if and only if $u \equiv v \pmod{k}$.

In particular, $a^u \equiv 1 \pmod{m}$ if and only if $k \mid u$.

In fact, the condition is obviously sufficient. To prove the necessity, we can assume that $u > v$ and let $l = u - v$. Then from $a^u \equiv a^v \pmod{m}$ and $(a, m) = 1$, we easily know that $a^l \equiv 1 \pmod{m}$. By the division with remainder, $l = kq + r$, where $0 \leq r < k$, so $a^{kq} \cdot a^r \equiv 1 \pmod{m}$, that is, $a^r \equiv 1 \pmod{m}$. From $0 \leq r < k$ and the definition of k, there must be $r = 0$, so $u \equiv r \pmod{k}$.

From (1) we can immediately derive the following result.

(2) Let $(a, m) = 1$ and let the order of a modulo m be k. Then the sequence $a, a^2, a^3, \ldots$, modulo m is periodic, and the minimum positive period

is k. In addition, the k numbers $a, a^2, \ldots, a^k$ are not congruent to each other modulo m.

The following (3) can be derived from (1) combined with Euler's theorem ((13) in Chapter 8).

(3) Let $(a, m) = 1$. Then the order of a modulo m exactly divides the Euler function value $\varphi(m)$. In particular, if m is a prime p, then the order of a modulo p exactly divides $p - 1$.

Remark 1. It is very useful to find the order of a modulo m. By using (1) and the order of a modulo m, we can derive the divisibility relation from the exponent of some integer powers, which is another basic method of generating exact division in number theory.

Remark 2. On the other hand, it is usually difficult to determine the order of a modulo m, which can be realized only when the problem has some particularity (see Example 2 and Example 3 below). For certain a and m, we can calculate the remainder of $a, a^2, a^3, \ldots$ modulo m one by one to obtain the order; if we use (3), then this procedure can be slightly simplified.

Remark 3. From (3), we know that the order of a modulo m cannot exceed $\varphi(m)$. If there is an integer g with $(g, m) = 1$ such that the order of g modulo m is exactly $\varphi(m)$, then it is said to **have a primitive root modulo m**, and g is said to be a **primitive root modulo m**.

If there exists a primitive root g modulo m, then the reduced system modulo m has a very simple structure. The reason is: We derive from (3) that $g, g^2, \ldots, g^{\varphi(m)}$ are not congruent to each other modulo m and all of them are coprime with m, so these $\varphi(m)$ numbers, which are terms in a geometric sequence, form a reduced system modulo m. This brings a convenience to deal with some problems.

However, there doesn't exist a primitive root for every $m \geq 2$. It can be proved that if and only if $m = 2, 4, p^l$, and $2p^l$ (p is an odd prime number), there exists a primitive root modulo m.

Chinese Remainder Theorem

From (12) in Chapter 8, we can easily get a basic result of the following congruence equation.

(4) Let $(a, m) = d$. Then the linear congruence equation

$$ax \equiv b(\mathrm{mod}\, m) \tag{1}$$

has solutions if and only if $d|b$.

Suppose (1) has a solution $x = c$. Then $m|(ac - b)$, thus $d|(ac - b)$. And $d|a$, so $d|b$. Conversely, if $d|b$, then the congruence equation (1) has the same solution as

$$\frac{a}{d}x \equiv \frac{b}{d}\left(\mathrm{mod}\,\frac{m}{d}\right) \tag{2}$$

Because $\left(\frac{a}{d}, \frac{m}{d}\right) = 1$, we deduce that (2) has solutions from (12) in Chapter 8, and these solutions form a congruence class modulo $\frac{m}{d}$. In fact, it is not difficult to prove that when $d|b$, the congruence equation (1) (modulo m) has exactly d solutions.

However, even if every (linear) congruence equation has solutions, the system of congruence equations made from them may not have solutions. Such an example is easy to find. The following result, known as the Chinese remainder theorem or Sun Tzu theorem, gives a sufficient condition for the existence of solutions of a system of congruence equations and is widely used.

(5) (**Chinese remainder theorem**) Let $m_1, \ldots, m_k$ be positive integers that are pairwise coprime. Then for any integers $b_1, \ldots, b_k$, there must be a solution to the system of the linear congruence equations

$$x \equiv b_1(\mathrm{mod}\, m_1), \ldots, x \equiv b_k(\mathrm{mod}\, m_k) \tag{3}$$

and all the solutions constitute one congruent class modulo $m_1 \cdots m_k$. To be exact, the solutions to the system of congruence equations (3) are

$$x \equiv M_1 M_1^{-1} b_1 + \cdots + M_k M_k^{-1} b_k (\mathrm{mod}\, m_1 \cdots m_k), \tag{4}$$

where M_i and M_i^{-1} $(1 \leq i \leq k)$ are determined by the following conditions:

$$m_1 \cdots m_k = M_i m_i, M_i M_i^{-1} \equiv 1(\mathrm{mod}\, m_i).$$

In fact, as known by the condition $(m_i, M_i) = 1$, there is an integer M_i^{-1} such that $M_i M_i^{-1} \equiv 1 \pmod{m_i}$. In addition, $m_j | M_i$ for $j \neq i$, so the number $d_i = M_i M_i^{-1}$ satisfies

$$d_i \equiv 1 \pmod{m_i}, d_i \equiv 0 \pmod{m_j} \quad \text{(for } j \neq i). \tag{5}$$

We note that the congruence equations (5) are k special cases of (3). It is easy to verify that for $1 \leq i \leq k$, the "superposition" $b_1 d_1 + \cdots + b_k d_k$ of solutions of (5) satisfies the system (3) of congruence equations, so (4) gives the solution of (3).

On the other hand, if x and x' both are solutions of (3), then $x \equiv x' \pmod{m_i}$ for $1 \leq i \leq k$.

Hence from the condition that $m_1, \ldots, m_k$ are pairwise coprime, $x \equiv x' \pmod{m_1 \cdots m_k}$ ((10) in Chapter 8). Therefore, the solutions of (3) compose a congruence class modulo $m_1 \cdots m_k$.

Remark 4. The main strength of the chinese remainder theorem is that when the moduli are pairwise coprime, the system (3) of congruence equations must have solutions.

In addition, the above derivation of the solutions (4) is similar to the derivation of Lagrange's interpolation formula in Chapter 14 (refer to (1) in Chapter 14). For specific problems, we can follow this spirit without applying formula (4).

Lucas' Theorem

Let $n \geq k$ be a positive integer and p be a prime number. Lucas' theorem gives a relation between the combination number $\binom{n}{k}$ modulo p and the base-p representations of n and k. This result provides a way to deal with some problems of combination numbers modulo p.

(6) (**Lucas' theorem**) Let $n \geq k$ be a positive integer and p be a prime number. Let the base-p representations of n and k be

$$n = (a_d \cdots a_0)_p, \ k = (b_d \cdots b_0)_p,$$

where $0 \leq a_i, b_i \leq p - 1 (i = 0, \ldots, d)$ and $a_d \neq 0$. Then

$$\binom{n}{k} \equiv \prod_{i=0}^{d} \binom{a_i}{b_i} \pmod{p}.$$

Therefore, $p | \binom{n}{k}$ if and only if there exists an i such that $a_i < b_i$.

To prove it, we first prove a lemma:

Lemma Let p be a prime number, $n = pA + a$, and $k = pB + b$, where $A \geq B \geq 0$ and $0 \leq a, b \leq p-1$. Then

$$\binom{n}{k} \equiv \binom{A}{B}\binom{a}{b} \pmod{p}.$$

Proof of the lemma: Since p is a prime number, $p \mid \binom{p}{i}$ $(1 \leq i \leq p-1)$, thus

$$(1+x)^p = \sum_{i=0}^{p} \binom{p}{i} x^i \equiv 1 + x^p \pmod{p}.$$

Therefore

$$(1+x)^n = (1+x)^{pA}(1+x)^a$$
$$\equiv (1+x^p)^A (1+x)^a$$
$$\equiv \sum_{i=0}^{A} \binom{A}{i} x^{pi} \sum_{j=0}^{a} \binom{a}{j} x^j$$
$$\equiv \sum_{i=0}^{A} \sum_{j=0}^{a} \binom{A}{i}\binom{a}{j} x^{pi+j} \pmod{p}.$$

Because the coefficient of x^k in $(1+x)^n = \sum_{i=0}^{n} \binom{n}{i} x^i$ is $\binom{n}{k}$, we consider the coefficient of $x^k = x^{pB+b}$ on the right side of the above formula.

Since $0 \leq a, b \leq p-1$, for $0 \leq i \leq A$ and $0 \leq j \leq a$, when $b \leq a$,

$$pi + j = pB + b(= k)$$

has a unique group of solutions $j = b$ and $i = B$. This implies that $\binom{n}{k} \equiv \binom{A}{B}\binom{a}{b} \pmod{p}$, so the conclusion is true.

When $b > a$, the equation $pi+j = k$ has no solutions, so $\binom{n}{k} \equiv 0 \pmod{p}$. Since $\binom{a}{b} = 0$, the conclusion is also true.

Proof of Lucas' theorem Prove by induction on the number d of digits in the base-p representation. When $d = 1$, the conclusion is obviously true. Assume that the theorem is true when the number of digits is d. When the number of digits is $d + 1$, take $a = a_0, A = a_1 + \cdots + a_d p^{d-1}, b = b_0$, and $B = b_1 + \cdots + b_d p^{d-1}$. (Note that the number of digits of A in the base-p representation is d). Then $n = pA + a$ and $k = pB + b$. Obviously, $A \geq B$ and $\leq a, b \leq p - 1$. Hence from the lemma, we obtain

$$\binom{n}{k} \equiv \binom{A}{B}\binom{a}{b}$$

$$\equiv \prod_{i=1}^{d} \binom{a_i}{b_i} \cdot \binom{a}{b}$$

$$\left(\text{use the inductive hypothesis for } \binom{A}{B}\right)$$

$$\equiv \prod_{i=0}^{d} \binom{a_i}{b_i} \pmod{p}.$$

This shows that the conclusion is also true when the number of digits is $d + 1$. $\square$

Illustrative Examples

Example 1. Let a and m be positive integers with $a > 1$. Prove: $m \mid \varphi(a^m - 1)$.

Proof. Obviously, a and $a^m - 1$ are coprime, and clearly we know that the order of a modulo $a^m - 1$ is m, so $m \mid \varphi(a^m - 1)$ from (3). (Refer to Remark 1.) $\square$

Example 2. Let p be an odd prime number. Prove that any prime factor of $2^p - 1$ has the form $2px + 1$, where x is a positive integer.

Proof. Let q be any prime factor of $2^p - 1$. Then $q \neq 2$. Let the order of 2 modulo q be k. Then $k \mid p$ from $2^p \equiv 1 \pmod{q}$, so $k = 1$ or p (note that p is prime, which is the main factor to determine the order k). Obviously, $k \neq 1$, since that otherwise $2^1 \equiv 1 \pmod{q}$, which is impossible, so $k = p$.

Now, from Fermat's little theorem $2^{q-1} \equiv 1 \pmod{q}$, we deduce that $k \mid q - 1$, namely $p \mid q - 1$ (refer to (3)). Because p and q are both odd numbers, $q - 1 = 2px$ (x is some positive integer). $\square$

Example 3. Let n be a positive integer. Prove that any prime factor of $2^{2^n} + 1$ has the form $2^{n+1}x + 1$, where x is a positive integer.

Proof. Let p be any prime factor of $2^{2^n} + 1$. Then $p \neq 2$. Let the order of 2 modulo p be k. From

$$2^{2^n} \equiv -1(\text{mod}\, p), \tag{6}$$

we obtain that $2^{2^{n+1}} \equiv 1(\text{mod}\, p)$, so $k | 2^{n+1}$, therefore k is a power of 2. Suppose $k = 2^l$, where $0 \leq l \leq n+1$. If $l \leq n$, then square both sides of $2^{2^l} \equiv 1(\text{mod}\, p)$ for several times to generate $2^{2^n} \equiv 1(\text{mod}\, p)$. Combining it with (6), we obtain $1 \equiv -1(\text{mod}\, p)$, which is impossible. Thus, there must be $l = n + 1$, i.e., the order $k = 2^{n+1}$. Similar to the previous problem we know that $2^{n+1} | p - 1$, so $p - 1 = 2^{n+1}x$ (x is a positive integer). $\square$

Remark 5. From Example 2 and Example 3, we see that any prime factor of a Mersenne number $M_p = 2^p - 1$ is greater than p, and any divisor of M_p must be $\equiv 1(\text{mod}\, p)$. Any divisor of a Fermat number $F_n = 2^{2^n} + 1$ is $\equiv 1(\text{mod}\, 2^{n+1})$. (Refer to Remark 4 in Chapter 7.)

In addition, Example 3 also shows that for any fixed $l \geq 2$, there is an infinite number of prime numbers of the form $2^l x + 1$.

This is because for $n \geq l$, all prime factors of F_n are $\equiv 1(\text{mod}\, 2^{n+1})$, which of course $\equiv 1(\text{mod}\, 2^l)$. Also $F_n (n \geq 1)$ are pairwise coprime, so these prime factors are mutually different (refer to Example 4 in Chapter 6).

Example 4. Suppose m, a, and b are all positive integers and $m > 1$. Then

$$(m^a - 1, m^b - 1) = m^{(a,b)} - 1.$$

Proof. Let $d = (m^a - 1, m^b - 1)$. Because $(a, b) | a$ and $(a, b) | b$, it's easy to know that $m^{(a,b)} - 1 | m^a - 1$ and $m^{(a,b)} - 1 | m^b - 1$, thus $m^{(a,b)} - 1 | d$ ((9) in Chapter 6).

On the other hand, m and d are obviously coprime. Let the order of m modulo d be k. Then from

$$m^a \equiv 1(\text{mod}\, d), \quad m^b \equiv 1(\text{mod}\, d),$$

we deduce that $k | a$ and $k | b$, so $k | (a, b)$. Therefore, $m^{(a,b)} \equiv 1(\text{mod}\, d)$, i.e., $d | m^{(a,b)} - 1$. It follows that $d = m^{(a,b)} - 1$. $\square$

This problem is Example 5 in Chapter 6. The above solution applied the property (1) of the order, which depends on the division with remainder. Therefore, the essences of the two solutions are the same.

Example 5. Let $n > 1$. Prove: $n \nmid 2^n - 1$.

Proof. Prove by contradiction. If there exists $n > 1$ such that $n | 2^n - 1$, then n must be odd. Let k be the order of 2 modulo n. Then from

$$2^n \equiv 1 (\mathrm{mod}\, n) \tag{7}$$

we know $k|n$. Since $2^k \equiv 1 \ (\mathrm{mod}\, n)$, we have $2^k \equiv 1 \ (\mathrm{mod}\, k)$, i.e.,

$$k|2^k - 1. \tag{8}$$

But the order k satisfies $1 \le k < n$, and obviously $k \ne 1$ (otherwise, $n = 1$ is deduced), so $1 < k < n$. By repeating the above argument from (8), it is found that there is an infinite number of integers $k_i (i = 1, 2, \ldots)$ that satisfy $k_i | 2^{k_i} - 1$, and $n > k > k_1 > k_2 > \cdots > 1$, which is obviously impossible.

This proof can also be expressed (more simply) as follows: Taking $n > 1$ as the minimum integer such that $n | 2^n - 1$. The above argument generates an integer k such that $k | 2^k - 1$ and $1 < k < n$, which contradicts the minimality of n.

In addition, this problem can also be solved in a slightly different way: Suppose there exists $n > 1$ such that $n | 2^n - 1$. If p is any prime factor of n, then $p \ne 2$. Let k be the order of 2 modulo p. Then it is easy to know that

$$2^n \equiv 1 (\mathrm{mod}\, p), \tag{9}$$

so $k|n$. The fact $k|p - 1$ is derived from Fermat's little theorem $2^{p-1} \equiv 1 (\mathrm{mod}\, p)$, so $k|(n, p - 1)$.

Now we take (a parameter) p as the minimum prime factor of n. Then $(n, p - 1) = 1$, so $k|1$, that is, $k = 1$, hence $2^1 \equiv 1 (\mathrm{mod}\, p)$, which is impossible. □

Remark 6. The method of Example 5 is the so-called method of infinite descent. Its main point is to assume (from the proof by contradiction) that there is a solution and to derive a "smaller" solution. For this, please refer to Remark 4 in Chapter 20.

Remark 7. In the second solution, we converted the congruence (7) modulo n into the congruence (9) modulo p (p is a prime factor of n), which weakens the hypothesis of proof by contradiction, but still characterizes $n > 1$ (refer to Remark 1 in Chapter 7).

There are many good properties of congruence modulo p (refer to the second section in Chapter 8, the fourth section in Chapter 11, and the second section in Chapter 12), but they are not particularly beneficial to this problem.

Example 6. Prove that for any positive integer n, there exist n consecutive positive integers, each of which is exactly divisible by a square number greater than 1.

Proof. Because there is an infinite number of primes, we may choose n distinct prime numbers $p_1, \ldots, p_n$, and $p_1^2, \ldots, p_n^2$ are pairwise coprime. From the Chinese remainder theorem, the system of congruence equations

$$x \equiv -1 (\bmod\, p_1^2), \ldots, x \equiv -n (\bmod\, p_n^2)$$

has an integer solution $x > 0$. Then $x + 1, \ldots, x + n$ are divisible by $p_1^2, \ldots, p_n^2$ respectively.

This problem can also be solved by the following method: Let $F_k = 2^{2^k} + 1 (k \geq 0)$. Then since $F_1, \ldots, F_n$ are pairwise coprime, the system of congruence equations

$$x \equiv -1 (\bmod\, F_1^2), \ldots, x \equiv -n (\bmod\, F_n^2)$$

has an integer solution $x > 0$. For any positive integer solution x, the numbers $x + 1, \ldots, x + n$ are the n consecutive positive integers that meet the requirement.

Note that this solution does not need to apply the concept of prime numbers. $\square$

Remark 8. The Chinese remainder theorem can often convert the problem of "finding n consecutive integers with some properties" into "finding n pairwise coprime numbers with some properties," and the latter is easy to solve.

Example 7. Let n and k be given integers with $n > 0$. Suppose $k(n - 1)$ is even. Prove: There exist x and y such that $(x, n) = (y, n) = 1$ and $x + y \equiv k (\bmod\, n)$.

Proof. First we prove that the conclusion holds when n is a power p^α of prime p. In fact, we can prove that there exist x and y such that $p \nmid xy$ and $x + y = k$.

If $p = 2$, then the condition shows that k is even, so we can take $x = 1$ and $y = k - 1$. If $p > 2$, then one of the pairs $x = 1$, $y = k - 1$ and $x = 2$, $y = k - 2$ can meet the requirements.

In general, let $n = p_1^{\alpha_1} \cdots p_r^{\alpha_r}$ be the standard factorization of n. It has been proved above that for each p_i, there are integers x_i and y_i such that $p_i \nmid x_i y_i$ and $x_i + y_i = k (i = 1, 2, \ldots, r)$. The Chinese remainder theorem shows that the system of congruence equations

$$x \equiv x_1 (\bmod\, p_1^{\alpha_1}), \ldots, x \equiv x_r (\bmod\, p_r^{\alpha_r})$$

has a solution x; similarly

$$y \equiv y_1 (\bmod\, p_1^{\alpha_1}), \ldots, y \equiv y_r (\bmod\, p_r^{\alpha_r})$$

also has a solution y. Now it is easy to verify that x and y meet the requirement of the problem: Because $p_i \nmid x_i y_i$, then $p_i \nmid xy (i = 1, \ldots, r)$, so $(xy, n) = 1$. Also

$$x + y \equiv x_i + y_i = k (\bmod\, p_i^{\alpha_i})\ (i = 1, \ldots, r),$$

thus $x + y \equiv k (\bmod\, n)$. $\square$

Remark 9. The demonstration of Example 7 shows the most basic effect of the Chinese remainder theorem: A problem related to any positive integer is reduced to the case of a power of a prime number (or prime number), which is easier to deal with (please compare it with Remark 5 in Chapter 7).

The following Example 8 is a basic result about the arithmetic property of the binomial coefficients.

Example 8. Let p be prime, n be a positive integer, and

$$n = (a_m \cdots a_1 a_0)_p$$

be its base-p representation, where a_i satisfy $0 \le a_i < p (i = 0, 1, \ldots, m)$ and $a_m \neq 0$.

Prove that there are exactly

$$N = (a_0 + 1)(a_1 + 1) \cdots (a_m + 1) \tag{10}$$

numbers in the binomial coefficients $\binom{n}{0}$, $\binom{n}{1}$, ..., $\binom{n}{n}$ that are not divisible by p, and determine for which n there will be $N = n + 1$.

Proof. The result can be deduced directly from Lucas' theorem. For $\leq$ $k \leq n$, let the p-ary representation of k be

$$k = (b_m \cdots b_1 b_0)_p.$$

From Lucas' theorem (see (6) in this chapter), we know that

$$\binom{n}{k} \equiv \prod_{i=0}^{m} \binom{a_i}{b_i} \pmod{p}.$$

Hence $p \nmid \binom{n}{k}$ if and only if for each $i = 0, 1, \ldots, m$, there is $0 \leq b_i \leq a_i$ (it means that there are $a_i + 1$ options for each b_i). Therefore, there are $N = (a_0 + 1)(a_1 + 1) \cdots (a_m + 1)$ such numbers k.

Then we prove $N = n + 1$ (that is, none of the coefficients of an nth degree binomial is divisible by p) if and only if $a_i = p - 1 (i = 0, 1, \ldots, m - 1)$ and $a_m \neq 0$.

In fact, from (10),

$$N = a_m \prod_{i<m}(a_i + 1) + a_{m-1} \prod_{i<m-1}(a_i + 1) + \cdots + (a_0 + 1). \qquad (11)$$

In addition, because $a_i \leq p - 1$ $(i = 1, \ldots, m)$, for $r \geq 1$,

$$\prod_{i<r}(a_i + 1) = (a_0 + 1) \cdots (a_{r-1} + 1) \leq p^r. \qquad (12)$$

From (11) and (12),

$$n + 1 = N \leq \sum_{i=0}^{m} a_i p^i + 1.$$

Comparing it with $N = n + 1 = \sum_{i=0}^{m} a_i p^i + 1$, we know that for all r, the inequality (12) must hold as equality, i.e., $a_i = p - 1 (i = 0, 1, \ldots, m - 1)$. $\square$

In particular, it is deduced from this problem that there are exactly 2^s odd numbers in $\binom{n}{0}$, $\binom{n}{1}$, ..., $\binom{n}{n}$, where s is the number of digit 1 in the binary representation of n; Therefore, all the coefficients of an nth degree binomial are odd if and only if n has the form $2^k - 1$ (k is a positive integer).

Exercises

Group A

1. Let p be an odd prime number. Prove that the prime factor of $2^p + 1$ either is 3 or has the form $2px + 1$ (x is a positive integer).
2. Let $a > 1$ be an integer and p be an odd prime. Prove that the prime factor of $a^p - 1$ can either be divisible by $a - 1$ or has the form $2px + 1$ (x is a positive integer).
3. Solve the system of congruence equations

$$x \equiv 1 \pmod 4, x \equiv 2 \pmod 5, x \equiv 4 \pmod 7.$$

4. Let n be a given positive integer. Prove that there are n consecutive positive integers, none of which is a power of a prime.

Group B

5. Let $n \geq 1$. Prove that the order of 2 modulo 5^n is $\varphi(5^n) = 4 \times 5^{n-1}$.
6. Let m and k be integers greater than 1. Prove that there is a reduced system modulo m such that all the prime factors of these numbers are greater than k.
7. In the Cartesian plane, for any given positive integer n, prove that there exists an integer point whose distance from any irreducible integer point is greater than n (an integer point refers to a point whose coordinates are integers in the Cartesian plane, and an irreducible integer point is an integer point whose coordinates are coprime).
8. Let A be an infinite subset of the set of positive integers, n be a given integer, and $n > 1$. It is known that for any prime number p that doesn't exactly divide integer n, there is an infinite number of elements in the set A that are not divisible by the integer p.

    Prove that for any integer $m > 1$ with $(m, n) = 1$, there is a finite number of different elements in the set A, whose sum S satisfies $S \equiv 1 \pmod m$ and $S \equiv 0 \pmod n$.
9. Let p be a given odd prime number. Prove that there is an infinite number of prime numbers $\equiv 1 \pmod p$.

Chapter 20

Indeterminate Equations (2)

By showing examples, this chapter introduces several kinds of special indeterminate equations and solutions that are often involved in (higher level) mathematics competitions.

Example 1. Prove that the indeterminate equation

$$(x+1)^y - x^z = 1, x, y, z > 1 \tag{1}$$

has only one group of positive integer solutions $x = 2$, $y = 2$, and $z = 3$.

Proof. First of all, simplify (1) modulo $x + 1$ (refer to Remark 3 in Chapter 9). Then we get

$$(-1)^{z+1} \equiv 1 \pmod{x+1},$$

so z is odd. Decompose (1) into

$$(x+1)^{y-1} = x^{z-1} - x^{z-2} + \cdots - x + 1.$$

Thus x must be even; otherwise the parities of the two sides of the above equality are different. Similarly, convert (1) into

$$(x+1)^{y-1} + (x+1)^{y-2} + \cdots + (x+1) + 1 = x^{z-1},$$

and then y is also even.

Now let $x = 2x_1$ and $y = 2y_1$. Then from (1),

$$((x+1)^{y_1} - 1)((x+1)^{y_1} + 1) = x^z. \tag{2}$$

Because x is even, the greatest common divisor of $(x + 1)^{y_1} - 1$ and $(x + 1)^{y_1} + 1$ is 2, and obviously $x \mid (x + 1)^{y_1} - 1$. From these and (2), it follows that

$$(x + 1)^{y_1} - 1 = 2x_1^z \quad (x + 1)^{y_1} + 1 = 2^{z-1}.$$

This implies $2^{z-1} > 2x_1^z$, so $x_1 = 1$, i.e., $x = 2$. Hence $y = 2$ and $z = 3$. □

Remark 1. Equation (1) is a special case of the famous **Catalan's conjecture**. Catalan conjectured that 3^2 and 2^3 are the only positive integer powers with a difference of 1, that is, the indeterminate equation

$$x^a - y^b = 1, \quad a, b > 1 \tag{3}$$

has only a group of positive integer solutions $x = 3$, $y = 2$, $a = 2$, and $b = 3$. With a deeper method, this problem has been completely solved recently, and the elementary method can deal with some special cases of (3). Also see Exercise 1 and Exercise 2 in Chapter 9.

Example 1 can also be proved by the following method: We first prove that x has no odd prime factor. Suppose p is an odd prime number such that $x = p^a x_1$, where a and x_1 are positive integers and $p \nmid x_1$. By the binomial theorem, rewrite (1) as

$$xy + \sum_{i=2}^{y} \binom{y}{i} x^i = x^z. \tag{4}$$

Hence $x \mid y$, so $p \mid y$. Let $y = p^b y_1$, where $b \geq a$ and $p \nmid y_1$. The main point of the argument is to compare the powers of prime p on both sides of formula (4).

For $2 \leq i \leq y$, suppose $p^c \| i$ (i.e., $p^c \mid i$ but $p^{c+1} \nmid i$). Then in

$$\binom{y}{i} x^i = \frac{y}{i} \binom{y-1}{i-1} x^i = \frac{p^b y_1}{i} \binom{y-1}{i-1} (p^a x_1)^i,$$

the power of prime p is at least $d = b + ai - c$. If $c = 0$, then $d > a + b$; if $c > 0$, since $p \geq 3$, then $i \geq p^c > c + 1$, so $d > b + a + c(a - 1) \geq a + b$, i.e., $d \geq a + b + 1$. In a word, $p^{a+b+1} \mid \binom{y}{i} x^i$, hence $p^{a+b+1} \mid \sum_{i=2}^{y} \binom{y}{i} x^i$. Also $p^{a+b} \| xy$, thus the power of p on the left side of formula (4) is $a + b$. But $p^{az} \| x^z$, and from (1) we know $z > y$, from which $az > ay \geq ap^b \geq a(b+1) \geq a+b$, i.e., $az > a+b$. Hence the powers of p on both sides of the formula are not equal, which is impossible. Therefore, x does not have an odd prime factor, i.e., x is a power of 2. Suppose $x = 2^r$. Then equation (1) will be reduced to Exercise 2 in Chapter 9, which is easy to solve.

Remark 2. The essence of the above argument is to derive a contradiction from (4) modulo p^{a+b+1}. Because this is based on the calculation and estimation of the power of p in each term of (4), sometimes it is called the **method of comparing the powers of prime numbers**. This technique is widely used.

Example 2. Prove that the indeterminate equation

$$x^4 + y^4 = z^2 \tag{5}$$

has no positive integer solutions.

Proof. Prove by contradiction. Suppose (5) has positive integer solutions. We choose a group of solutions that minimize z among all such solutions. The idea of the argument is to create another group of solutions (r, s, t) so that $0 < t < z$. Since the selected solutions give the minimum, this leads to a contradiction.

First of all, there must be $(x, y) = 1$, because if $d = (x, y)$ and $d > 1$, then from (5) we know $d^4 \mid z^2$, that is, $d^2 \mid z$, and $\left(\frac{x}{d}, \frac{y}{d}, \frac{z}{d^2}\right)$ is also a group of positive integer solutions of (5), but $\frac{z}{d^2} < z$, which is contradictory to the selection of z.

Rewrite (5) as

$$(x^2)^2 + (y^2)^2 = z^2.$$

Since $(x, y) = 1$, we have $(x^2, y^2) = 1$, so (x^2, y^2, z) is a group of primitive Pythagorean numbers. From the Pythagorean numbers theorem in the second section of Chapter 9, we see that there exist integers $a > b > 0$ with $(a, b) = 1$, one of which is even and the other is odd, such that (suppose y is even)

$$x^2 = a^2 - b^2, \quad y^2 = 2ab, \quad z = a^2 + b^2. \tag{6}$$

From $x^2 + b^2 = a^2$ and $(a, b) = 1$, we know that a is odd and b is even. By applying the above theorem, there exist integers $p > q > 0$ with $(p, q) = 1$, one of p and q being even and the other being odd, such that

$$x = p^2 - q^2, \quad b = 2pq, a = p^2 + q^2.$$

From (6), we see that $y^2 = 4pq(p^2 + q^2)$, i.e.,

$$\left(\frac{y}{2}\right)^2 = pq(p^2 + q^2).$$

Because $(p, q) = 1$, the positive integers p, q, and $p^2 + q^2$ are pairwise coprime. The above formula shows that their product is a square of integers, so there are positive integers r, s, and t such that

$$p = r^2, q = s^2, p^2 + q^2 = t^2,$$

thus

$$r^4 + s^4 = t^2.$$

Hence we get a group of positive integer solutions (r, s, t) of (5). But

$$0 < t = \sqrt{p^2 + q^2} = \sqrt{a} < \sqrt[4]{z} \le z,$$

which contradicts the minimality of z.   □

Remark 3. From Example 2, we immediately derive that the Fermat equation

$$x^n + y^n = z^n, n > 2,$$

has no positive integer solutions when $n = 4$. For the other two simple corollaries of Example 2, please refer to Exercise 1 in this chapter.

Remark 4. The proof method of Example 2 is called the **infinite descent method**, which is an important method first proposed by Fermat.

Here are the main steps to prove that an indeterminate equation doesn't have positive integer solutions by the infinite descent method. Start from the opposite conclusion, and suppose there is a group of positive integer solutions we try to construct another group of positive integer solutions to this equation, such that the new solution is "strictly smaller" than the original one. Here, strictly smaller eyes refers to that a quantity taking positive integer values related to solutions strictly decreases. If the above process can go on infinitely, then there will be a contradiction because there is only a finite number of terms in a strictly decreasing positive integer sequence.

The infinite descent method can also take different forms (the essence is the same as above): Suppose there exist positive integer solutions and choose a group of "minimum" solutions. The "minimum" solution here means that a quantity taking positive integer values and related to solutions reaches the minimum. The core of the argument is still to try to create a new solution to the equation, which is "strictly smaller" than the selected solution, thus generating a contradiction (refer to Example 5 in Chapter 19).

How to construct a smaller solution depends on the specific problem of course. Example 5 in Chapter 19 applied the property of the order, which depends on the division with remainder (refer to (1) in Chapter 19). In Example 2, Pythagorean theorem was applied twice, and in Example 3 below, (the more elementary) Vieta's theorem will be used.

Remark 5. The infinite descent method is not limited to dealing with indeterminate equations. If a problem can generate another state from one state, and there is a strict reduction in a quantity of positive integer values related to the state, the descent method can be used to generate a contradiction (i.e., the problem has no solution) or a solution. For example, Euclidean algorithm in (7) of Chapter 6 is essentially an infinite descent method: After a finite number of divisions with remainder, it must reach the "minimum state" and produce the greatest common divisor. The argument in (5) of Chapter 11 is also a descending method: From one state to another, it corresponds to a strictly decreasing positive integer (the degree of polynomials related to the state), so the process must stop after a finite number of steps. Thus the representation with remainder is given.

Example 3. Prove that if the equation

$$x^2 + y^2 + 1 = xyz \tag{7}$$

has positive integer solutions (x, y, z), then $z = 3$.

Proof. Prove by contradiction. If there is a positive integer $z \neq 3$ such that the equation (7) has a positive integer solution (x, y), then $x \neq y$, since otherwise we get $2x^2 + 1 = x^2 z$, so $x = 1$ and $z = 3$, a contradiction.

Since x and y are symmetric, we may assume that $x > y$. Select a solution (x_0, y_0) from these solutions to minimize x_0. Consider the quadratic equation in one variable x

$$x^2 - y_0 z x + y_0^2 + 1 = 0. \tag{8}$$

From Vieta's theorem, the other root of (8) is

$$x_1 = y_0 z - x_0,$$

which is an integer, and

$$0 < x_1 = \frac{y_0^2 + 1}{x_0} \leq \frac{y_0^2 + 1}{y_0 + 1} \leq y_0.$$

Therefore, there is another positive integer solution (y_0, x_1) of (7), which satisfies $y_0 > x_1$ (from the previous discussion, $y_0 \neq x_1$) and $y_0 < x_0$. This contradicts the minimality of x_0.   □

There are some indeterminate equations, which are difficult to deal with by the decomposition method of (as mentioned in Chapter 9) and the estimation method of but need to apply some in-depth congruence knowledge: quadratic residue (or called the square residue). For the purpose of this book, we will not discuss it in depth, but only introduce one of the most basic and commonly used results:

Lemma Let p be a prime number of the form $4k + 3$. Then

$$x^2 + 1 \equiv 0 \pmod{p} \tag{9}$$

has no integer solutions.

In other words, for any integer x, a prime factor of $x^2 + 1$ is either 2, or $\equiv 1 \pmod 4$.

The proof is quite simple. Assume that there exists x satisfying (9). Then

$$x^2 \equiv -1 \pmod{p},$$

so $x^{p-1} \equiv (-1)^{\frac{p-1}{2}} \pmod{p}$. But $p \nmid x$, thus from Fermat's little theorem, $x^{p-1} \equiv 1 \pmod{p}$, and hence $(-1)^{\frac{p-1}{2}} \equiv 1 \pmod{p}$, that is, $-1 \equiv 1 \pmod{p}$. This implies that $p = 2$, which is impossible.

Remark 6. It is not difficult to prove that there is an infinite number of prime numbers of the form $4k + 1$ from the above lemma and Euclid's idea of proving that there is an infinite number of prime numbers (see (4) in Chapter 7) (try to compare it with Exercise 5 in Chapter 7):

Suppose there is only a finite number of prime numbers of the form $4k + 1$. Let all of them be $p_1, \ldots, p_k$. Consider

$$N = (2p_1 \cdots p_k)^2 + 1.$$

Because $N > 1$, N has a prime factor p. Because N is odd, we see that $p \neq 2$. Then from the above lemma, $p \equiv 1 \pmod 4$. Since p is obviously different from $p_1, \ldots, p_k$ (otherwise $p \mid 1$, which is impossible), we obtain a new prime number of the form $4k+1$, which is contradictory to the previous assumption.

Example 4. Let p be a prime number of the form $4k + 3$, and let n be a positive integer satisfying $p \mid n$. If n can be expressed as the sum of squares of two integers, that is, there are integers x and y, such that

$$n = x^2 + y^2, \tag{10}$$

then the power of p in n is even.

Proof. First prove that x and y are both divisible by p. If not, assume that $p \nmid x$. Then there is an integer x_1 such that $xx_1 \equiv 1 \pmod p$. Multiply the both sides of (10) by x_1^2, and then modulo p to get (notice that $p \mid n$)

$$(x_1 y)^2 + 1 \equiv 0 \pmod p.$$

Since $p \equiv 3 \pmod 4$, this contradicts the lemma above, so $p \mid (x, y)$. Let $p^\alpha \| (x, y)$. Then $p^{2\alpha} \mid n$. We prove that $p^{2\alpha} \| n$, that is, the power of p in n is even.

Let $n = p^{2\alpha} m$, $x = p^\alpha x_1$ and $y = p^\alpha y_1$. Then m, x_1, and y_1 are all integers, and $p \nmid (x_1, y_1)$. From (10), we obtain

$$m = x_1^2 + y_1^2.$$

Suppose $p \mid m$. Then by the same argument as above, $p \mid (x_1, y_1)$, which leads to a contradiction. Therefore the proposition is proved. $\square$

Remark 7. There are many important and interesting results about the sum of squares of positive integers, which are not discussed in this book. To express a positive integer as the sum of the squares of two integers has a complete characterization, which is the following **Gauss' theorem**:

A necessary and sufficient condition for a positive integer n to be expressed as the sum of the squares of two integers is that the power of any prime factor of the form $4k + 3$ in n is even.

Example 4 proves the necessity of the theorem, while the proof of the sufficiency of Gauss' theorem relies on an important fact: Prime numbers of the form $4k + 1$ can be expressed as the sum of the squares of two integers. It is not easy to prove this proposition. This book does not introduce it.

Remark 8. The demonstration of Example 4 shows that if p is a prime number of the form $4k + 3$, and integers x and y satisfy

$$x^2 + y^2 \equiv 0 \pmod{p},$$

then both x and y can be divisible by p. This version of the lemma is sometimes more convenient for applications.

Example 5. Prove that the equation

$$y^2 = x^3 - 5 \tag{11}$$

has no integer solutions.

Proof. First of all, if x is even, take (11) modulo 4 to get $y^2 \equiv 3 \pmod 4$, which is impossible. If x is an odd number, when $x \equiv 3 \pmod 4$, from (11), $y^2 \equiv 2 \pmod 4$, which is still impossible. Hence $x \equiv 1 \pmod 4$, so $x^2 + x + 1 \equiv 3 \pmod 4$, and $x^2 + x + 1$ must have a prime factor p of the form $4k + 3$. Deform (11) into

$$y^2 = (x - 1)(x^2 + x + 1) - 4,$$

and take modulo p. We obtain $y^2 + 2^2 \equiv 0 \pmod p$. From Remark 8, $p|(2, y)$, that is, $p = 2$, which is impossible. Thus, equation (11) has no integer solutions.

In the above argument, some necessary conditions (of having solutions) are derived by congruence. The purpose is to generate a prime factor of the form $4k + 3$ on the right side of (11), to apply the lemma or the fact in Remark 8.

Any integer that is $\equiv 3 \pmod 4$ must have a prime factor of the form $4k + 3$. This simple conclusion is fundamental in this kind of argument.

$\square$

Example 6. Prove that the equation

$$y^2 = x^5 - 4 \tag{12}$$

has no integer solutions.

Proof. First, if x is even, then y is even. Thus the left side of (12) $\equiv 0, 4, 16 \pmod{32}$ while the right side $\equiv -4 \pmod{32}$, which is impossible. Therefore, x is odd. If $x \equiv -1 \pmod 4$, then from (12), we obtain $y^2 \equiv -5 \equiv 3 \pmod 8$, which is also impossible, so $x \equiv 1 \pmod 4$. Deform (12) into

$$y^2 + 6^2 = (x + 2)(x^4 - 2x^3 + 2^2 x^2 - 2^3 x + 2^4). \tag{13}$$

Since $x + 2 \equiv 3 \pmod 4$, the number $x + 2$ has a prime factor p of the form $4k + 3$. From (13), we know that $y^2 + 6^2 \equiv 0 \pmod p$, so $p \mid (6, y)$, thus $p = 3$. Hence $3 \mid x + 2$, i.e., $x \equiv 1 \pmod 3$.

On the other hand, from $x \equiv 1 \pmod 4$, the second factor on the right side of (13) is also $\equiv 3 \pmod 4$, so it has a prime factor of the form $4k + 3$. Similar to the above case, it can only be 3, namely $x^4 - 2x^3 + 2^2 x^2 - 2^3 x + 2^4$ is divisible by 3.

However, since (already proven) $x \equiv 1 \pmod 3$,

$$x^4 - 2x^3 + 2^2 x^2 - 2^3 x + 2^4 \equiv 1 \pmod 3,$$

a contradiction! So equation (12) has no integer solutions.

The following solution can also be used: Therefore, Take (12) modulo 11. On one hand, it is known from Fermat's little theorem that for any integer x, there is $x^{10} \equiv 0, 1 \pmod{11}$, since 11 is a prime number. Thus we know $x^5 \equiv 0, \pm 1 \pmod{11}$, and the right side of (12) is congruent to $-4, -3, -5 \pmod{11}$.

On the other hand, it is easy to verify $y^2 \equiv 0, 1, 3, 4, 5, 9 \pmod{11}$. Therefore, (12) modulo 11 has no solutions, so it has no integer solutions.

$\square$

Example 7. Let p be prime and $p \equiv 3 \pmod 4$. Integers x, y, z, and t satisfy $x^{2p} + y^{2p} + z^{2p} = t^{2p}$. Prove: $p \mid xyzt$.

Proof. Because the equation in the problem is homogeneous, we can assume that $(x, y, z, t) = 1$.

First of all, consider the original equation modulo 4. Then from this fact that a square number modulo 4 is or 1, and the above assumptions, it is easy to know that there must be two even and one odd numbers in x, y, and z (so t is an odd number); suppose x is odd, and y and z are even.

The equation can be decomposed into

$$y^{2p} + z^{2p} = (t^2 - x^2)M, \tag{14}$$

where $M = t^{2(p-1)} + t^{2(p-2)} x^2 + \cdots + x^{2(p-1)}$. Since x and t are odd numbers,

$$M = (t^{p-1})^2 + (t^{p-2}x)^2 + \cdots + (x^{p-1})^2 \equiv p \equiv 3 \pmod 4.$$

Therefore, M must have a prime factor q of the form $4k + 3$, and q has an odd power in M.

From $q \mid M$ and (14), we know that $(y^p)^2 + (z^p)^2 \equiv 0 \pmod q$. Since $q \equiv 3 \pmod 4$, we can deduce that $q \mid y$ and $q \mid z$ from the lemma (see Remark 8), so $q^{2p} \mid (y^{2p} + z^{2p})$. On the other hand, q appears odd times in M, and from

(14), $q|(t^2 - x^2)$, i.e., $t^2 \equiv x^2 \pmod{q}$, thus
$$M = (t^2)^{p-1} + (t^2)^{p-2}x^2 + \cdots + x^{2(p-1)} \equiv px^{2(p-1)} \pmod{q}.$$
Since $q|M$, this implies that $q|px^{2(p-1)}$. If $q \neq p$, then $q|x^{2(p-1)}$, so the
prime number $q|x$, which also shows that $q|t$. Combining it with (the
proven) $q|y$ and $q|z$, we obtain that q is a common divisor of x, y, z,
and t, which contradicts the previous assumption. Therefore, $q = p$, so $p|y$
and $p|z$, which proves the conclusion.

Exercises

Group A

1. Prove: The area of a right triangle with integer side lengths cannot be
   twice a square number; the area of an isosceles triangle with integer side
   lengths cannot be a square number.
2. Prove: A positive integer of the form $4k + 3$ cannot be expressed as the
   sum of the squares of two integers; there is an infinite number of positive
   integers of the form $4k + 1$, which cannot be expressed as the sum of the
   squares of two integers.
3. Prove: There doesn't exist an even number a and an odd number $x \geq 5$
   such that
   $$\sum_{k=2}^{\frac{x-1}{2}} (-1)^k \binom{x}{2k} a^{2k-2} = \frac{x(x-1)}{2}.$$
4. Prove: The equation $y^2 = x^3 + 7$ has no integer solutions.
5. Prove: The equation $y^2 = x^3 + 11$ has no integer solutions.

Group B

6. Let a and b be positive integers such that $ab + 1 \,|\, a^2 + b^2$. Prove that
   $\frac{a^2 + b^2}{ab + 1}$ is a square number.
7. Prove: The area of a right triangle with integer side lengths cannot be
   a square number.
8. Prove: The equation $x^4 - y^4 = z^2$ has no positive integer solutions x, y,
   and z.
9. (i) Prove: The equation $x^4 + y^4 = 2z^2$ has no positive integer solutions
   x, y, and z, where $x \neq y$.
   (ii) Prove: The equation $x^4 - y^4 = 2z^2$ has no positive integer solutions
   x, y, and z.

Comprehensive Exercises

1. Three of eight boys and five girls are chosen as representatives, including at least one girl. How many choices are there in all?

2. Six people stand in a row, where A is not at the left end and B is not at the right end. How many different arrangements are there?

3. Prove: Among n-digit numbers composed of digits 1 and 2, we cannot choose more than $\left[\frac{2^n}{n+1}\right]$ distinct numbers that have at least three digits different from each other.

4. Seventeen pairs of men and women are arranged in a row. It is known that the height difference between each pair of man and woman does not exceed 10 cm. Prove: If both men and women are arranged from high to low, then the height difference between each pair is still not greater than 10 cm.

5. In a chess round robin, each player is known to score up to k points (1 point for the winner, for the loser, and $\frac{1}{2}$ for each draw). Prove: There is a player who plays no more than $2k$ games.

6. There are $n(\geq 3)$ players in a table tennis round robin, and no one beats everyone else. Prove: There must be three players A, B, and C such that A beats B, B beats C, and C beats A.

7. Let $a_1, \ldots, a_n$ be integers such that $a_1 + \cdots + a_n$ is a multiple of 6. Prove: $a_1^3 + \cdots + a_n^3$ is also a multiple of 6.

8. Starting from the second term in a sequence, each term is equal to the sum of the previous term and all its digits, and the first term is 1. Is there a term equal to 123456 in this sequence?

9. Prove: $\frac{1}{n+1}\binom{2n}{n}$ is an integer.

10. Let n be a positive integer, Prove: $\binom{2n}{0} + \binom{2n}{2} \cdot 3 + \cdots + \binom{2n}{2n} \cdot 3^n$ is divisible by 2^n.

11. Let $(m, n) = 1$. Prove $\frac{(m+n-1!)}{m!n!}$ is an integer.

12. Let m and n be coprime positive integers. Prove:

$$\sum_{x=1}^{n-1} \left[\frac{xm}{n}\right] = \frac{(m-1)(n-1)}{2}.$$

13. Let $n > 1$ be an integer. Prove: $n^5 + n^4 + 1$ is not prime.

14. Let $a_1 = 1$, $a_2 = 2$, and for $n \geq 1$ define

$$a_{n+2} = \begin{cases} 5a_{n+1} - 3a_n, & \text{if } a_n \cdot a_{n+1} \text{ is even;} \\ a_{n+1} - a_n, & \text{if } a_n \cdot a_{n+1} \text{ is odd.} \end{cases}$$

Prove: For $n \geq 1$, we have $a_n \neq 0$.

15. Let $H = \{h \in \mathbb{C} | h^n = -1\}$, and n be a positive integer such that $n \geq 2$. Evaluate $\sum_{h \in H} (1 - h)^{-2}$.

16. Prove: Any positive integer n can be expressed in the form

$$n = \varepsilon_1 1^2 + \varepsilon_2 2^2 + \varepsilon_3 3^2 + \cdots + \varepsilon_m m^2,$$

where m is some positive integer and $\varepsilon_i = 1$ or $-1 (i = 1, 2, \ldots, m)$.

17. Prove: For any positive integer $a_1 > 1$, there exists an increasing positive integer sequence $a_1, a_2, a_3, \ldots$ such that $a_1^2 + \cdots + a_k^2$ is divisible by $a_1 + \cdots + a_k$ for all $k \geq 1$.

18. Let n be a given positive integer. Prove that there exist n consecutive positive integers, none of which can be expressed as the sum of the squares of two integers.

19. Let $n \geq 2$. Prove: In any $(4n - 3) \times (4n - 3)$ numerical table composed of 0 and 1, there must be a $2 \times n$ sub-table in which elements are all 1 or all. (The n columns of this table need not be adjacent.)

20. Let $B(n)$ denote the number of partitions of n as the sum of powers of 2. Prove: When $n \geq 2$, the number $B(n)$ is even. (Here the partition is an unordered partition; refer to Exercise 3 in Chapter 3.)

21. Let k be a positive integer, $m = 2^k + 1$, and $r \neq 1$ is a root of $z^m - 1 = 0$. Prove: There exist $f(x), g(x) \in \mathbb{Z}[x]$ such that $f^2(r) + g^2(r) = -1$.

22. Let $p(x)$ be an integer coefficient polynomial and $p(n) > n$ for $n \geq 1$. Define $x_1 = 1, x_2 = p(x_1), \ldots, x_n = p(x_{n-1})$, and suppose that for any positive integer N, there is a term divisible by N in the sequence $\{x_n\}$. Prove: $p(x) = x + 1$.

23. Suppose $a > 1$ is a given odd number. Prove that there are only finitely many numbers m such that $2^m | a^m - 1$.

24. Let $v(k)$ denote the number of 1 in the binary representation of k, and n be a positive integer. Evaluate $\sum_{k=1}^{2^n-1}(-1)^{v(k)}k^i(i=1,2,\ldots,n)$.

25. Let $a_1,\ldots,a_n$ be distinct positive integers such that for any positive integer k, the number $(a_1+k)\cdots(a_n+k)$ is divisible by $a_1\cdots a_n$. Prove: $a_1,\ldots,a_n$ is a permutation of $1,2,\ldots,n$.

26. Determine all positive integers n such that there is an integer m satisfying that 2^n-1 exactly divides m^2+9.

27. Prove: There do not exist integers $a,b,c>1$ that are pairwise coprime, such that $b|2^a+1$, $c|2^b+1$, and $a|2^c+1$.

28. Let $n \geq 1$ and $k \geq 0$ both be integers. Suppose $a_1,\ldots,a_n$ are integers, and their remainders on the division by $n+k$ have at least $k+1$ different values. Prove: The sum of some of these n numbers is divisible by $n+k$. (Note that when $k=0$, this question becomes a well-known result.)

29. Let r and s be nonzero integers and k be a positive real number. A sequence $\{a_n\}$ is defined as: $a_1=r, a_2=s$, and $a_{n+2}=\frac{a_{n+1}^2+k}{a_n}$ for $n \geq 1$. Prove: A necessary and sufficient condition for all a_n to be integers is that $\frac{r^2+s^2+k}{rs}$ is an integer.

30. Prove: For any odd number, a multiple of it can be found such that every digit in the decimal representation of the multiple is odd.

31. Let $n>1$ be an odd number and $c_1,\ldots,c_n$ be any integers. For any permutation $a^*=(a_1,\ldots,a_n)$ of $\{1,2,\ldots,n\}$, define $S(a^*)=\sum_{i=1}^n c_i a_i$. Prove: There exist two different permutations a^* and b^* of $\{1,2,\ldots,n\}$ such that $n!$ exactly divides $S(a^*)-S(b^*)$.

32. Let a_i, b_i, and c_i be integers $(1 \leq i \leq n)$, and for each i, at least one of the three numbers a_i, b_i, and c_i is an odd number. Prove: There exist integers A, B and C, such that among the n numbers

$$Aa_i+Bb_i+Cc_i \quad (i=1,\ldots,n),$$

there are at least $\frac{4}{7}n$ odd numbers.

33. Let $a_1<a_2<\cdots<a_{2000}<10^{100}$ be given positive integers. Prove: There are two non-empty and disjoint subsets A and B of the set $\{a_1,a_2,\ldots,a_{2000}\}$ such that:

   (i) the numbers of elements of A and B are equal;
   (ii) the sums of the elements of A and B are equal;
   (iii) the sums of the squares of elements of A and B are equal.

34. Let S_0 be a finite set of positive integers. For $n \geq 1$, (recursively) define the finite sets $S_1,S_2,\ldots$ of positive integers: Integer $a \in S_n$ if and only if exactly one of a and $a-1$ belongs to S_{n-1}. Prove: There is an infinite

number of positive integers N such that $S_N = S_0 \cup N + a$: $a \in S_0\}$ (i.e., S_N is the union of S_0 and an integer translation of S_0).

35. Suppose we are given 31 distinct positive integers, where there are at least 94 pairs of numbers that are coprime. Prove: There are four integers a, b, c, and d in the given integers such that $(a, b) = (b, c) = (c, d) = (d, a) = 1$.

36. Prove: There is an infinite sequence of positive integers $\{a_n\}(n \geq 1)$ such that for $n = 1, 2, \ldots$, the number $a_1^2 + \cdots + a_n^2$ is always a perfect square.

37. Let $a_1, \ldots, a_n$ be integers, which satisfy $\sum_{i=1}^{n} i^l a_i = 0$ for all $l = 1, \ldots, k-1$. Prove: $k! \,|\, \sum_{i=1}^{n} i^k a_i$.

38. Let $f(x)$ be a non-constant integer coefficient polynomial. Prove: The sequence $\{f(n)\}(n = 0, 1, \ldots)$ has an infinite number of (different) prime factors.

39. Prove: For each $n \geq 2$, there exists an n-degree polynomial $f(x) = x^n + a_1 x^{n-1} + \cdots + a_n$ with integer coefficients, which satisfies:

   (1) none of $a_1, a_2, \ldots, a_n$ is zero;
   (2) $f(x)$ can't be decomposed into the product of two non-constant integer coefficient polynomials;
   (3) for any integer x, the number $|f(x)|$ is not prime.

40. Prove: There is an infinite sequence of positive integers $\{a_n\}(n \geq 0)$ such that $(a_i, a_j) = 1$ for any $i \neq j$, and for all positive integers n, the polynomial $a_n x^n + \cdots + a_1 x + a_0$ cannot be decomposed into the product of two non-constant polynomials with integer coefficients.

41. Let $f(x)$ be an integer coefficient polynomial and p be prime. Suppose $f(0) = 0$ and $f(1) = 1$. If $f(k)$ is congruent to either 0 or 1 modulo p for any integer k, prove that the degree of $f(x)$ is at least $p - 1$.

42. Let $f(x)$ be a polynomial of degree n, in which the coefficients are all ± 1, and has the form of

$$f(x) = (x-1)^m g(x),$$

where $g(x)$ is a polynomial with integer coefficients and m is a positive integer. Prove: If $m \geq 2^k$ ($k \geq 2$ is an integer), then $2^{k+1} \,|\, n+1$.

Solutions

Solution 1

Permutations and Combinations

1. (i) P_6^3; (ii) 6^3.

2. The number equals to the number of the 7-permutations of $1, 2, \ldots, 9$ minus the number of permutations where 8 and 9 are adjacent. That is $P_9^7 - 2! \times 6 \times P_7^5 = 151200$. (the number of 7-permutations where 8 and 9 are adjacent can be determined as follows: At step 1, take the 5-permutations of $1, 2, \ldots, 7$ and there are P_7^5 ways; at step 2, insert 8 and 9 as a whole to the head, the tail, or between any two numbers of the 5-permutations and there are 6 ways; at last, there are 2! ways to make a full permutation of 8 and 9. According to the multiplication principle, the number of such permutations is $2! \times 6 \times P_7^5$.)

3. There are 6! arrangements for 6 singing programs. Each arrangement produces 7 "gaps". Insert 4 dance programs into these 7 "gaps". There are P_7^4 ways in total. Hence, there are $6!P_7^4$ arrangements of the program list.

4. According to the full permutation with repetition formula, there are $\frac{9!}{2!3!4!}$ plans.

5. It is easy to know that the "words" can be divided into three categories: 2 numbers of a, 3 numbers of c; 2 numbers of a, 1 number of b, 2 numbers of c; 1 number of a, 1 number of b, 3 numbers of c. Therefore, according to the addition principle and the full permutation with repetition formula, the number is $\frac{5!}{2!0!3!} + \frac{5!}{2!1!2!} + \frac{5!}{1!1!3!} = 60$.

6. First, let the boys sit in a circle. Then from the cyclic permutation formula, there are $(n-1)!$ ways. There is a space between every two adjacent boys; there is a total of n ways. Now let n girls arrange

in these n spaces, so there are $n!$ ways. Overall there are $(n-1)!n!$ arrangements. (It should be noted that after the arrangement of boys, the arrangement of girls is a (straight line) permutation rather than a cyclic permutation.)

7. Obviously, there are two balls in one box, and one ball in each of the other boxes. Step 1: Take two out of the $n+1$ balls, and there are $\binom{n+1}{2}$ ways. Step 2, Regard these two balls as one, and put them in n (different) boxes with the rest $n-1$ balls, so there are $n!$ methods. Therefore, there are $\binom{n+1}{2} n! = \frac{n}{2}(n+1)!$ ways in total.

8. The result is a k-combination with repetition of $1, 2, \ldots, 6$. Thus, the number is equal to

$$\binom{k+6-1}{k} = \binom{k+5}{k}.$$

9. The numbers are divided into three categories according to their remainders on division by 3: $A = \{3, 6, \ldots, 300\}$, $B = \{2, 5, \ldots, 299\}$, and $C = \{1, 4, \ldots, 298\}$. There are four cases when the sum of the three numbers is divisible by 3: Three numbers all belong to one of A, B and C or three numbers are taken from A, B, and C respectively. According to the addition principle, there are $3\binom{100}{3} + 100^3$ different ways.

10. Let these five points be A_1, A_2, A_3, A_4, and A_5. Connect any two points, so there are $\binom{5}{2} = 10$ straight lines. Every three points form a triangle, so there are $\binom{5}{3} = 10$ triangles. Any four points can generate $\binom{4}{2} = 6$ straight lines, and from the fifth point we can draw a perpendicular line to each of these 6 straight lines, so there are a total of $5 \times 6 = 30$ perpendicular lines. These perpendicular lines have at most $\binom{30}{2} = 435$ intersection points. However, for each of the above 10 lines, there are three perpendicular lines to it that are parallel to each other (without intersection points), so 30 intersection points should be subtracted. In addition, the three altitudes of any of the 10 triangles intersect at one point, so there are 20 intersection points to be subtracted. Each point A_i is the intersection point of 6 perpendicular lines, so we need to subtract $5\binom{6}{2} = 75$ intersections points. Therefore, there are at most 310 intersection points.

11. Let $|A| = k$. Then $|B| = 12 - k$ according to condition (1). Because A and B are not empty, $k \neq 0, 1$ and 2. From condition (3), we know that $k \in A$ and $12 - k \notin B$. Therefore, from (1), $k \in B$ and $12 - k \in A$.

Obviously, $k \neq 6$; otherwise $6 \in B$ and $6 = 12 - 6 \in A$, which doesn't meet condition (2).

For each $k(1 \leq k \leq 11$ and $k \neq 6)$, we have determined that $12 - k \in A$, and $k \in B$, so the remaining $k - 1$ elements of A are selected from the remaining 10 elements of the set $A \cup B$ excluding k and $12 - k$, and there are $\binom{10}{k-1}$ selections. Then condition (1) uniquely determines B.

Therefore, $N = \sum_{k=1}^{11} \binom{10}{k-1} - \binom{10}{6-1} = 2^{10} - \binom{10}{5} = 772$. (where the equality $\sum_{k=1}^{11} \binom{10}{k-1} = \sum_{k=0}^{10} \binom{10}{k} = 2^{10}$ can be calculated directly, or refer to (8) in Chapter 2.)

12. Because the intersection of any two of A, B, and $\overline{A \cup B}$ is the empty set, and their union is set S (where $\overline{A \cup B}$ is the complement of $A \cup B$ with respect to S), any element of S is in A, B, or $\overline{A \cup B}$, so there are three situations. Then the number of ordered subset pairs (A, B) satisfying $A \cap B = \emptyset$ in S is 3^{10} (refer to Example 4 in this chapter).

The number obtained above includes the case that A and B are empty sets, which should be excluded: If $A = \emptyset$, then each element of S is either in B or in $\overline{B}$, so there are 2^{10} sets B. Similarly, when $B = \emptyset$, there are 2^{10} sets A. And because there is only one set pair where A and B are empty sets, there are at least $2^{10} + 2^{10} - 1 = 2^{11} - 1$ set pairs in which at least one of A and B is an empty set. Therefore, there are $3^{10} - (2^{11} - 1)$ ordered non-empty subset pairs (A, B) that satisfy $A \cap B = \emptyset$, and the number of the desired unordered pairs is $\frac{1}{2}(3^{10} + 1) - 2^{10} = 28501$.

Solution 2

Binomial Coefficients

1. Induct on k. When $k = 1$, the conclusion is obviously true. Assume that the equality is true for $k - 1\,(k \geq 2)$. Then from the inductive hypothesis and (5),

$$\sum_{i=0}^{k}(-1)^i \binom{n}{i} = (-1)^{k-1}\binom{n-1}{k-1} + (-1)^k \binom{n}{k}$$

$$= (-1)^{k-1}\left(\binom{n-1}{k-1} - \binom{n}{k}\right)$$

$$= (-1)^{k-1}\left(-\binom{n-1}{k}\right)$$

$$= (-1)^k \binom{n-1}{k}.$$

2. Let $a_n = \sum_{k=0}^{n}\binom{n+k}{k}\frac{1}{2^k}\,(n \geq 1)$. Then $a_1 = 2$. From (5), we obtain that

$$a_{n+1} = \sum_{k=0}^{n+1}\binom{n+1+k}{k}\frac{1}{2^k} = \sum_{k=0}^{n+1}\binom{n+k}{k}\frac{1}{2^k} + \sum_{k=0}^{n+1}\binom{n+k}{k-1}\frac{1}{2^k}$$

$$= a_n + \binom{2n+1}{n+1}\frac{1}{2^{n+1}} + \frac{1}{2}\sum_{k=1}^{n+2}\binom{n+1+k-1}{k-1}\frac{1}{2^{k-1}}$$

$$- \binom{2n+2}{n+1}\frac{1}{2^{n+2}} = a_n + \frac{1}{2}a_{n+1}.$$

271

Therefore, $a_{n+1} = 2a_n$ is true for all $n \geq 1$, thus from this and $n \geq 1$, we see that $a_n = 2^n$.

3. Let a_n be the sum. Then $a_0 = a_1 = 1$. For $n \geq 2$, we have (using (5))

$$a_n = \sum_{k=0}^{\left[\frac{n}{2}\right]} \frac{(-1)^k}{4^k} \binom{n-k-1}{k} + \sum_{k=1}^{\left[\frac{n}{2}\right]} \frac{(-1)^k}{4^k} \binom{n-k-1}{k-1}.$$

In the first term, if n is odd, then $\left[\frac{n}{2}\right] = \left[\frac{n-1}{2}\right]$; if n is even, then $\binom{n-k-1}{k} = 0$ and $\left[\frac{n}{2}\right] - 1 = \left[\frac{n-1}{2}\right]$ when $k = \frac{n}{2}$. Hence the first term of the above formula is

$$\sum_{k=0}^{\left[\frac{n-1}{2}\right]} \frac{(-1)^k}{4^k} \binom{n-1-k}{k} = a_{n-1}.$$

Similarly, the second term is

$$\sum_{k=1}^{\left[\frac{n}{2}\right]} \frac{(-1)^k}{4^k} \binom{n-1-k}{k} = -\frac{1}{4} \sum_{k=0}^{\left[\frac{n-2}{2}\right]} \frac{(-1)^k}{4^k} \binom{n-2-k}{k}.$$

Hence, we can deduce that $a_n = a_{n-1} - \frac{1}{4} a_{n-2}$ $(n \geq 2)$. That is,

$$a_n - \frac{1}{2} a_{n-1} = \frac{1}{2} \left(a_{n-1} - \frac{1}{2} a_{n-2} \right).$$

Easily we know that $a_n - \frac{1}{2} a_{n-1} = \frac{1}{2^n}$, so for $n \geq 1$,

$$a_n = \frac{1}{2^n} + \frac{1}{2} a_{n-1} = \frac{1}{2^n} + \frac{1}{2} \left(\frac{1}{2^{n-1}} + \frac{1}{2} a_{n-2} \right)$$

$$= \cdots = \frac{n}{2^n} + \frac{a_0}{2^n} = \frac{n+1}{2^n}.$$

4. Using $\frac{1}{k+1} \binom{n}{k} = \frac{1}{n+1} \binom{n+1}{k+1}$ and the binomial theorem, we can easily get the result. Refer to the deformation method of (i) in Example 1.

5. Let a_n be the sum. Using

$$\frac{1}{k} \binom{n}{k} = \frac{1}{k} \left(\binom{n-1}{k-1} + \binom{n-1}{k} \right) = \frac{1}{n} \binom{n}{k} + \frac{1}{k} \binom{n-1}{k}$$

and the binomial theorem, we get

$$a_n = \sum_{k=1}^{n-1} (-1)^{k+1} \frac{1}{k} \binom{n}{k} + (-1)^{n+1} \frac{1}{n}$$

$$= \sum_{k=1}^{n-1} (-1)^{k+1} \frac{1}{k} \binom{n-1}{k} + \frac{1}{n} \sum_{k=1}^{n-1} (-1)^{k+1} \binom{n}{k} + (-1)^{n+1} \frac{1}{n}$$

$$= a_{n-1} + \frac{1}{n} \sum_{k=1}^{n} (-1)^{k+1} \binom{n}{k} = a_{n-1} + \frac{1}{n}.$$

Since $a_1 = 1$, the result is obtained from the above formula and mathematical induction.

6. When $n < m$ and $n = m$, the conclusion is obviously true. When $n > m$, using (12) and (9), we obtain that

$$\sum_{k=m}^{n} (-1)^{k+m} \binom{n}{k} \binom{k}{m}$$

$$= \sum_{k=m}^{n} (-1)^{k+m} \binom{n}{m} \binom{n-m}{k-m} \qquad (\text{let } i = k - m)$$

$$= \binom{n}{m} \sum_{i=0}^{n-m} (-1)^{i} \binom{n-m}{i} = 0.$$

7. Let a_n be the sum and $b_n = \sum_{k=n+1}^{2n} \binom{2n}{k}$, which is the sequence "complementary" to a_n. Then $a_n + b_n = 2^{2n}$. From (4),

$$b_n = \sum_{k=n+1}^{2n} \binom{2n}{2n-k} = \sum_{k=0}^{n-1} \binom{2n}{k} = a_n - \binom{2n}{n}.$$

Thus the result is obtained.

8. Let

$$a_n = \sum_{k=0}^{\left[\frac{n}{2}\right]} \left\{ \binom{n}{k}^2 + \binom{n}{k-1}^2 \right\} \quad \text{and} \quad b_n = 2 \sum_{k=0}^{\left[\frac{n}{2}\right]} \binom{n}{k} \binom{n}{k-1}.$$

We prove that $a_n - b_n = \frac{1}{n+1} \binom{2n}{n}$.

When n is even, from (4) and the Vandermonde identity (refer to Remark 4)

$$a_n = \sum_{k=0}^{n} \binom{n}{k}^2 = \binom{2n}{n},$$

$$b_n = \sum_{k=0}^{n} \binom{n}{n-k}\binom{n}{k-1} = \binom{2n}{n-1} = \binom{2n}{n+1}.$$

Hence $a_n - b_n = \binom{2n}{n}\left(1 - \frac{n}{n+1}\right) = \frac{1}{n+1}\binom{2n}{n}$. If n is odd, then

$$a_n = \binom{2n}{n} - \left(\frac{n}{\frac{n+1}{2}}\right)^2, \quad b_n = \binom{2n}{n-1} - \left(\frac{n}{\frac{n-1}{2}}\right)^2.$$

Thus $a_n - b_n = \binom{2n}{n} - \binom{2n}{n-1} = \frac{1}{n+1}\binom{2n}{n}$.

9. Suppose there exist n and k such that the four numbers form an arithmetic sequence. Then from

$$2\binom{n}{k+1} = \binom{n}{k} + \binom{n}{k+2},$$

we obtain that

$$\frac{k+1}{n-k} + \frac{n-k-1}{k+2} = 2. \tag{1}$$

Further, from $2\binom{n}{k+2} = \binom{n}{k+1} + \binom{n}{k+3}$, we deduce that when k is replaced by $k+1$, (1) also holds. In addition, if we replace k by $n-k-2$, then the left side of (1) remains unchanged. The above discussion shows that the quadratic equation has four roots: $k, k+1, n-k-2$ and $n-(k+1)-2$. It follows that $k = n-k-3$ and $k+1 = n-k-2$. Thus $n = 2k+3$, so n is odd and $k = \frac{n-3}{2}$. Hence the four numbers mentioned in the problem are the middle four terms of the degree-n binomial coefficients. From the unimodal property of the binomial coefficients, we know that they can't form an arithmetic sequence.

10. There are $\binom{2n}{2}$ ways to choose 2 elements from $2n$ different elements. On the other hand, this can also be counted as follows: If $2n$ elements are (arbitrarily) divided into two n-element sets, then two elements can be taken from the same set or we can take one element from each set. The former has $2\binom{n}{2}$ methods, and (from the multiplication principle)

the latter has n^2 methods. Obviously, there is no repetition between these two kinds of methods, so there are $2\binom{n}{2} + n^2$ kinds of methods, and we get the result.

11. From (11), (4), and the Vandermonde identity,

$$\sum_{k=1}^{n} k \binom{n}{k}^2 = \sum_{k=1}^{n} n \binom{n-1}{k-1}\binom{n}{k} = n\sum_{k=1}^{n}\binom{n-1}{n-k}\binom{n}{k}$$
$$= n\binom{2n-1}{n} = n\binom{2n-1}{n-1}.$$

Considering the following counting problem, we can derive a combinatorial proof of the equality: Select n people from n men and n women to form a committee and specify that the chairman of the committee must be a woman.

First of all, there are n ways to choose one of the n women as chairman. After the chairman being selected, $n-1$ people (as members of the committee) are selected from the remaining $2n-1$ people, with a total of $\binom{2n-1}{n-1}$ options. From the multiplication principle, there are $n\binom{2n-1}{n-1}$ ways of selecting the n-member committee.

On the other hand, for any k $(1 \leq k \leq n)$, there are $\binom{n}{k}$ ways to select k from the n women as members of the committee first, and there are k ways to choose one of them as chairman; There are $\binom{n}{n-k}$ ways to select $n-k$ from the n men as members, therefore, the n member committee with k women (one of them is the chairman) has $\sum_{k=1}^{n}$ according to the principle of addition, sum for $k = 1, 2, \ldots, n$, then there are $k\binom{n}{k}\binom{n}{n-k} k\binom{n}{k}^2$ selection methods options. Combining these two counting results, we get the equality.

Solution 3

Counting: Correspondence and Recursion

1. Let f be any mapping that satisfies the requirements, and suppose $1, 2, \ldots, n$ appear in $f(1), f(2), \ldots, f(n)$ (from left to right) $x_1 x_2, \ldots, x_n$ times, respectively. Then $(x_1, \ldots, x_n)$ is a set of ordered nonnegative integer solutions to the equation $x_1 + \cdots + x_n = n$. On the other hand, from any set of nonnegative integer solutions $(x_1, \ldots, x_n)$ to the equation, a satisfactory mapping f can be made (taking $f(1) = \cdots = f(x_1) = 1, f(x_1 + 1) = \cdots = f(x_1 + x_2) = 2, \ldots, f(x_1 + \cdots + x_{n-1} + 1) = \cdots = f(x_1 + \cdots + x_n) = n$). Thus, from Example 1, we know that the number is $\binom{2n-1}{n}$.

2. The sending method of the ith postcard is equal to the number of (ordered) nonnegative integer solutions of $x_1 + \cdots + x_n = a_i$, which is $\binom{a_i + n - 1}{n - 1}$. Hence the number of methods is

$$\binom{a_1 + n - 1}{n - 1} \cdots \binom{a_k + n - 1}{n - 1}.$$

3. Use the solution of Example 1: Arrange n identical balls in a row. There is a gap between every two balls. There are $n - 1$ gaps in total. Each gap has two options: "insert" and "not insert" a partition. One way of inserting partitions corresponds to an ordered partition of n, and an ordered partition also corresponds to a way of inserting partitions. Therefore, the number of partitions is 2^{n-1}.

4. Every four points on the circumference form a convex quadrilateral, and its diagonals (two chords under consideration) intersect at one point.

Therefore, the set of every four points corresponds to an intersection point. Because there are no three chords intersecting at one point, different sets of four points correspond to different intersection points. Conversely, for any intersection point, it is easy to know that there is a set of four corresponding points (as mentioned above). Therefore, the number of intersection points is the number of 4-element subsets of n points, i.e., $\binom{n}{4}$.

5. Suppose n straight lines divide the plane into a_n parts. Then $a_1 = 2$ and $a_{n+1} = a_n + n + 1$, so $a_n = \frac{n(n+1)}{2} + 1$.

6. Let a_n be the number of the desired subsets. Then $a_1 = 2$ and $a_2 = 3$. The subsets that meet the requirement are divided into two categories: The first kind does not contain n, and there are a_{n-1} such subsets; the second kind contains n, so it can not contain $n - 1$, and there are a_{n-2} such subsets. Hence $a_n = a_{n-1} + a_{n-2}$. This is the same as the recurrence formula of the Fibonacci sequence. Considering the initial value of a_n, we obtain $a_n = f_{n+2}$, where the definition of f_n is $f_1 = f_2 = 1$ and $f_n = f_{n-1} + f_{n-2}(n \geq 3)$. Its general formula is easy to find.

Please note that if we classify the required subsets into 0-element set, 1-element sets, ... , n-element sets, and apply the result of Example 8, then we get $a_n = \sum_{k=0}^{n} \binom{n-k+1}{k}$, and by combining the above result, we generate an interesting identity

$$\sum_{k=0}^{n} \binom{n - k + 1}{k} = f_{n+2}.$$

7. Let b_n be the numbers. Then $b_1 = 2, b_2 = 3$ and $b_3 = 5$. When n is odd, the said sequence can be divided into two types: If $a_n = 1$, then $a_{n-1} = 1$, so the number of the sequences is b_{n-2}. If $a_n = 0$, then the number of the sequences is b_{n-2} when $a_{n-1} = 1$, and $a_{n-2} = 0$ when $a_{n-1} = 0$, so the number of such sequences is b_{n-3}. Therefore,

$$b_n = 2b_{n-2} + b_{n-3}(n \geq 4). \tag{1}$$

This recurrence formula is still valid when n is even. Finally, it is not difficult to prove by induction and (1) that $b_n = b_{n-1} + b_{n-2}$. Thus $b_n = f_{n+2}$, where f_n is the Fibonacci sequence.

8. Let the k numbers taken out be $a_1 < a_2 < \cdots < a_k$, and we prove that this group of numbers corresponds to $a_1 < a_2 - 1 < a_3 - 2 < \cdots < a_k - (k - 1)$ one by one, and the latter is a k-combination of

the $n - (k - 1)$ elements $1, 2, \ldots, n - (k - 1)$. Therefore, the number is $\binom{n - k + 1}{k}$.

9. Let $b_i = a_i - i + 1 (i = 1, \ldots, n)$. Then $1 \le b_1 \le b_2 \le \cdots \le b_k \le n - k + 1$, and each b_i is an odd number. Conversely, each such sequence b_i determines a satisfactory sequence a_i. Therefore, the number is equal to the number of k-combinations with repetition of all odd numbers in $1, 2, \ldots, n - k + 1$, which is $\binom{m - k + 1}{k}$, where $m = \left[\frac{n - k + 2}{2}\right]$.

10. Suppose there are x_n sequences with an even number of zeros and y_n sequences with an odd number of zeros. Then $x_n + y_n = k^n$.

When there is an even number of zeros in $a_1, a_2, \ldots, a_{n-1}$, take $a_n \ne$, and otherwise take $a_n = 0$. This generates a sequence with n terms and an even number of zeros. It is easy to see that each n-term sequence with an even number of zeros can be generated in this way. Hence $x_n = (k - 1)x_{n-1} + y_{n-1}$, and combining it with $x_{n-1} + y_{n-1} = k^{n-1}$ (see the previous results), we obtain that

$$x_n = (k - 2)x_{n-1} + k^{n-1}. \tag{2}$$

To solve (2), we first use (2) to derive that

$$
\begin{aligned}
x_n - kx_{n-1} &= (k - 2)x_{n-1} + k^{n-1} - k(k - 2)x_{n-2} - k^{n-1} \\
&= (k - 2)x_{n-1} - k(k - 2)x_{n-2},
\end{aligned}
$$

or equivalently $x_n - (2k - 2)x_{n-1} + k(k - 2)x_{n-2} = 0$. Therefore $x_n = \frac{1}{2}[k^n + (k - 2)^n]$. This can also be derived from (2) recursively.

11. Using the same numerical table as Example 2, we see that $S_1 \subseteq S_2 \subseteq \cdots \subseteq S_k$ is equivalent to that the numbers (from top to bottom) in any column in the numerical table is a monotone non-decreasing sequence composed of 0 and 1. There are exactly $k + 1$ such sequences. Therefore, it is concluded that there are $(k + 1)^n$ numerical tables in total, which is the number of subset groups.

In addition, it is not difficult to derive the recurrence formula for the problem. Let $f(n, k)$ be the number. For $\le i \le n$ and a fixed i-element subset S_k, the number of the corresponding subsets $(S_1, \ldots, S_{k-1})$ is obviously

$f(i, k-1)$, and because there are $\binom{n}{i}$ such S_k,

$$f(n, k) = \sum_{i=0}^{n} \binom{n}{i} f(i, k-1), \tag{3}$$

and $f(0, k) = 1$. It is not easy to solve $f(n, k)$ out directly from this recurrence relation, but if the answer to the question is guessed out, we can easily prove it by using (3) and by induction.

12. There are $\binom{2n}{n}$ sequences composed of n ones and n zeros. Then calculate the number of sequences that don't meet the condition (2), which we call bad sequences. In a bad sequence S, there must be a moment (counting from left to right), when the number of zeros exceeds the number of ones for the first time, and at this moment, the number of zeros exceeds the number of ones by 1. If one exchanges 0 and 1 at this moment and before (replace 0 with 1 while replace 1 with 0), then the new sequence T has $n+1$ ones and $n-1$ zeros.

It is not difficult to prove that the above exchanging rule of 0 and 1 gives a one-to-one correspondence from all bad sequences to sequences composed of $n+1$ ones and $n-1$ zeros.

In fact, this correspondence is an injection. This is because for another bad sequence S', either the moment when the number of zeros exceeds the number of ones is different for S and S', or it is the same but the following terms are not identical since $S \neq S'$. In either case, the corresponding T is different from T'.

This correspondence is a surjection: For any sequence T composed of $n+1$ ones and $n-1$ zeros, since the number of ones is more than the number of zeros, when T is counted from left to right, there must be a moment when the number of ones is more than the number of zeros for the first time. Exchanging 0 and 1 at this time and before produces a bad sequence S. Obviously, this S corresponds to T in the previous setting. Therefore, the correspondence is a surjection.

Because the correspondence is a one-to-one correspondence, the number of bad sequences is the number of sequences composed of $n+1$ ones and $n-1$ zeros, that is $\binom{2n}{n+1}$.

Therefore, the number of sequences in the problem is $\binom{2n}{n} - \binom{2n}{n+1} = \frac{1}{n+1}\binom{2n}{n}$.

Remark The number $C_n = \frac{1}{n+1}\binom{2n}{n}$ is called the nth Catalan number. Many counting problems are related to this.

Solution 4

Counting: Inclusion-Exclusion Principle

1. Let S_k be the set of positive integers not exceeding 1000 and divisible by k. Use Theorem 2 to find $|S_2 \cup S_3 \cup S_5|$. Answer: 734.

2. Let S_1 be the number of permutations that contain abc and S_2 be the number of permutations that contain ef. Then the desired number is $|\overline{S}_1 \cap \overline{S}_2| = 6! - 5! - 4! + 3! = 582$.

3. Let S_k be the number of permutations where k is adjacent to $k+n$ ($k = 1, 2, \ldots, n$). For any $r(1 \le r \le n)$ and $1 \le i_1 < \cdots < i_r \le n$, we can easily know that $|S_{i_1} \cap \cdots \cap S_{i_r}| = 2^r(2n-r)!$. Thus, from the inclusion-exclusion principle (Theorem 2),

$$\left| S_1 \bigcup \cdots \bigcup S_n \right| = \sum_{r=0}^{n} (-2)^r \binom{n}{r} (2n-r)!.$$

4. Using the notation in Remark 3 and from the above result, we know that

$$\sigma_r = 2^r \binom{n}{r} (2n-r)!(r = 1, 2, \ldots, n).$$

From Theorem 4, we obtain that

$$\left| S_1 \bigcup \cdots \bigcup S_n \right| < \sigma_1 - \sigma_2 + \sigma_3 = \frac{2}{3} \cdot (2n)!$$

and

$$\left| S_1 \bigcup \cdots \bigcup S_n \right| > \sigma_1 - \sigma_2 + \sigma_3 - \sigma_4 > \frac{5}{8} \cdot (2n)!.$$

5. Make the substitution $y_1 = x_1 - 1$, $y_2 = x_2$, $y_3 = x_3 - 4$ and $y_4 = x_4 - 2$. Then the problem is reduced to finding the number of integer solutions to the equation

$$y_1 + y_2 + y_3 + y_4 = 13 \qquad (1)$$

satisfying $0 \le y_1 \le 5$, $0 \le y_2 \le 7$, $0 \le y_3 \le 4$ and $0 \le y_4 \le 4$.

Let S be the set of nonnegative integer solutions of equation (2); S_1, S_2, S_3 and S_4 are subsets of S satisfying $y_1 > 5$, $y_2 > 7$, $y_3 > 4$, and $y_4 > 4$ respectively. Then the number of solutions is $|\overline{S_1} \cap \overline{S_2} \cap \overline{S_3} \cap \overline{S_4}|$.

From Example 1 of Chapter 3, we know that $|S| = \binom{16}{3} = 560$. In order to find $|S_1|$, let $y_1' = y_1 - 6$. Then $|S_1|$ is equal to the number of non-negative integer solutions of $y_1' + y_2 + y_3 + y_4 = 7$, i.e., $\binom{4+7-1}{4} = 120$. Similarly,

$$|S_2| = \binom{4+5-1}{4} = 56,$$

$$|S_3| = |S_4| = \binom{4+8-1}{4} = 165,$$

and

$$|S_1 \cap S_3| = |S_1 \cap S_4| = \binom{4+2-1}{4} = 5,$$

$$|S_2 \cap S_3| = |S_2 \cap S_4| = 1,$$

$$|S_3 \cap S_4| = \binom{4+3-1}{4} = 15,$$

$$|S_1 \cap S_2| = 0.$$

In addition, $|S_i \cap S_j \cap S_k|$ and $|S_1 \cap S_2 \cap S_3 \cap S_4|$ are both 0. Finally, by the inclusion-exclusion principle, we conclude that the number of solutions is 81.

6. From Example 1 of Chapter 3, the number of nonnegative integer solutions $(x_1, \ldots, x_n)$ of the equation

$$x_1 + \cdots + x_n = n \qquad (2)$$

is $\binom{2n-1}{n}$. If there is an $x_i \ge 3$ in the equation (1), then replace x_i with $x_i - 3$, and the number of non-negative integer solutions of (1) with $x_i \geq 3$ is the same as the number of non-negative integer solutions of

$$x_1 + \cdots + x_n = n - 3,$$

which is $\begin{pmatrix} 2n - 4 \\ n - 3 \end{pmatrix} = \begin{pmatrix} 2n - 4 \\ n - 1 \end{pmatrix}$.

In general, for $1 \leq k \leq n$, if there are k numbers $x_i \geq 3$ in the equation (1), then replace these x_i with $x_i + 3$, and the number of nonnegative integer solutions of (1) is equal to the number of nonnegative integer solutions of the equation

$$x_1 + \cdots + x_n = n - 3k,$$

which is $\begin{pmatrix} 2n - 3k - 1 \\ n - 3k \end{pmatrix} = \begin{pmatrix} 2n - 3k - 1 \\ n - 1 \end{pmatrix}$.

Note that there are $\begin{pmatrix} n \\ k \end{pmatrix}$ selection methods of k numbers in $x_1, \ldots, x_n$ satisfying $x_i \geq 3$, so by the inclusion-exclusion principle, the number is

$$\sum_{k=0}^{n} (-1)^k \begin{pmatrix} n \\ k \end{pmatrix} \begin{pmatrix} 2n - 3k - 1 \\ n - 1 \end{pmatrix} = \sum_{0 \leq k \leq \frac{n}{3}} (-1)^k \begin{pmatrix} n \\ k \end{pmatrix} \begin{pmatrix} 2n - 3k - 1 \\ n - 1 \end{pmatrix}.$$

Solution 5

Combinatorial Problems

1. Let the numbers written on the blackboard be $a_1, \ldots, a_n$. Since $(1+a)(1+b) = 1 + (a+b+ab)$, it is easy to know by the rule of the operation that the product

$$(1+a_1)(1+a_2)\cdots(1+a_n)$$

remains unchanged under the operation. In the case of this problem, let the last remaining number be x. Then

$$1 + x = (1+1)\left(1 + \frac{1}{2}\right)\cdots\left(1 + \frac{1}{100}\right)$$

$$= 2 \cdot \frac{3}{2} \cdot \frac{4}{3} \cdot \ldots \cdot \frac{100}{99} \cdot \frac{101}{100} = 101,$$

so $x = 100$.

2. Let $f(x) = ax^2 + bx + c$. Then

$$x^2 f\left(\frac{1}{x} + 1\right) = a(x+1)^2 + bx(x+1) + cx^2$$

$$= (a+b+c)x^2 + (2a+b)x + a,$$

where the discriminant is $(2a+b)^2 - 4a(a+b+c) = b^2 - 4ac$. Similarly, we can verify that the discriminant of $(x-1)^2 f\left(\frac{1}{x-1}\right)$ is also $b^2 - 4ac$. Therefore, under such an operation, the discriminant of $f(x)$ remains unchanged. Therefore, we can't get $x^2 + 10x + 9$ from $x^2 + 4x + 3$ (because the discriminants of the two are not equal).

3. (i) Yes. It is not difficult to specify the deformation method. (ii) No. Suppose the three numbers in the first row (from left to right) of the

table are a_1, a_2 and a_3, the second row is a_4, a_5 and a_6, and the third row is a_7, a_8 and a_9. Then

$$S = (a_1 + a_3 + a_5 + a_7 + a_9) - (a_2 + a_4 + a_6 + a_8)$$

is unchanged under the deformation.

4. Considering the two points A and B with the largest distance in the given points, we see that the rest of the points are on the circumference with AB as a diameter, thus deducing that $n \leq 4$.

5. If $A = \varnothing$, then $B = \varnothing$, so $|A| = |B|$. If $A \neq \varnothing$, then suppose $(a, b) \in A$ and let $f((a, b)) = (a + b, a)$. Since $a + b \in S$ and $(a + b) - a \in S$, we have $(a + b, a) \in B$. That is, f is a mapping from A to B.

We prove that this is an injection, because if there are two elements $(a, b) \neq (a', b')$ in A while $f((a, b)) = f((a', b'))$, then $(a + b, a) = (a' + b', a')$, so $a = a'$, and $b = b'$, a contradiction. Therefore f is an injection from A to B, so that $|A| \leq |B|$.

Similarly, for $(a, b) \in B$, let $g((a, b)) = (b, a - b)$. Then g is a mapping from B to A, which is also an injection, and $|B| \leq |A|$. Combining the above results, we know that $|A| = |B|$.

6. Note that there are only 2^n numbers of n-digit integers consisting of 1 and 2. We prove the conclusion by induction on n. When $n = 2$, the conclusion is obviously true. For $n \geq 2$, assume that conclusion is true for n and suppose $C_1, C_2, \ldots, C_{2^n}$ are the numbers arranged counterclockwise starting from a certain point. Then the 2^{n+1} numbers $\overline{1C_1}, \overline{1C_2}, \ldots, \overline{1C_{2^n}}, \overline{2C_{2^n}}, \ldots, \overline{2C_2}, \overline{2C_1}$ (arranged counterclockwise) meet the requirement.

7. Let $1, 2, \ldots, 6$ denote these 6 questions. Obviously, no one has worked out the 6 questions. If someone A has worked out 5 questions, suppose the question that he hasn't worked out is question 6. Then for any other student B, the question that neither of them has worked out can only be question 6, so no one has worked out question 6, which is inconsistent with the condition. Therefore, everyone can work out at most 4 questions.

(i) Suppose someone worked out 4 questions, assumed to be $1, 2, 3$, and 4. It is easy to know that no one can work out both questions 5 and 6. Since there are 100 people who answered each of question 5 and question 6 correctly, the total number of people is $\geq 200 + 1 = 201$.

(ii) If each student answered at most 3 questions correctly, then because the sum of the numbers of the questions that all students answered correctly equals 600, the number of students is greater than or equal to $\frac{600}{3} = 200$.

The following example shows that there can be exactly 200 students. If there are 50 students who worked out questions $(1, 2, 3), (1, 5, 6), (2, 4, 5)$ and $(3, 4, 6)$ respectively then there will be 200 students, and each question is answered correctly by exactly 100 people.

Therefore, there are at least 200 students.

8. Let $a_{ij}(1 \le i, j \le n)$ be the number in the cross grid of row i and column j, and s_i be the sum of the numbers in row i. Because there are only n columns, and there is only a finite number of ways to exchange the numbers in each column, only a limited number of new tables can be generated, so there must be a numerical table such that the corresponding $|s_1| + |s_2| + \cdots + |s_n|$ is minimal. We will prove that there must be $|s_k| \le 2(k = 1, 2, \ldots, n)$ in this case.

Proof by contradiction: Suppose there is one k that makes $|s_k| > 2$. We may assume that $s_k > 2$. The sum of the numbers in the numerical table is 0, so $s_1 + \cdots + s_n = 0$. Hence there exists one i, such that $s_i < 0$. In addition, there is obviously one j such that $a_{kj} > a_{ij}$ (otherwise $s_i > s_k$). By exchanging a_{kj} and a_{ij} in the jth column, we shall prove that the corresponding $|s_1'| + \cdots + |s_n'|$ in the table obtained after the exchange will be strictly decreased (here s_m' is the sum of the numbers of the mth row in the new numerical table), which leads to a contradiction (because we have chosen $|s_1| + \cdots + |s_n|$ as the minimum), hence all $|s_k| \le 2$.

In order to prove the assertion just mentioned, we note that for $m \ne i$ and k, it is obvious that $s_m' = s_m$ and

$$s_k' = s_k - a_{kj} + a_{ij}, \quad s_i' = s_i + a_{kj} - a_{ij}.$$

Since $|a_{kj}| \le 1$ and $|a_{ij}| \le 1$, we have $s_k > 2 \ge a_{kj} - a_{ij}$, i.e., $s_k' > 0$. Also since $s_i < 0$ and $a_{kj} - a_{ij} > 0$, there is the strict inequality

$$|s_i'| = |s_i + a_{kj} - a_{ij}| < |s_i| + (a_{kj} - a_{ij}).$$

Hence

$$|s_k'| + |s_i'| < (s_k - a_{kj} + a_{ij}) + (|s_i| + a_{kj} - a_{ij})$$
$$= |s_k| + |s_i|.$$

Thus, $|s_1'| + \cdots + |s_n'| < |s_1| + \cdots + |s_n|$, and the above conclusion is proved.

$\square$

9. Induct on n. When $n = 1$ and 2, the conclusion is obviously true. Let $n \geq 3$, and assume that the conclusion is true for n. Now we consider $2(n+1)$ pairs of numbers (a_i, b_i). By symmetry, suppose $a_1 \geq a_2 \geq \cdots \geq a_{2n+2}$. According to the inductive hypothesis, the $2n$ pairs $(a_3, b_3), (a_4, b_4), \ldots, (a_{2n+2}, b_{2n+2})$ can be divided into two groups. Let A_1 and A_2 denote the sums of a_i in the first group and the second group respectively, and B_1 and B_2 denote the sums of b_i in the first group and the second group respectively. Then $|A_1 - A_2| \leq a_2$ (because $a_3, \ldots, a_{2n+2} \leq a_2$, $\displaystyle\max_{3 \leq i \leq 2n+2} a_i \leq a_2$), and $|B_1 - B_2| \leq \displaystyle\max_{3 \leq i \leq 2n+2} b_i$.

We may assume that $B_1 \leq B_2$. If $b_1 \leq b_2$, then put (a_2, b_2) into the first group and (a_1, b_1) into the second group. In this case, the sums of the first group and the second group are $A_1 + a_2$ and $A_2 + a_1$ respectively, and

$$|(A_1 + a_2) - (A_2 + a_1)| \leq |A_1 - A_2| + |a_1 - a_2|$$

$$\leq a_2 + (a_1 - a_2)$$

$$= a_1 = \max_{1 \leq i \leq 2n+2} a_i.$$

The sums of b_i in the two groups are $B_1 + b_2$ and $B_2 + b_1$ respectively. Because $B_1 - B_2 \leq 0$ and $b_2 - b_1 \geq 0$,

$$|(B_1 + b_2) - (B_2 + b_1)| = |(B_1 - B_2) + (b_2 - b_1)|$$

$$\leq \max\{B_2 - B_1, b_2 - b_1\}$$

$$\leq \max_{1 \leq i \leq 2n+2} b_i.$$

(if x and y have opposite signs, then $|x+y| \leq \max\{|x|, |y|\}$) Therefore, the above grouping method meets the requirement.

If $b_1 > b_2$, then put (a_1, b_1) into the first group, and put (a_2, b_2) into the second group. Then the same proof as above shows that this grouping method also meets the requirement the (details are omitted here).

Solution 6

Exact Division

1. Because the multiple of k has the form kx, where x is an integer, the number of the integers divisible by k in $1, 2, \ldots, n$ is the largest integer x satisfying $kx \leq n$. By definition, this is $\left[\frac{n}{k}\right]$.
2. Because $(ad + bc) - (ab + cd) = (a - c)d - (a - c)b$ can be divisible by $a - c$, from the condition $a - c | ab + cd$, the conclusion $a - c | ad + bc$ is derived.
3. Use $b(a^2 + ab + 1) - a(b^2 + ab + 1) = b - a$ (refer to (ii) in Remark 3).
4. Let $(a, b) = d$. Then $a = a_1 d$ and $b = b_1 d$, where a_1 and b_1 are coprime integers (property (11)). The fact $d | a_1 + b_1$ can be deduced from the known conditions, so $d \leq a_1 + b_1$ (property (3)). Thus $d^2 \leq a + b$, that is, $d \leq \sqrt{a + b}$.
5. Let $d = (2^m - 1, 2^n + 1)$. Then $2^m - 1 = du$ and $2^n + 1 = dv$, where u and v are integers. From $(du + 1)^n = (dv - 1)^m$, and expanding both sides (note that m is an odd number), we can get $dA + 1 = dB - 1$ (A and B are integers), from which $d = 1$.

Another Solution: Since m is an odd number, formula (6) gives $(2^n + 1) | (2^{mn} + 1)$. In addition, $2^{mn} - 1$ is a multiple of $2^m - 1$. Let $2^{mn} - 1 = (2^m - 1)q$. Then $2^n + 1$ exactly divides $(2^m - 1)q + 2$. Thus we obtain the result (refer to Remark 10).

6. By the division with remainder, $n = mq + r$ with $0 \leq r < m$, and $q \geq 0$. If q is even then from $a^n + 1 = (a^{mq} - 1)a^r + (a^r + 1)$, there must be $a^m + 1 | a^r + 1$, but $0 < a^r + 1 < a^m + 1$, which can't be true. Hence, q is an odd number. From $a^n + 1 = (a^{mq} + 1)a^r - (a^r - 1)$, and noticing

that $a^m + 1 | a^{mq} + 1$, we deduce that $a^m + 1 | a^r - 1$, but $0 \leq r < m$, so there must be $a^r - 1 = 0$, i.e., $r = 0$. Therefore $n = mq$.

7. Easily we know that the condition is necessary and sufficient. It's easy to verify that $x = x_0 + \frac{b}{(a,b)}t$ and $y = y_0 - \frac{a}{(a,b)}t$ give a solution of the equation. We only need to prove that any solution (x', y') to the equation is in this form. By subtracting $ax' + by' = c$ from $ax_0 + by_0 = c$, we get $a(x_0 - x') + b(y_0 - y') = 0$. Thus, $a | b(y_0 - y')$, so $\frac{a}{(a,b)} | \frac{b}{(a,b)}(y_0 - y')$. But $\frac{a}{(a,b)}$ and $\frac{b}{(a,b)}$ are coprime (property (11)), so $\frac{a}{(a,b)} | (y_0 - y')$. Therefore $y_0 - y' = \frac{a}{(a,b)}t$, where t is some integer, and thus $x' = x_0 + \frac{b}{(a,b)}t$.

   Please note that in order to find a group of integer solutions of equation (1), we can first find a group of solutions of the equation $ax + by = (a, b)$, which can be found by Euclidean algorithm. But when $|a|$ and $|b|$ are not too large, we can also find it by the trial-and-error method.

8. Because $(a, b) = 1$, there are integers x and y that make $ax + by = 1$. By the division with remainder, $x = bq + r$ with $0 \leq r < b$. Let $aq + y = -s$. Then $ar - bs = 1$. Since $b \neq 1$, we have $r \neq 0$, i.e. $0 < r < b$. Thus, $0 < s < a$. The proof of the uniqueness can refer to the demonstration of (4).

9. Because $(a, b) = 1$, the equation $ax + by = n$ must have a group of integer solutions. According to the method in the previous question, we may assume that $0 \leq x < b$. In this way, when $n > ab - a - b$,

$$by = n - ax > ab - a - b - ax \geq ab - a - b - a(b - 1) = -b,$$

so $y > -1$ is also a non-negative integer.

   When $n = ab - a - b$, if the equation has a nonnegative integer solution (x, y), then $ab = (x + 1)a + (y + 1)b$, so $a | (y + 1)b$. Since $(a, b) = 1$, we have $a | y + 1$. Since $y + 1 > 0$, we see that $y + 1 \geq a$. Similarly, $x + 1 \geq b$. (Refer to Remark 3) Therefore, $ab = (x + 1)a + (y + 1)b \geq ab + ab = 2ab$, a contradiction.

10. (i) From $(a, b) = 1$ and property (12), we deduce that $(a^2, b) = 1, \ldots, (a^m, b) = 1$. Thus $(a^m, b^2) = 1, \ldots, (a^m, b^n) = 1$.

    (ii) Let $d = (a, b)$. If $d = 1$, then using (i) we know that $(a^n, b^n) = 1$. Therefore,

    $$d^n = d^n \left(\left(\frac{a}{d}\right)^n, \left(\frac{b}{d}\right)^n \right)$$

$$= \left(\left(\frac{a}{d} \right)^n \cdot d^n, \left(\frac{b}{d} \right)^n \cdot d^n \right)$$

$$= (a^n, b^n).$$

(Use properties (10) and (11).)

11. Let $ab = x^k$ and $x > 0$. Since $(a, b) = 1$, we see that $(a^{k-1}, b) = 1$. Suppose $(a, x) = d$. Then from the previous question

$$d^k = (a^k, x^k) = (a^k, a^b) = a(a^{k-1}, b) = a,$$

so a is the kth power of an integer. Similarly, $b = (b, x)^k$, and b is also the kth power of an integer.

12. Let the rational number $x = \frac{p}{q}$, (p and q are coprime integers), and since x^k is an integer, $q^k | p^k$. But $(p, q) = 1$, so $(p^k, q^k) = 1$. Therefore $q^k = \pm 1$, i.e., $q = \pm 1$, and $x = \pm p$ is an integer.

13. The uniqueness is obvious: Let $d | mn$ and $d \geq 1$. If $d = m_1 n_1 = m_2 n_2$, where m_1 and m_2 are positive divisors of m, and n_1 and n_2 are positive divisors of n, then $m_2 | m_1 n_1$. But $(m, n) = 1$, so $(m_2, n_1) = 1$, thus $m_2 | m_1$. Similarly, $m_1 | m_2$, so $m_1 = m_2$, therefore $n_1 = n_2$.

In order to prove that there exists such a representation, let $d | mn$ with $d \geq 1$. Then

$$mn = dx, \text{ for some positive integer } x. \tag{1}$$

Let $(d, m) = r$. Then $d = d'r$ and $m = m'r$, where $(m', d') = 1$. Plugging into the formula (1), we obtain $m'n = d'x$. Hence $d' | m'n$, but $(d', m') = 1$, thus $d' | n$. And r is obviously a positive divisor of m, so $d = d'r$ is the product of a positive divisor of m and a positive divisor of n.

14. (i) Use q to do the division with remainder repeatedly, and then n can be expressed as:

$$n = n_1 q + a_0, 0 \leq a_0 \leq q - 1,$$

$$n_1 = n_2 q + a_1, 0 \leq a_1 \leq q - 1,$$

The obtained quotients n_i are strictly decreasing, and finally reaching 0. Let k be the maximum subscript that makes $a_k \neq 0$. Then $n = a_0 + n_1 q, n_1 = a_1 + n_2 q, \ldots, n_{k-1} = a_{k-1} + n_k q, n_k = a_k$, so $n = a_0 + a_1 q + \cdots + a_k q^k$, where $0 \leq a_i \leq q - 1 (i = 0, 1, \ldots, k)$ and $a_k \neq 0$.

Suppose n has another satisfactory representation

$$n = a_0' + a_1'q + \cdots + a_l'q^l, \quad 0 \le a_i' \le q-1 \quad (i = 0, 1, \ldots, l).$$

Then $q \mid (a_0 - a_0')$, but $0 \le |a_0 - a_0'| \le \max(a_0, a_0') < q$, so $a_0 = a_0'$. Thus,

$$a_1 + a_2 q + \cdots + a_k q^{k-1} = a_1' + a_2' q + \cdots + a_l' q^{l-1}.$$

Similarly, we obtain $a_1 = a_1'$. In this way, we know that that there must be $k = l$, and $a_i = a_i' (i = 0, 1, \ldots, k)$.

(ii) Easily we know that $n_{i+1} = \left[\frac{n_i}{q}\right]$ (refer to Remark 5). Therefore, it is not difficult to generalize that $n_i = \left[\frac{n}{q^i}\right]$ (by using that $\left[\frac{[x]}{m}\right] = \left[\frac{x}{m}\right]$ for the positive integer m), so (ii) is obvious.

15. When $1 \le k \le 2^n - 1$, from the uniqueness of the binary representation of a positive integer (see the previous question), the 2^n numbers $a_0 + a_1 \cdot 2 + \cdots + a_{n-1} \cdot 2^{n-1}$ ($a_i = 0$ or 1) are exactly $0, 1, \ldots, 2n - 1$. Hence k is the sum of the different divisors of 2^{n-1}, so it is the sum of the different divisors of m.

When $2^n - 1 < k \le m$, by the division with remainder, $k = (2^n - 1)t + r$ with $0 \le r < 2^n - 1$. Obviously, $t \le 2^{n-1}$, so from the above conclusion, both r and t are the sums of different divisors of 2^{n-1}, and $(2^n - 1)t$ can be expressed as the sum of different divisors of $m = 2^{n-1}(2^n - 1)$ (these divisors are obviously different from the divisors in the representation of r).

16. Let S denote the set of qualified numbers. First, we prove that if $m \in S$, then m and a are coprime. In fact, for any $n \ge 1$,

$$(a, A_n) = \left(a, 1 + a \sum_{k=0}^{n-1} a^k\right) = (a, 1) = 1.$$

Therefore, if $m \in S$, then there exists $n \ge 1$ such that $m \mid A_n$, which implies that $(m, a) \le (A_n, a) = 1$, so m and a are coprime.

Obviously, $m = 1 \in S$. Then we prove that if $m > 1$ and $(m, a) = 1$, then $m \in S$.

Consider the $m + 1$ numbers $A_1, A_2, \ldots, A_{m+1}$. Because there are m possible values for the remainder of an integer on division by m, two of the $m + 1$ numbers get the same remainder on division by m, so the difference between the two numbers is a multiple of m, which we assume to be A_i and $A_j (1 \le i < j \le m+1)$. Because

$$A_j - a_i = a^{i+1} \sum_{k=0}^{j-i-1} a^k,$$

which is divisible by m, and $(a, m) = 1$, we have $(a^{i+1}, m) = 1$. Therefore, from the above formula, we deduce that $A_{j-i-1} = \sum_{k=0}^{j-i-1} a_k$ is divisible by m (notice that $m > 1$, so $j - i - 1$ must be a positive integer). Hence $m \in S$.

To summarize, the desired numbers are all the positive integers that are coprime with a.

Solution 7

Prime Numbers

1. When $n = 1$, the conclusion is obvious. Suppose $n = p_1^{\alpha_1} \cdots p_k^{\alpha_k}$ is the standard decomposition of $n > 1$, and assume that $\alpha_1, \ldots, \alpha_l$ are even and $\alpha_{l+1}, \ldots, \alpha_k$ are odd. Then

$$\alpha_i = 2a_i (i = 1, \ldots, l), \alpha_j = 2b_j + 1 (j = l+1, \ldots, k).$$

Hence n can be expressed as

$$n = (p_1^{a_1} \cdots p_l^{a_l} p_{l+1}^{b_{l+1}} \cdots p_k^{b_k})^2 p_{l+1} \cdots p_k,$$

which has the form $q^2 r$, where r has no square factor.

In order to prove the uniqueness, let $n = q^2 r = q_1^2 r_1$, where r and r_1 are integers without square factors. We prove that $r = r_1$, so $q = q_1$.

To show $r = r_1$, we prove that the prime factors of r and r_1 are exactly the same. Suppose this assertion is wrong and we may assume that there exists a prime p that exactly divides r but does not exactly divide r_1. Let the power of p in q and q_1 be a and a_1, respectively. Because r has no square factors, p appears only once in r. Therefore, the powers of p in $q^2 r$ and $q_1^2 r_1$ are $2a + 1$ and $2a_1$, respectively. One of the two powers is odd and the other is even, so they are obviously not equal, but it is impossible. This proves the above assertion.

2. (i) This can be derived from the second proof of (8). If $\sqrt{n}$ is not an integer, then the number of the positive divisors of n that are less than

$\sqrt{n}$ is $\leq [\sqrt{n}]$, so the number of the divisors greater than $\sqrt{n}$ is also $\leq [\sqrt{n}]$, and

$$\tau(n) \leq 2[\sqrt{n}] < 2\sqrt{n}.$$

If $\sqrt{n}$ is an integer, then the number of the positive divisors less than $\sqrt{n}$ is $\leq \sqrt{n} - 1$, thus

$$\tau(n) \leq 2(\sqrt{n} - 1) + 1 = 2\sqrt{n} - 1 < 2\sqrt{n}.$$

(ii) Note that if d is a positive divisor of n, then $\frac{n}{d}$ is also a positive divisor of n. Suppose $1 = d_1 < \cdots < d_k = n$ are all the positive divisors of n, where $k = \tau(n)$. Then $\frac{n}{d_k} < \cdots < \frac{n}{d_1}$ are also all the positive divisors of n. Therefore

$$d_1 \cdots \cdots d_k = \frac{n}{d_k} \cdots \cdots \frac{n}{d_1},$$

so the result is obtained.

3. From Example 5 in Chapter 6, we know that $(M_p, M_q) = 2^{(p,q)} - 1 = 2 - 1 = 1$ (for any prime numbers $p \neq q$).

4. For $n \geq 2$,

$$2^{2^n} + 2^{2^{n-1}} + 1 = (2^{2^{n-1}} + 1)^2 - (2^{2^{n-2}})^2$$
$$= (2^{2^{n-1}} - 2^{2^{n-2}} + 1)(2^{2^{n-1}} + 2^{2^{n-2}} + 1).$$

Thus, let $a_n = 2^{2^n} + 2^{2^{n-1}} + 1$, and for $n \geq 2$,

$$a_n = (2^{2^{n-1}} - 2^{2^{n-2}} + 1)a_{n-1}. \tag{1}$$

It's simple to verify that a_{n-1} is coprime with $2^{2^{n-1}} - 2^{2^{n-2}} + 1$. Obviously, $a_1 > 1$ has (at least one) prime factors. Induct on n, and suppose a_{n-1} has at least $n - 1$ different prime factors ($n \geq 2$). Because $2^{2^{n-1}} - 2^{2^{n-2}} + 1 > 1$ has a prime factor p, and p does not exactly divide a_{n-1} (since $(a_{n-1}, 2^{2^{n-1}} - 2^{2^{n-2}} + 1) = 1$), from (1), we know that a_n has at least n different prime factors.

5. Suppose there is only a finite number of prime numbers of the form $4k - 1$, assumed to be $p_1, \ldots, p_n$. Consider the odd number $N = 4p_1 \cdots p_n - 1$. Easily we know that $N > 1$, so N has prime factors. If all these prime factors are of the form $4k + 1$, then their product is also in this form. But N is of the form $4k - 1$, so N must have a prime factor p of the form $4k - 1$, and obviously p is different from $p_1, \ldots, p_n$, which contradicts the above assumption.

The same idea can be used to prove that there is an infinite number of prime numbers of the form $6k - 1$.

6. Take these n numbers as $(n+1)!+2, (n+1)!+3, \ldots, (n+1)!+(n+1)$. For $2 \le k \le n+1$, because $(n+1)!+k$ has a proper divisor k, it is not a prime number.

7. It can be proved by dividing n into three cases: $3k, 3k+1$, or $3k+2$. Perhaps a more essential solution is to use Exercise 9 in Chapter 6. It is easy to know that if two positive integers a and b are coprime, then n can be expressed in the form of $ax + by$ when $n > ab + a + b$, where $x \ge 2$ and $y \ge 2$. In this problem, we take $a = 2$ and $b = 3$.

8. Let $n = 2^{k-1}m$, where $k \ge 2$ and $2 \nmid m$. According to the formula (5) in Chapter 7, $2^k m = 2n = \sigma(n) = \frac{2^k-1}{2-1} \cdot \sigma(m)$, so $\sigma(m) = m + \frac{m}{2^k-1}$. But both m and $\frac{m}{2^k-1}$ are divisors of m, which are not equal, and $\sigma(m)$ is the sum of all positive divisors of m, so m only has these two divisors, that is, m is prime, and $\frac{m}{2^k-1} = 1$.

9. Let $2^k \le n < 2^{k+1}$. Then $k \ge 1$. We prove that 2^k appears exactly once in (the standard decompositions of) $1, 2, \ldots, n$. If there is $m \le n$ with $m \ne 2^k$ such that $2^k | m$, then $n \ge m = 2^k \cdot l \ge 2^k \cdot 2 > n$, a contradiction! Let M denote the product of all the odd numbers not exceeding n. Then in

$$M \cdot 2^{k-1}\left(1 + \frac{1}{2} + \cdots + \frac{1}{n}\right) = M \cdot 2^{k-1} + \frac{M \cdot 2^{k-1}}{2} + \cdots + \frac{M \cdot 2^{k-1}}{n},$$

every term is an integer except $\frac{M \cdot 2^{k-1}}{2^k}$, so $M \cdot 2^{k-1}\left(1 + \frac{1}{2} + \cdots + \frac{1}{n}\right)$ is not an integer. Therefore, $1 + \frac{1}{2} + \cdots + \frac{1}{n}(n > 1)$ is not an integer.

10. We prove it by a way similar to the previous problem. Let $3^k \le 2n-1 < 3^{k+1}$. Then $k \ge 1$. Show that 3^k appears exactly once in $1, 3, \ldots, 2n-1$.

11. Because $2^k + 1$ is an odd number greater than 1, it has an odd prime factor p. Thus, $2|p^k + 1$. It shows that any prime factor p of $2^k + 1$ and $q = 2$ meet the requirement.

12. For any prime p, let $p^{\alpha_i} | a_i$ for $i = 1, \ldots, n$. We see from (9) in this chapter that it suffices to prove that

$$\max\{\alpha_1, \ldots, \alpha_n\} \ge \sum_{i=1}^n \alpha_i - \sum_{i \le i < j \le n} \min\{\alpha_i, \alpha_j\}.$$

This inequality is very easy to verify (suppose $\alpha_1 \ge \alpha_2 \ge \cdots \ge \alpha_n$, and look at the left and right sides of the above inequality respectively).

13. Since $(m,n)[m,n] = mn$ (see (16) in Chapter 6), from Vieta's theorem, both the number pair (m,n) and the number pair $((m,n),[m,n])$ are two roots of the quadratic equation $x^2 - (m+n)x + mn = 0$. Hence these two pairs of numbers are exactly the same. If $m = (m,n)$, then $m|n$; if $n = (m,n)$, then $n|m$.

14. Since $(ax + by)(ay + bx) = ab(x^2 + y^2) + xy(a^2 + b^2)$, from the condition, we know that $(a^2 + b^2)|ab(x^2 + y^2)$. If $a^2 + b^2$ and $x^2 + y^2$ are coprime, then $(a^2 + b^2)|ab$. But $a^2 + b^2 > ab > 0$, which will generate a contradiction. Therefore $a^2 + b^2$ and $x^2 + y^2$ are not coprime.

Solution 8

Congruence (1)

1. An integer can be expressed as one of the three forms: $3k$, $3k+1$, and $3k+2$. Therefore, its square is congruent to 0 or 1 modulo 3. Similarly, the other conclusion can be proved.
2. It is not necessary to classify the integer modulo 8; modulo 2 is enough (note that the product of adjacent integers is even).
3. It is sufficient to classify integers modulo 3.
4. Classify integers modulo 2.
5. Induct on n. Easy to verify for $n = 1$. Assume that the conclusion is true for $n - 1$, i.e., $a^{2^{n-1}} = 1 + 2^{n+1}x$ (x is an integer). Square both sides, and then $a^{2^n} = 1 + 2^{n+2}x'$ (x' is an integer).

   Another Proof: Use the identity (see the proof of Example 1 in Chapter 6)
   $$a^{2^n} - 1 = (a-1)(a+1)\left(a^2+1\right)\left(a^{2^2}+1\right)\cdots\left(a^{2^{n-1}}+1\right).$$
   Because the right side of the above formula is the product of $n+1$ terms, each term is even (because a is odd); one of $a-1$ and $a+1$ must be divisible by 4, so $2^{2^n} - 1$ can be divisible by 2^{n+2}.

6. Notice that $k \equiv -(p-k)(\bmod p)$, so from Wilson's theorem,
   $$1^2 \cdot 3^2 \cdot \ \cdots \ \cdot (p-2)^2$$
   $$\equiv 1 \cdot (p-1) \cdot 3 \cdot (p-3) \cdots \cdot (p-2) \cdot 2 \cdot (-1)^{\frac{p-1}{2}}$$
   $$= (-1)^{\frac{p-1}{2}} \cdot (p-1)!$$
   $$\equiv (-1)^{\frac{p+1}{2}} (\bmod p).$$
   The other congruence can be proved in the same way.

301

7. We have $2^{pq-1} - 1 = \left(2^{(p-1)q} - 1\right) 2^{q-1} + 2^{q-1} - 1$, and by exchanging p and q, the other equality is obtained. Hence the both aspects of the problem hold (using Fermat's little theorem).

8. Let $n = a_k \times 10^k + \cdots + a_1 \times 1^0 + a_0$ ($0 \le a_i \le 9$, for $i = 0, 1, \ldots, k$ and $a_k \ne 0$). Then from $10^i \equiv 1(\mathrm{mod}\,9)$ (for all $i \ge 0$), we know that $S(n) \equiv n(\mathrm{mod}\,9)$. From $10^i \equiv 1(\mathrm{mod}\,11)$ (for odd number $i \ge 1$), we see that $T(n) \equiv n(\mathrm{mod}\,11)$.

9. Let $1 = a_1 < \cdots < a_k = m-1$ be all the positive integers not exceeding m that are coprime with m, where $k = \varphi(m)$ with $m \ge 2$. Because $(m - a_i, m) = (a_i, m) = 1$, the set $\{a_1, \ldots, a_k\}$ is the same as $\{m - a_k, \ldots, m - a_1\}$. Therefore $a_1 + \cdots + a_k = (m - a_k) + \cdots + (m - a_1)$, from which we obtain the result.

10. Take $n = m = p - 1$ in equation 6 of Chapter 4.

11. Because $(a, 10) = 1$, we have $a^{\varphi(25)} \equiv 1(\mathrm{mod}\,25)$, that is, $a^{20} \equiv 1(\mathrm{mod}\,25)$. Also $a^2 \equiv 1(\mathrm{mod}\,4)$, so $a^{20} \equiv 1(\mathrm{mod}\,4)$. Therefore $a^{20} \equiv 1(\mathrm{mod}\,100)$, i.e., the last two digits of a^{20} are 01 (refer to (6) and (10)).

12. If there is a group of a_i ($1 \le i \le n$) and b_i ($1 \le i \le n$) such that $a_1 + b_1, \ldots, a_n + b_n$ form a complete system modulo n, then

$$1 + 2 + \cdots + n \equiv (a_1 + b_1) + \cdots + (a_n + b_n)$$
$$\equiv (a_1 + \cdots + a_n) + (b_1 + \cdots + b_n)$$
$$\equiv 2(1 + 2 + \cdots + n)(\mathrm{mod}\,n),$$

i.e., $n \left| \frac{n(n+1)}{2} \right.$. This can't be true because n is even.

13. According to Wilson's theorem, the product of numbers in any reduced system modulo p is $\equiv (p-1)\cdots 2 \cdot 1 = (p-1)! \equiv -1(\mathrm{mod}\,p)$.

14. Let $m = p_1^{\alpha_1} \cdots p_k^{\alpha_k}$ be the standard decomposition of m. We just need to prove that for any a that is coprime with m, there is $a^{\varphi(m)} \equiv 1(\mathrm{mod}\,p_i^{\alpha_i})(i = 1, \ldots, k)$ (refer to Remark 5 in Chapter 7 and (10)).

From (6) and the calculation formula of $\varphi(m)$, we only need to prove that $a^{p^{\alpha-1}(p-1)} \equiv 1(\mathrm{mod}\,p^\alpha)$ for any prime number p and integer $\alpha \ge 1$. It is not difficult to prove by induction on α. The case of $\alpha = 1$ is Fermat's little theorem (refer to the proof of Exercise 5).

15. We prove that $a^p \equiv a(\mathrm{mod}\,p)$ for any integer a (see Remark 3). Since, $a^p \equiv b^p(\mathrm{mod}\,p)$ when $a \equiv b(\mathrm{mod}\,p)$, it suffices to show the conclusion when $a = 0, 1, \ldots, p - 1$.

For any integer i, from the binomial theorem and Example 7 in Chapter 7,

$$(i+1)^p - i^p = \sum_{k=0}^{p-1} \binom{p}{k} i^k \equiv \binom{p}{0} = 1 \,(\bmod \, p).$$

Summing over $i = 0, 1, \ldots, a - 1$, we obtain the result.

16. If $p = 2$, then all the positive even numbers n meet the requirements. If $p > 2$, take $n = (p-1)^{2k}$, where k is any positive integer. Then it is obtained by Fermat's little theorem that

$$2^n - n = 2^{(p-1)^{2k}} - (p-1)^{2k} \equiv \left(2^{p-1}\right)^{(p-1)^{2k-1}} - 1$$

$$\equiv 1 - 1 = 0 (\bmod \, p).$$

17. For any given positive integer m, take a sufficiently large n_0 to make $2^{n_0} + 3^{n_0} - m > 1$. Then $2^{n_0} + 3^{n_0} - 1$, $2^{n_0} + 3^{n_0} - 2, \ldots, 2^{n_0} + 3^{n_0} - m$ are all greater than 1, so they have prime factors $p_1, p_2, \ldots, p_m$, respectively. Take

$$n_k = n_0 + k\,(p_1 - 1)\,(p_2 - 1)\cdots(p_m - 1),$$

where k is any positive integer.

If $p_i \neq 2$, then from Fermat's little theorem, $2^{p_i - 1} \equiv 1 (\bmod \, p_i)$, so

$$2^{n_k} = 2^{n_0} \cdot 2^{k(p_1-1)(p_2-1)\cdots(p_m-1)} \equiv 2^{n_0} \cdot 1 \equiv 2^{n_0} (\bmod \, p_i).$$

If $p_i = 2$, then the above formula obviously holds as well.
Similarly, $3^{n_k} \equiv 3^{n_0} (\bmod \, p_i)$, we have

$$2^{n_k} + 3^{n_k} - i \equiv 2^{n_0} + 3^{n_0} - i \equiv 0 \,(\bmod \, p_i) \,\, (i = 1, 2, \ldots, m).$$

Since $2^{n_k} + 3^{n_k} - i > 2^{n_0} + 3^{n_0} - i \geq p_i$, from the above formula, $2^{n_k} + 3^{n_k} - i$ is always a composite number for all $i = 1, 2, \ldots, m$. Since k is an arbitrary positive integer, there is an infinite number of n_k.

Solution 9

Indeterminate (Diophantine) Equations (1)

1. Obviously x is odd. If y is odd, then the equation is decomposed into

$$(x+1)(x^{y-1} - x^{y-2} + \cdots - x + 1) = 2^z.$$

The second factor on the left side of the above formula is the sum of an odd number of odd integers, which can only be 1 (because the right side is the power of 2). This leads to $x + 1 = 2^z$, i.e., $y = 1$, a contradiction. When y is an even number, x^y is the square of an odd number. Consider the equation modulo 4, and there must be $z = 1$. This also leads to $x = 1$, a contradiction. Therefore, the equation has no solution.

2. This integer x is odd. If y is odd, then the equation has no solution (using the solution of Exercise 1); if y is even, refer to the solution of equation (9) in the text, so the only solution is $x = 3$, $y = 2$, and $z = 3$.

3. Let $(x-1)x(x+1) = y^k$, where x and y are positive integers with $x \geq 2$. The equation is converted into $(x^2 - 1)x = y^k$. Since $(x, x^2 - 1) = 1$, both x and $x^2 - 1$ are the kth power of positive integers, i.e., $x^2 = (u^2)^k$ and $x^2 - 1 = v^k$, where u and v are positive integers. It can be verified easily that this is impossible (refer to Example 1).

4. Let $n = x^2 - y^2$, that is, $n = (x - y)(x + y)$. Because $x - y$ and $x + y$ have the same parity, either n is odd or $4 \mid n$ (this can also be derived from the original equation modulo 4).

Conversely, if n is an odd number, then we can take $x - y = 1$ and $x + y = n$; if $4 \mid n$, take $x - y = 2$ and $x + y = \frac{n}{2}$.

5. Modulo 9 (refer to Exercise 3 in Chapter 8).
6. Modulo 16 (refer to Exercise 4 in Chapter 8).
7. The equation can be rearranged into $(2x + 1)^2 = 4(y^4 + y^3 + y^2 + y) + 1$. Because

$$4\left(y^4 + y^3 + y^2 + y\right) + 1 = \left(2y^2 + y + 1\right)^2 - y^2 + 2y$$
$$= \left(2y^2 + y\right)^2 + 3y^2 + 4y + 1,$$

when $y > 2$ or $y < -1$, there holds that $(2y^2 + y) < (2x + 1)^2 < (2y^2 + y + 1)^2$. Thus, the equation has no solution when $y > 2$ or $y < -1$ (refer to Example 4(ii)). When $-1 \leq y \leq 2$, it's easy to verify that the equation has 6 groups of integer solutions.

8. Suppose $a > b > c >$. If $a + b = 2^u$, $b + c = 2^v$, and $c + a = 2^w$, then $u > w > v$, so $u \geq w + 1$ and $u > v + 1$, which implies that $(c + a) + (b + c) < 2^{u-1} + 2^{u-1} = 2^u = a + b$, a contradiction.

9. Let integers a, b, and c be the side lengths of a triangle. By Heron's formula, the problem is equivalent to finding all positive integer solutions of the equation

$$4(a + b + c) = (a + b - c)(a + c - b)(b + c - a).$$

In the above formula, the parities of the three factors on the right side are the same, and the left side are divisible by 2, so the factors on the right side are all even. Let

$$x = \frac{1}{2}(b + c - a), \ y = \frac{1}{2}(a + c - b), \ z = \frac{1}{2}(a + b - c).$$

We may assume that $a \leq b \leq c$. Then $x \geq y \geq z$ and $x + y + z = xyz$. Hence $yz \leq 3$, and then we find $x = 3$, $y = 2$ and $z = 1$, i.e., $a = 3$, $b = 4$ and $c = 5$.

10. Considering the equation modulo 3 and modulo 4, we see that x and z are even numbers. Let $x = 2m$ and $z = 2n$. Then $(3^m, 2^y, 5^n)$ is a group of primitive Pythagorean numbers. Therefore, there are positive integers a and b, $(a, b) = 1$, and one of a and b is odd and the other is even, such that

$$2^y = 2ab, \quad 3^m = a^2 - b^2.$$

From $(a, b) = 1$ and $a > b$, we know that $a = 2^{y-1}$ and $b = 1$. Thus

$$3^m + 1 = 2^{2y-2}.$$

If $y \geq 3$, then the right side of the formula is $\equiv 0 \pmod{8}$, and the left side is $\equiv 2, 4 \pmod{8}$. Thus $y = 2$ and hence $x = z = 2$.

11. Suppose (x, y, z) is a group of primitive solutions of the equation

$$x^2 + y^2 = z^2 \tag{1}$$

Because x and z are odd, both $\frac{z-x}{2}$ and $\frac{z+x}{2}$ are integers, and

$$\left(\frac{z-x}{2}, \frac{z+x}{2} \right) = \left(\frac{z-x}{2} + \frac{z+x}{2}, 2 \cdot \frac{z+x}{2} \right)$$

$$= (z, z+x) = (z, x) = 1.$$

Also from (1),

$$\frac{z-x}{2} \cdot \frac{z+x}{2} = \left(\frac{y}{2} \right)^2.$$

Thus the positive integers $\frac{z-x}{2}$ and $\frac{z+x}{2}$ are both perfect squares that is, there exist integers $a > b > 0$ with $(a, b) = 1$ such that $\frac{z+x}{2} = a^2$, $\frac{z-x}{2} = b^2$ and $y = 2ab$, i.e.,

$$x = a^2 - b^2, \quad y = 2ab, \quad z = a^2 + b^2. \tag{2}$$

Because x is odd, one of a and b is odd and the other is even.

On the other hand, when a and b satisfy the above conditions, we show that (x, y, z) given by (2) is a primitive solution of (1), and $2|y$.

First, it is easy to verify that the (x, y, z) given by (2) satisfies (1), and obviously $2 \mid y$. We just need to prove that $(x, y) = 1$. Let $(x, y) = d$. Then $d|x$, and $d|z$ i.e., $d \mid (a^2 - b^2)$ and $d|(a^2 + b^2)$, so $d|2a^2$ and $d|2b^2$, which implies that $d \mid (2a^2, 2b^2)$, i.e., $d \mid 2(a^2, b^2)$. Since $(a, b) = 1$, we have $(a^2, b^2) = 1$. Hence $d \mid 2$. But one of a and b is odd and the other is even, which implies that x is odd, hence $d = 1$.

12. When $m \leq 4$, the solutions are $(m, n) = (1, 1)$ and $(3, 3)$. When $m \geq 5$, $1! + 2! + 3! + 4! \equiv 3 \pmod{5}$, and $k! \equiv 0 \pmod{5}$ when $k \geq 5$. Hence, $1! + 2! + \cdots + m! \equiv 3 \pmod{5}$ when $m \geq 5$, but $n^2 \equiv 0, \pm 1 \pmod{5}$. Therefore, the equation has no solution when $m \geq 5$.

13. Suppose $1 \leq x_1 < \cdots < x_n$ satisfy $x_1 \cdots x_n = x_1 + \cdots + x_n$. We have

$$x_n \cdot (n-1)! \leq x_1 \cdots x_n = x_1 + \cdots + x_n < n x_n,$$

i.e., $(n-1)! < n$, so $n = 2$ or 3. Therefore, the only solution is $x_1 = 1$, $x_2 = 2$ and $x_3 = 3$.

14. There is at least one even number in x, y, and z. Then taking modulo 4, we deduce that x, y, and z are all even numbers. We suppose $x = 2x_1$, $y = 2y_1$, and $z = 2z_1$. Plug them into the original equation and repeat the above argument. Then we know that x_1, y_1, and z_1 are all even numbers. The above process is operated repeatedly, and then we deduce that x, y, and z are all divisible by any positive integer powers of 2. Therefore the only solution is $x = y = z = 0$.

15. Let $m = 2^k r$, where r is odd. Then the congruence equation has the solution $x = \frac{r+1}{2}$ and $y = \frac{2^{2k+1}+1}{3}$.

16. Let $2^p - 1 = q^s$. Then according to the conclusion of Exercise 1, there must be that $s = 1$.

Solution 10

Problems in Number Theory

1. The key of the demonstration is to establish the following recurrence relation:

$$a_n = a_{n-1}a_{n-2}\cdots a_1 + k(n \geq 2). \tag{1}$$

It is not difficult to prove it by induction. The conclusion is obviously true for $n = 2$. Assume that it is true for n. Then from the known recurrence formula,

$$a_{n+1} = a_n(a_n - k) + k = a_n a_{n-1} \cdots a_1 + k.$$

That is, the conclusion is also true for $n + 1$.

For any m and n $(1 \leq m < n)$, it can be seen from (1) that $a_m \mid a_n - k$. By mathematical induction, a_m and k are coprime, so a_m and $(a_n - k) + k = a_n$ are also coprime.

2. Note that if $d \mid n$, then $\frac{n}{d} \mid n$, so

$$S = \sum_{i=1}^{k-1} d_i d_{i+1} = n^2 \sum_{i=1}^{k-1} \frac{1}{d_i d_{i+1}}$$

$$\leq n^2 \sum_{i=1}^{k-1} \left(\frac{1}{d_i} - \frac{1}{d_{i+1}} \right) < \frac{n^2}{d_1} = n^2.$$

3. If the common difference of the sequence is d, then the sequence is $a_1, a_1 + d, \ldots, a_1 + (p-1)d$. Because $(p, d) = 1$, these p numbers form a complete system modulo p, so there is exactly an a_i divisible by p, and the rest $p - 1$ terms form a permutation of $1, 2, \ldots, p - 1$ in the

sense of modulo p. Therefore, according to Wilson's theorem (see (15) in Chapter 8),

$$\frac{a_1 a_2 \cdots a_p}{a_i} \equiv (p-1)! \equiv -1 (\text{mod } p).$$

From the above formula and the fact $p \mid a_i$, we obtain that $a_1 a_2 \cdots a_p + a_i \equiv 0 \ (\text{mod } p^2)$.

4. The known equation can be rearranged into

$$(a-b)(c-b) = b^2. \tag{2}$$

If there exists a prime p such that $p \mid (a-b, c-b)$, then from the above formula, $p \mid b^2$, so $p \mid b$. Also from $p \mid c-b$, we deduce that $p \mid c$; From $p \mid a-b$, we know that $p \mid a$, thus $p \mid (a, b, c)$, which contradicts the known condition. Therefore $(a-b, c-b) = 1$, and it follows from (2) that $a-b$ and $c-b$ are both perfect squares (note that from the known equation and the condition that a, b, and c are positive numbers, we can deduce that $a-b > 0$, and $c-b > 0$) follows from (2).

5. There do not exist such two powers of 2. If such numbers exist, assume that m and n are powers of 2 that meet the requirement of the problem, and suppose $m > n$. Then $m < 10n$. Since both m and n are powers of 2, $m = 2n, 4n$, or $8n$.

    Let $S(k)$ denote the sum of the digits of k (in the decimal representation). Then, from Exercise 8 of Chapter 8, we know that $S(m) \equiv m(\text{mod } 9)$ and $S(n) \equiv n(\text{mod } 9)$. Also $S(m) = S(n)$, so $m \equiv n(\text{mod } 9)$, i.e., n, $3n$, or $7n$ is divisible by 9, which is impossible (since n is a power of 2).

6. Take the equation modulo 4. Because the right side $\equiv 2(\text{mod } 4)$, we have $x = 1$ (otherwise, the left side will be divisible by 4, which leads to a contradiction).

    If $y > 1$, then the left side of the equation $\equiv 0(\text{mod } 9)$, and 5^z modulo 9 is $5, 7, 8, 4, 2, 1$ periodically, so $z = 6k + 3$. Hence $5^3 + 1$ exactly divides $5^z + 1 = (5^3)^{2k+1} + 1$, thus $5^3 + 1$ exactly divides $2 \cdot 3^y$. In particular, $7 \mid 2 \cdot 3^y$, which is impossible, so $y = 1$, thus $z = 1$. Therefore, the solution is $x = y = z = 1$.

7. Suppose $a \geq 0$. If $b = 0$, then $a = 0$. Below we suppose $b \neq 0$. Because $2^{2k}a + b$ is always a square number for all $k \geq 1$, we see that $4(2^{2k-2}a + b) = 2^{2k}a + 4b$ is a square number. Suppose

$$2^{2k}a + b = x_k^2, \ 2^{2k}a + 4b = y_k^2, \ (x_k \text{ and } y_k \text{ are positive integers}).$$

From this (eliminating 2^{2k}), we get

$$x_k + y_k \le (x_k + y_k)|x_k - y_k| = |x_k^2 - y_k^2| = |3b|.$$

Hence $2^{2k}a + b = x_k^2 \le (x_k + y_k)^2 \le 9b^2$ is valid for any k, so $a = 0$. (Since the right side of the inequality is a bounded quantity, if $a > 0$, then the left side can be arbitrarily large with k.)

8. By mathematical induction, we know that $a_n \equiv 2 \pmod{3} (n = 1, 2, \ldots)$, so $a_r \cdot a_s \equiv 2 \cdot 2 \equiv 1 \pmod 3$. Therefore $a_r \cdot a_s$ is not a term in the sequence $\{a_n\} (n \ge 1)$.

9. Each a_i has at least two prime factors p_i and q_i (p_i may be the same as q_i), so it can be expressed as $a_i = p_i q_i b_i$.

For $i \ne j$, there is $(a_i, a_j) = 1$, so p_i and q_i are not the same as p_j and q_j. Let N be the largest number of all p_i and q_i. Then

$$\sum_{i=1}^{n} \frac{1}{a_i} \le \sum_{i=1}^{n} \frac{1}{p_i q_i} \le \sum_{i=1}^{n} \frac{1}{\min(p_i, q_i)^2} \le \sum_{i=1}^{n} \frac{1}{k^2} < 1.$$

10. Suppose there are positive integers n and k ($1 < k < n - 1$) such that $\binom{n}{k} = p^\alpha$, where p is a prime number and $\alpha \ge 1$. Let $p^\beta \le n < p^{\beta+1}$ (β is a non-negative integer). Then the power of p in $\binom{n}{k}$ is

$$\sum_{i=1}^{\infty} \left(\left[\frac{n}{p^i} \right] - \left[\frac{k}{p^i} \right] - \left[\frac{n-k}{p^i} \right] \right)$$

$$= \sum_{i=1}^{\beta} \left(\left[\frac{n}{p^i} \right] - \left[\frac{k}{p^i} \right] - \left[\frac{n-k}{p^i} \right] \right) - \sum_{i=\beta+1}^{\infty} \left[\frac{k}{p^i} \right]$$

$$\le \sum_{i=1}^{\beta} 1 - \sum_{i=\beta+1}^{\infty} \left[\frac{k}{p^i} \right] \le \beta.$$

(The inequality $[x + y] \le [x] + [y] + 1$ was applied here.)

Because the power of p in $\binom{n}{k}$ is α, we have $p^\alpha \le p^\beta \le n$, which means that $\binom{n}{k} \le n = \binom{n}{1}$.

However, due to $1 < k < n-1$, and from the unimodality of the binomial coefficients, $\binom{n}{k} > \binom{n}{1}$, which is a contradiction.

Solution 11

Operations and Exact Division of Polynomials

1. The quotient is $x^4 - x^2 - x + 1$ and the remainder is $2x^2 - 2$.
2. Refer to Example 1. The polynomial

$$x^4 + x^3 + x^2 + x + 1 = \left(x^2 + \frac{1}{2}x + 1 \right)^2 - \left(\frac{1}{2}\sqrt{5} \right)^2,$$

   is the only solution.
3. Using the property (20) repeatedly we can get the result.
4. Note that all the constant terms of the polynomials in S are even. If such $d(x)$ exists in S, then $d(x) \mid x$ and $d(x) \mid 2$, and it is easy to know that this can't be true.
5. (i) If the conclusion is not true, let the degrees of $f(x)$ and $g(x)$ modulo p be n and m respectively. Then the coefficient of x^{n+m} in $f(x)g(x)$ is not divisible by p, thus, $f(x)g(x) \not\equiv 0 \pmod{p}$.
   (ii) Let $m = ab$, where a and b are integers greater than 1. Then $f(x) = ax$ and $g(x) = bx$ are both $\not\equiv 0 \pmod m$. But obviously $f(x)g(x) = abx^2 = mx^2 \equiv 0 \pmod m$.
6. Let $f(x) = \left(1 + x^2 - x^3 \right)^{100}$ and $g(x) = \left(1 - x^2 + x^3 \right)^{100}$. Because $(-x)^{20} = x^{20}$, the coefficients of x^{20} in $f(x)$ and $g(x)$ are the same as the coefficients of x^{20} in $f(-x)$ and $g(-x)$, respectively. Because $f(-x) = \left(1 + x^2 + x^3 \right)^{100}$ is a polynomial with non-negative coefficients, the coefficients of all the terms containing x^{20} in its expansion are all 1. Suppose there are n terms of x^{20} in total. Then the combined coefficient of x^{20} in $f(-x)$ is n. In the expansion of $g(-x) = \left(1 - x^2 - x^3 \right)^{100}$, the number of x^{20}-terms is also n. However, the coefficient of each x^{20} is ± 1, and it's easy to know that there must be -1 in it, so the final coefficient

of x^{20} in $g(-x)$ is less than n. Therefore, the coefficient of x^{20} in $f(x)$ is greater than that of x^{20} in $g(x)$.

7. According to the condition in the problem, the polynomial

$$x^{2k+1} + x + 1 - (x^k + x + 1) = x^k(x^{k+1} - 1)$$

is divisible by $x^k + x + 1$. Note that when k is odd,

$$x^k = (x^k + 1) - 1 = (x + 1)(x^{k-1} - x^{k-2} + \cdots - x + 1) - 1;$$

when k is even,

$$x^k = (x^k - 1) + 1 = ((x^2)^{\frac{k}{2}} - 1) + 1$$
$$= (x + 1)(x - 1)(x^{2k-2} + \cdots + x^2 + 1) + 1.$$

Therefore

$$(x^k, x^k + x + 1) = (x^k, x + 1) = ((-1)^k, x + 1) = 1.$$

Combining it with the above $(x^k + x + 1) \mid x^k(x^{k+1} - 1)$, we see that $x^{k+1} - 1$ is divisible by $x^k + x + 1$. Because

$$x^{k+1} - 1 = x(x^k + x + 1) - (x^2 + x + 1),$$

$x^2 + x + 1$ is also divisible by $x^k + x + 1$. Thus $k = 1$ or 2. But obviously $k = 1$ does not meet the requirement, so $k = 2$.

Solution 12

Zeros of Polynomials

1. Use the method of Example 1. Answer: $x^3 - 1$.
2. Use the method of Example 7. Answer: $n = 6k \pm 1$ (k is a positive integer).
3. The condition implies that $f(x) = (x - a)(x - b)(x - c) - 1$. If there is an integer d such that $f(d) = 0$, then $(d - a)(d - b)(d - c) = 1$, which is impossible.
4. Suppose $f(x)$ has an integer root a. Then $f(x) = (x - a)g(x)$, where $g(x)$ is an integer coefficient polynomial. Since $f(k)$ and $f(l)$ are both odd numbers, $(k - a)g(k)$ and $(l - a)g(l)$ are both odd numbers, but k and l have different parities, so one of $k - a$ and $l - a$ must be even. This is a contradiction.
5. Because the coefficients of the polynomial $f(x)$ are all nonnegative, all their roots are non-positive. We suppose $f(x) = (x + \alpha_1) \cdots (x + \alpha_n)$ with $\alpha_i \geq 0$. Because $2 + \alpha_i = 1 + 1 + \alpha_i \geq 3\sqrt[3]{\alpha_i}$, $(i = 1, \ldots, n)$, by Vieta's theorem, $\alpha_1 \cdots \alpha_n = 1$. Hence $f(2) = (2 + \alpha_1) \cdots (2 + \alpha_n) \geq 3^n \sqrt[3]{\alpha_1 \cdots \alpha_n} = 3^n$ (the arithmetic-geometric mean inequality).
6. Take $x_0 = 0$ and $x_{n+1} = x_n^2 + 1$. Then it is easy to use mathematical induction to prove that $f(x_n) = x_n$ ($n = 0, 1, \ldots$). But x_n are strictly increasing, so they are different from each other, and hence the equation $f(x) = x$ has an infinite number of different roots. Therefore from (5), we know that $f(x) = x$.
7. By plugging α into the equation and taking conjugation, we know that $f(\overline{\alpha}) = 0$.
8. The leading coefficient of $f(x)$ must be positive, and the same linear factors in its standard decomposition must appear even times. Since the quadratic factors of $f(x)$ are all of the form $(x - b)^2 + c^2$, these are

the sums of the squares of two real polynomials. At last, the identity

$$\left(x_1^2 + x_2^2\right)\left(y_1^2 + y_2^2\right) = \left(x_1 y_1 + x_2 y_2\right)^2 + \left(x_1 y_2 - x_2 y_1\right)^2$$

shows that the product of the above two factors is still the sums of the squares of two real polynomials (refer to Remark 7).

9. Let $k = (p-1)q + r$. Then since $(p-1) \nmid k$, we have $0 < r < p-1$. Because $a^k - 1 \equiv a^r - 1 \pmod{p}$, we can suppose $0 < k < p-1$. Since the congruence equation $x^k \equiv 1 \pmod{p}$ has at most k different solutions (see (6)), and there are $p-1$ reduced congruence classes modulo p, such a must exist.

10. If $|\alpha| < 1$, then there is no need to prove. Suppose $|\alpha| > 1$, and let $M = \max_{0 \le k \le n-1}\left|\frac{a_k}{a_n}\right|$. From $a_0 + \cdots + a_{n-1}\alpha^{n-1} = -a_n\alpha^n$,

$$|\alpha|^n = \left|\frac{a_0}{a_n} + \cdots + \frac{a_{n-1}}{a_n}\alpha^{n-1}\right|$$

$$\le M(1 + |\alpha| + \cdots + |\alpha|^{n-1})$$

$$< M\frac{|\alpha|^n}{|\alpha| - 1}.$$

Therefore $|\alpha| < M + 1$.

11. If $\mathrm{Re}(\alpha) \le 0$ or $|\alpha| \le 1$, then there is no need to prove. For $\mathrm{Re}(\alpha) > 0$ and $|\alpha| > 1$, we have $\mathrm{Re}\left(\frac{1}{\alpha}\right) > 0$, so from

$$0 = \left|\frac{f(\alpha)}{\alpha^n}\right| \ge \left|a_0 + \frac{a_1}{\alpha}\right| - \frac{a_2}{|\alpha|^2} - \cdots - \frac{a_n}{|\alpha|^n}$$

$$\ge \mathrm{Re}\left(a_0 + \frac{a_1}{\alpha}\right) - \frac{9}{|\alpha|^2} - \cdots - \frac{9}{|\alpha|^n}$$

$$> 1 - \frac{9}{|\alpha|^2 - |\alpha|},$$

we derive the conclusion.

12. For any $|x| \le 1$ and $x \ne 1$,

$$|(1-x)(a_0 + a_1 x + \cdots + a_n x^n)|$$

$$= |a_0 - (a_0 - a_1)x - (a_1 - a_2)x^2 - \cdots - (a_{n-1} - a_n)x^n - a_n x^{n+1}|$$

$$\ge a_0 - |(a_0 - a_1)x + (a_1 - a_2)x^2 + \cdots + a_n x^{n+1}|$$

$$> a_0 - (a_0 - a_1) - (a_1 - a_2) - \cdots - (a_{n-1} - a_n) - a_n$$

$$= 0.$$

This is because the arguments of $(a_0 - a_1)\, x$, $(a_1 - a_2)\, x^2, \ldots, a_n x^{n+1}$ can't be all equal (unless $x \geq 0$, but in this case, the conclusion is obviously true). Therefore, the moduli of the roots of $f(x)$ are all larger than 1.

13. Since all $c_i > 0$, for $x \geq 0$, there is $f(x) > 0$, so $f(x)$ has no non-negative real root. If $b_k \leq 0$ for a certain k, then $x^2 + a_k x + b_k$ has a non-negative root, and so does $f(x)$, which is impossible. Hence $b_k > 0$ for any k.

On the other hand, from the multiplication of polynomials, $c_1 = a_1 + \cdots + a_n$. Because $c_1 > 0$, there must be a certain k such that $a_k > 0$. This shows that there exists such a quadratic polynomial $x^2 + a_k x + b_k$.

14. There doesn't exist a set S satisfying the condition.
Suppose there exists such a set $S = \{a_1, \ldots, a_k\}$. Let

$$m = \min\{|a_1|, \ldots, |a_k|\},$$

$$M = \max\{|a_1|, \ldots, |a_k|\}.$$

Since S does not contain 0, clearly $M \geq m > 0$.

Suppose for any positive integer n, there is a polynomial of degree $N \geq n$

$$f(x) = b_N x^N + b_{N-1} x^{N-1} + \cdots + b_1 x + b_0,$$

such that all the coefficients $b_0, \cdots, b_N$ and N roots $x_1, \ldots, x_N$ belong to the set S. From Vieta's theorem,

$$x_1 + \cdots + x_N = -\frac{b_{N-1}}{b_N},$$

$$x_1 x_2 + x_1 x_3 + \cdots + x_1 x_N + \cdots + x_{N-1} x_N = \frac{b_{N-2}}{b_N},$$

so

$$x_1^2 + \cdots + x_N^2 = \left(-\frac{b_{N-1}}{b_N}\right)^2 - 2\frac{b_{N-2}}{b_N}.$$

Therefore

$$Nm^2 \leq x_1^2 + \cdots + x_N^2 = \frac{b_{N-1}^2}{b_N^2} - 2\frac{b_{N-2}}{b_N} \leq \frac{M^2}{m^2} + 2\frac{M}{m}.$$

Hence $N \leq \frac{M^2}{m^4} + 2\frac{M}{m^3}$, i.e., N is bounded. This is contradictory to $N \geq n$ and that n can be arbitrarily large.

15. Do the division with remainder in $Q[x]$:

$$f(x) = g(x)h(x) + r(x), \tag{1}$$

where $h(x)$ and $r(x)$ are rational coefficient polynomials, $r(x)$ is either zero or $\deg r < \deg g$. Let N be the absolute value of the common denominator of the coefficients of $h(x)$ and $r(x)$, and multiply both sides of (1) by N. Then the condition of the problem becomes: There is a monotone increasing sequence $\{a_n\}$ composed of positive integers such that

$$\frac{Nr(a_n)}{g(a_n)} \quad (n = 1, 2, \ldots) \tag{2}$$

are all integers. (Note that $g(x)$ has only a finite number of zeros. If there are zeros of then $g(x)$ in $\{a_n\}$, removing these terms does not change the problem.) If $r(x)$ is not zero then it can be seen from $\deg r < \deg g$ and $a_n \to +\infty$ that the sequence in (2) tends to 0, but these numbers are all integers, so there is an M such that when $n \geq M$, all the terms in (2) are zero, i.e., $r(a_n) = 0$ (for $n \geq M$). However, $r(x)$ is not zero, and there is only a limited number of zeros, which comes into conflict. Therefore $r(x)$ is the zero polynomial, that is $g(x) \mid f(x)$ (in $\mathbb{Q}[x]$).

Solution 13

Polynomials with Integer Coefficients

1. Use Eisenstein's criterion directly.
2. Let $f(x) = x^7 + 7x + 1$. Then we can apply Eisenstein's criterion to $g(x) = f(x-1)$ (refer to (5)).
3. First of all, the polynomial doesn't have a linear factor (just check whether $x = \pm 1$ and ± 5 is a root; refer to (8) of Chapter 12). Therefore, if the polynomial is reducible in $\mathbb{Q}[x]$ (and thus also in $Z[x]$), it must be the product of quadratic and cubic factors. With the method of undetermined coefficients, a contradiction can be derived.
4. If n is not prime, let $n = ab$ $(a, b > 1)$. Then the polynomial can be expressed as $\frac{x^{ab}-1}{x-1} = \frac{(x^a)^b-1}{x-1}$, which is easy to decompose into the product of two (non-constant) integer coefficient polynomials. Note that (5) implies that the converse of the proposition is true.
5. Let $f(x)$ be the polynomial. Then $f(-x) = \Phi_p(x)$ (refer to (5)).
6. Let $f(x)$ be the polynomial. First of all, $f(x)$ does not have a linear factor. This is because if $f(x)$ has an integer root a, then $5|k$ from $a^5 - a \equiv \pmod 5$. Assume that $f(x)$ has a quadratic factor $x^2 - ax - b$. It is easy to find that the remainder of $f(x)$ on division by $x^2 - ax - b$ is $\left(a^4 + 3a^2b + b^2 - 1\right)x + \left(a^3b + 2ab^2 + k\right)$; this must be the zero polynomial, so $a^4 + 3a^2b + b^2 - 1 = 0$ and $a^3b + 2ab^2 + k = 0$. Hence $a(a^4 + 3a^2b + b^2 - 1) - 3(a^3b + 2ab^2 + k) = 0$, that is, $a^5 - a - 5ab^2 = 3k$. Because the left side is a multiple of 5, we get $5 \mid k$, a contradiction.

7. First prove that the moduli of the complex roots of $f(x)$ are all greater than 1. In fact, let x_0 be a root of $f(x)$. If $|x_0| \leq 1$, then

$$p = |a_n x_0^n + \cdots + a_0 x|$$

$$\leq |a_n| \cdot |x_0|^n + \cdots + |a_1| \cdot |x_0|$$

$$\leq |a_n| + \cdots + |a_1|,$$

which contradicts the known condition (p is not required to be prime here). Now let $f(x) = g(x)h(x)$, where $g(x)$ and $h(x)$ are non-constant polynomials with integer coefficients. Then

$$p = |f(0)| = |g(0)| \cdot |h(0)|. \tag{1}$$

Let b be the leading coefficient of $g(x)$. Then it is known from Vieta's theorem that $\frac{g(0)}{b}$ is equal to the product of all roots of $g(x)$. But the roots of $g(x)$ are all the roots of $f(x)$, so their moduli are all greater than 1. Hence $\left| \frac{g(0)}{b} \right| > 1$, and of course $|g(0)| > 1$. Similarly, $|h(0)| > 1$. But p is prime, and $|g(0)|$ and $|h(0)|$ are integers greater than 1, so (1) can't be true.

8. Suppose $f(x)$ can be decomposed into the product of two positive degree polynomials (with integer coefficients) $g(x)$ and $h(x)$. Since $|f(m)|$ is prime, we see that one of $|g(m)|$ and $|h(m)|$ is 1. Suppose $|g(m)| = 1$.

On the other hand, it can be seen from Exercise 10 in Chapter 12 that the moduli of the roots of $f(x)$ are all less than $M + 1$. Therefore if

$$g(x) = a(x - \alpha_1) \cdots (x - \alpha_r)$$

(a is the coefficient of the first term of $g(x)$), then

$$|g(m)| \geq (m - |\alpha_1|) \cdots (m - |\alpha_r|)$$

$$> |m - (M + 1)|^r \geq 1,$$

which leads to a contradiction.

Solution 14

Interpolation and Difference of Polynomials

1. $ix^2 - (1+i)x + 1$.
2. Express the polynomial as (refer to (8))

$$a \binom{x}{4} + b \binom{x}{3} + c \binom{x}{2} + d \binom{x}{1} + e.$$

   By using the substitution method (take $x = 0, 1, 2, 3$, and 4 in turn), we obtain the undetermined coefficients $a = b = 1$, $c = -1$, $d = 0$, and $e = 1$. Therefore, the polynomial is an integer valued polynomial.
3. Refer to the proof of formula (5).
4. For any integer a, the polynomial $f(x)$ is integer valued if and only if $g(x) = f(x+a)$ is an integer valued polynomial. Hence we may assume that the $n+1$ consecutive integers are $0, 1, \ldots, n$. First, express $f(x)$ in the form of (8), and since $f(k)$ $(k = 0, 1, \ldots, n)$ are integers, it can be deduced that all coefficients of $f(x)$ are integers.

   The significance of this problem is mainly theoretical. It uses the values of the polynomial (at a finite number of consecutive integers) to give the characterization of the integer valued polynomial. For a specific polynomial, in order to judge whether it is an integer valued polynomial or not, we prefer to use the undetermined coefficient method in Exercise 2 to express it in the form of (8).
5. Answer: $n = 2$. Both two methods of Example 1 can be used to solve this problem.

6. Use the method of Example 2. Consider $g(x) = (x+1)f(x) - x$. It's 0 at $x = 0, 1, \ldots, n$ and $g(x) = \frac{(-1)^{n+1}}{(n+1)!} x(x-1)\cdots(x-n)$ and because $g(-1) = 1$, then we find that

$$f(n+1) = \begin{cases} 1, & n \text{ is odd}; \\ \dfrac{n}{n+2}, & n \text{ is even}. \end{cases}$$

7. From Lagrange's interpolation formula and the $2n+1$ values $f(k)$ $(-n \le k \le n)$, we see that $f(x)$ can be uniquely determined:

$$f(x) = \sum_{k=-n}^{n} f(k) \prod_{i \ne k, -n \le i \le n} \frac{x-i}{k-i}.$$

From the given conditions,

$$|f(x)| \le \sum_{k=-n}^{n} \prod_{i \ne k} \left| \frac{x-i}{k-i} \right|.$$

For any real number x such that $-n \le x \le n$, we have

$$\prod_{i \ne k, -n \le i \le n} |x-i| \le (2n)!.$$

In fact, when $x \ge k$,

$$\prod_{i \ne k} |x-i| = (|x-(k+1)| \cdots |x-n|)(|x-(k-1)| \cdots |x+n|)$$

$$\le (n-k)! \cdot (n-k+1) \cdots (2n) = (2n)!.$$

Similarly, the case of $x < k$ can also be proved. Thus,

$$\prod_{i \ne k} \left| \frac{x-i}{k-i} \right| \le (2n)! \prod_{i \ne k} \frac{1}{|k-i|} \le (2n)! \frac{1}{(n+k)!(n-k)!}.$$

Therefore $|f(x)| \le \sum_{k=0}^{2n} \binom{2n}{k} = 2^{2n}$.

8. The form of the sum is related to Lagrange's interpolation formula of the degree-k polynomial x^k at n points $a_1, \ldots, a_n$. But k may be larger than n, so we take an integer coefficient polynomial $r(x)$ with $\deg r \le n-1$, which satisfies $r(a_i) = a_i^k (i = 1, \ldots, n)$. For this purpose, let $g(x) = \prod_{i=1}^{n} (x - a_i)$. Note that $g(x)$ is an integer coefficient polynomial with the leading coefficient 1, so doing the division with remainder in $\mathbb{Z}[x]$, we obtain that

$$x^k = g(x)h(x) + r(x),$$

where both $h(x)$ and $r(x)$ are polynomials with integer coefficients, and $\deg r \leq n - 1$. From the above formula, we see that $r(a_i) = a_i^k (i = 1, \ldots, n)$. From Lagrange's interpolation formula,

$$r(x) = \sum_{i=1}^{n} r(a_i) \prod_{j \neq i} \frac{x - a_j}{a_i - a_j}. \tag{1}$$

The sum in the problem is the coefficient of x^{n-1} on the right side of formula (1), and $r(x)$ is an integer coefficient polynomial, hence the coefficient of x^{n-1} is an integer (possibly zero), which proves the proposition.

Solution 15

Roots of Unity and their Applications

1. Use the 5th roots of unity.

2. Let $f(x^5) = f(x)q(x) + r(x)$, where $r(x) = 0$ or $\deg r \le 3$. Let $\zeta \ne 1$ be a 5th root of unity. Then $r(\zeta) = r(\zeta^2) = r(\zeta^3) = r(\zeta^4) = 5$ and $\deg r \le 3$. Therefore, $r(x) = 5$, that is, the remainder is the constant 5.

3. We have $f(x^n) = (x-1)g(x)$. Take $\zeta = e^{\frac{2\pi i}{n}}$ as an nth root of unity, and from $f(1) = 0$, we know that $f(\zeta^{kn}) = 0$ $(k = 1, \ldots, n)$. Therefore, $f(x^n)$ is divisible by $(x - \zeta)(x - \zeta^2) \cdots (x - \zeta^n) = x^n - 1$.

4. Refer to the solution of Example 2. In the binomial expansion of $(x+1)^n$, the result can be obtained by plugging in $x = 1, -1$, i, and -i (the 4th roots of unity).

5. (i) Take $x = 1$ in (4). (ii) Refer to the proof of Example 4.

6. (i) In the expansion of $(1 + x + x^2)^n$, plug in $x = 1$ and -1 (2nd roots of unity) to get the result. The solution of (ii) is similar to that of Example 2.

7. Let $\zeta = e^{\frac{2\pi i}{m+1}}$ be an $(m+1)$st root of unity. Then all roots of $1 + x + \cdots + x^m$ are ζ^k $(k = 1, 2, \ldots, m)$ (refer to (4)). Therefore, a necessary and sufficient condition for $f(x) = 1 + x^n + \cdots + x^{mn}$ to be divisible by $1 + x + \cdots + x^m$ is $f(\zeta^k) = 0$ $(1 \le k \le m)$. Because

$$f(x) = \frac{x^{n(m+1)} - 1}{x^n - 1},$$

the above condition is equivalent to that ζ^k is a zero of $x^{n(m+1)} - 1$, but not a zero of $x^n - 1$ (note that these two polynomials have no multiple roots). The former is obviously true, while the latter is

$$\cos \frac{2kn\pi}{m+1} + i\sin \frac{2kn\pi}{m+1} \neq 1, \quad k = 1, 2, \ldots, m.$$

Therefore, a necessary and sufficient condition is that $m + 1$ and n are coprime.

8. Using the solution of Example 1 in Chapter 14, we obtain that

$$\sum_{k=0}^{3n+1} (-1)^k \binom{3n+1}{k} P(3n + 1 - k) = 0,$$

in other words,

$$729 + 2\sum_{j=0}^{n} (-1)^{3j+1} \binom{3n+1}{3j+1} + \sum_{j=0}^{n} (-1)^{3j} \binom{3n+1}{3j} = 0.$$

Use the method of Example 2 to find

$$\sum_{j=0}^{n} (-1)^j \binom{3n+1}{3j+1} = 2\left(\sqrt{3}\right)^{3n-1} \cos \frac{3n-1}{6}\pi,$$

$$\sum_{j=0}^{n} (-1)^j \binom{3n+1}{3j} = 2\left(\sqrt{3}\right)^{3n-1} \cos \frac{3n+1}{6}\pi.$$

Hence $729 - 4\left(\sqrt{3}\right)^{3n-1} \cos \frac{3n-1}{6}\pi + 2\left(\sqrt{3}\right)^{3n-1} \cos \frac{3n+1}{6}\pi = 0$.
By considering the parity of n, it is easy to get that $n = 4$ only.

Solution 16

Generating Function Method

1. Compare the coefficients of the both sides of
$$(1+x)^n = (1+x)(1+x)^{n-1}.$$

2. The method is the same as in Example 3. Verify that the left side of the equation is the constant term of
$$\sum_{k=0}^{n} \binom{n}{k} 2^{n-k}(1-x)(x^{-1}+x)^k.$$

3. For the first part of the problem one can use the method of Example 5 (refer to Remark 1). For the second part, the result is the coefficient of x^8 in
$$(x+x^2+x^3)(1+x+x^2+x^3)(x^2+x^3+x^4+x^5+x^6).$$

4. Refer to Remark 4 of this chapter.

5. The principle in (1) can be used conversely for this problem. The coefficient of x^k is the number of solutions to the equation
$$k = a_1 + a_2, \tag{1}$$
where $0 \le a_1, a_2 \le n-1$. For $0 \le k \le n-1$, then (1) has exactly $k+1$ solutions:
$$0+k, 1+(k-1), \ldots, (k-1)+1, k+0.$$
However, if $n \le k \le 2n-2$, then $0+k, 1+(k-1), \ldots, (k-n)+n$ and $k+0, (k-1)+1, \ldots, n+(k-n)$ in the above solutions do not satisfy $a_1, a_2 \le n-1$. Therefore, the equation (1) has a total of $k+1-2(k-n+1) = 2n-k-1$ solutions that meet the requirements. In both cases, the coefficients of x^k can be written uniformly as
$$n-|n-k-1|.$$

6. The problem is equivalent to finding positive integers $a_1, \ldots, a_6$ and $b_1, \ldots, b_6$, satisfying

$$(x^{a_1} + \cdots + x^{a_6})(x^{b_1} + \cdots + x^{b_6}) = (x + \cdots + x^6)(x + \cdots + x^6),$$

and the sets $\{a_1, \ldots, a_6\}$ and $\{b_1, \ldots, b_6\}$ are both different from $\{1, 2, 3, 4, 5, 6\}$. The above equation is

$$(x^{a_1} + \cdots + x^{a_6})(x^{b_1} + \cdots + x^{b_6})$$

$$= \left(x \frac{x^6 - 1}{x - 1} \right)^2 = \left[x \cdot \frac{x^3 - 1}{x - 1}(x^3 + 1) \right]^2$$

$$= x^2 \left(x^2 + x + 1 \right)^2 (x + 1)^2 \left(x^2 - x + 1 \right)^2.$$

Because a_i is a positive integer, $x^{a_1} + \cdots + x^{a_6}$ must contain the factor x, and also must contain the factors $x + 1$ and $x^2 + x + 1$ (because the value of $x^{a_1} + \cdots + x^{a_6}$ at $x = 1$ is 6). The same conclusion holds for $x^{b_1} + \cdots + x^{b_6}$. Hence only the factor $\left(x^2 - x + 1 \right)^2$ is optional. Note that $x^2 - x + 1$ is irreducible in $\mathbb{Z}[x]$, so we can only take

$$x^{a_1} + \cdots + x^{a_6} = x(x + 1)(x^2 + x + 1) = x + 2x^2 + 2x^3 + x^4,$$

$$x^{b_1} + \cdots + x^{b_6} = x\left(x + 1 \right)\left(x^2 + x + 1 \right)\left(x^2 - x + 1 \right)^2$$

$$= x + x^3 + x^4 + x^5 + x^6 + x^8,$$

i.e., $a_1, \ldots, a_6 = \{1, 2, 2, 3, 3, 4\}$ and $\{b_1, \ldots, b_6\} = \{1, 3, 4, 5, 6, 8\}$ form the unique group of solutions.

7. Differentiate the identity

$$(1 + x)^n = \sum_{k=0}^{n} \binom{n}{k} x^k \tag{2}$$

with respect to x, we obtain that

$$n(1 + x)^{n-1} = \sum_{k=1}^{n} k \binom{n}{k} x^{k-1}.$$

In the above formula, replace x with $\frac{1}{x}$, and then multiply both sides by x^n to get

$$n(1 + x)^{n-1} = \sum_{k=1}^{n} k \binom{n}{k} x^{n-k}.$$

Multiplying the above equality by (1) gives that

$$n \left(1 + x\right)^{2n-1} = \left(\sum_{k=1}^{n} k \binom{n}{k} x^{n-k}\right) \left(\sum_{k=0}^{n} \binom{n}{k} x^k\right).$$

By comparing the coefficients of x^n on both sides of the above formula, we get

$$\sum_{k=1}^{n} k \binom{n}{k}^2 = n \binom{2n-1}{n} = n \binom{2n-1}{n-1}.$$

8. Let R, S, and T denote the sets of the numbers colored in three colors in M respectively. Let

$$f(x) = \left(\sum_{r \in R} x^r\right)^3 + \left(\sum_{s \in S} x^s\right)^3 + \left(\sum_{t \in T} x^t\right)^3,$$

$$g(x) = \left(\sum_{r \in R} x^r\right) \left(\sum_{s \in S} x^s\right) \left(\sum_{t \in T} x^t\right).$$

Note that x, y and z can be the same for $(x, y, z) \in A$, so $n \mid x + y + z$ means that $x + y + z$ is n, $2n$, or $3n$. Then $x + y + z$ can only be n or $2n$ for $(x, y, z) \in B$. Therefore, $|A|$ is equal to the sum of the coefficients of the terms of x^n, x^{2n}, and x^{3n} in $f(x)$, while $|B|$ is equal to $3! = 6$ times the sum of the coefficients of the terms of x^n and x^{2n} in $g(x)$.

Let $\zeta = e^{\frac{2\pi}{n} i}$ be an nth root of unity. Then the sum of the coefficients of the terms with degrees being multiples of n in any polynomial $p(x)$ is $\frac{1}{n} \sum_{k=1}^{n} p(\zeta^k)$. The terms of degrees being multiples of n in $f(x)$, can only be x^n, x^{2n}, and x^{3n}, and the terms with degrees being multiples of n in $g(x)$ can only be x^n and x^{2n}. Therefore,

$$|A| = \frac{1}{n} \sum_{k=1}^{n} f(\zeta^k),$$

$$|B| = \frac{6}{n} \sum_{k=1}^{n} g(\zeta^k).$$

We need to prove that $\sum_{k=1}^{n} f(\zeta^k) - 3 \sum_{k=1}^{n} g(\zeta^k) \geq 0$.

Let $a_k = \sum_{r \in R} \zeta^{kr}$, $b_k = \sum_{s \in S} \zeta^{ks}$, and $c_k = \sum_{t \in T} \zeta^{kt}$ $(k = 1, \ldots, n)$. Because

$$a_k^3 + b_k^3 + c_k^3 - 3a_k b_k c_k = (a_k + b_k + c_k)(a_k^2 + b_k^2 + c_k^2 - a_k b_k - b_k c_k - c_k a_k)$$

and $R \cup S \cup T = M$,

$$a_k + b_k + c_k = \sum_{m=1}^{n} \zeta^{km} = \begin{cases} 0, & \text{if } 1 \le k \le n-1, \\ n, & \text{if } k = n. \end{cases}$$

Therefore,

$$\sum_{k=1}^{n} f(\zeta^k) - 3 \sum_{k=1}^{n} g(\zeta^k) = f(1) - 3g(1)$$

$$= |R|^3 + |S|^3 + |T|^3 - 3|R| \cdot |S| \cdot |T| \ge 0.$$

Solution 17

Sets and Families of Subsets

1. Suppose there are m political parties. If X denotes the set of all promises and A_i denotes the set of promises of the ith political party $(i = 1, \ldots, m)$, then $|X| = n$, $A_i \cap A_j \neq \varnothing$, and $A_i \neq A_j$ $(1 \leq i, j \leq m$ with $i \neq j)$. According to the proof of Example 1, $m \leq 2^{n-1}$. It is not difficult to construct an example to show that the equality can hold.

2. Consider any element a less than n. If a is in a subset A without n, then it is also in the subset $A \cup \{n\}$ with n, and vice versa. Therefore, if the contribution of a to the alternating sum of A is $+a$ (or $-a$), then the contribution of a to the alternating sum of $A \cup \{n\}$ is $-a$ (or $+a$), and vice versa. Therefore the contribution of a to the sum of all alternating sums is 0.

   On the other hand, the contribution of the element n to the alternating sum of every subset containing n is n, so the contribution of n to all the sums is $n \cdot 2^{n-1}$. To summarize, $S = n \cdot 2^{n-1}$. (Refer to Solution 1 of Example 2.)

3. Any subset with more than 3 elements in $B_1, \ldots, B_k$ generates a family of 3-element subsets of the set A, and it is known from the condition that there will be no common 3-element subsets in these families. Therefore, in order to maximize k, each B_i must satisfy the requirement of $|B_i| \leq 3$ $(i = 1, \ldots, k)$. Also it is easy to know that all the families of 1-element, 2-element, and 3-element subsets meet the requirement of the problem, so the maximum value of k is $\binom{10}{1} + \binom{10}{2} + \binom{10}{3} = 175$.

4. Use the method of Example 4 to obtain (refer to (2) and (3))

$$\frac{1}{10}\left(\sum_{i=1}^{k}|A_i|\right)^2 \le \sum_{i=1}^{k}|A_i| + 2\sum_{1\le i<j\le k}|A_i \bigcap A_j|.$$

Hence $\frac{1}{10}(5k)^2 \le 5k + 2\binom{k}{2}\times 2$, thus $k \le 6$. Please construct an example to show that the maximum k is 6.

5. Let n be the minimum number of locks that meet the requirement. Let A_i be the set of locks that the ith member can open $(i = 1, \ldots, 11)$, and A be the set of all locks. The given condition implies that the union of any 5 sets of $A_1, \ldots, A_{11}$ is not equal to A, and the union of any 6 sets of $A_1, \ldots, A_{11}$ is equal to A, so from the argument in the first paragraph of Example 3 (note that there is no need for $|A_1| = |A_2| = \cdots$), we get $n = |A| \ge \binom{11}{5}$.

The following points out that if $\binom{11}{5}$ locks are added, then the members can be assigned the keys as required in the problem:

We establish a one-to-one correspondence between the $\binom{11}{5}$ locks and the 5-element subsets of the set $\{1, 2, \ldots, 11\}$. If a lock corresponds to a subset $\{i_1, \ldots, i_5\}$, then its keys are given to all members whose label is not $i_1, \ldots, i_5$. It is easy to verify that this distribution meets the requirement.

6. If the element a_j belongs to m_j sets A_i $(1 \le i \le n)$, then by using the subordination dependence table of the sets and elements, it is easy to know that $m_1 + \cdots + m_n = 2n$.

On the other hand, if there is a certain $m_j > 2$, then there are at least three subsets containing a_j, two of which are A_s and A_t (both s and t are not equal to j). However, from the known condition, because $A_s \cap A_t$ is not empty (since both contain a_j), one of them must be $\{a_s, a_t\}$, but it does not contain a_j, a contradiction. Therefore, all $m_j \le 2$, so all $m_j = 2$.

7. If a 10-element subset S' has the property: for any $k \in S'$, there is $f(S'\backslash\{k\}) \ne k$, then S' is called a "good set," and it is called a "bad set" otherwise.

If S' is a bad set, then there exists $k_0 \in S'$ such that

$$f(S'\backslash\{k_0\}) = k_0.$$

Let $T = S'\backslash\{k_0\}$. Then $|T| = 9$ and $f(T) = k_0$, so $S' = T \cup \{f(T)\}$. This shows that every bad set is generated by a 9-element set (and f).

Moreover, any 9-element set can generate at most one bad set according to the above formula, so the number of bad sets does not exceed the number of 9-element subsets of S, i.e., $\binom{20}{9}$. But the number of 10- element subsets is $\binom{20}{10}$, and since $\binom{20}{10} > \binom{20}{9}$, good sets must exist. (The main point of the argument is to point out a single-valued correspondence from bad sets to 9-element subsets.)

8. We may suppose $|A_1| = m$. Also suppose there are s sets that do not intersect A_1 in $A_1, \ldots, A_n$, and mark them as $B_1, \ldots, B_s$; suppose there are t sets containing A_1, and mark them as $C_1, \ldots, C_t$.

Because $B_i \cup A_1$ contains A_1, according to the condition, it is a certain A_k, so $B_i \cup A_1 \in \{C_1, \ldots, C_t\}$. Because B_i and B_j do not intersect A_1, if $B_i \neq B_j$, then $B_i \cup A_1 \neq B_j \cup A_1$. Hence the number of $B_i \cup A_1 (i = 1, \ldots, s)$ does not exceed the number of $C_1, \ldots, C_t$, that is, $s \leq t$.

Let $A_1 = \{a_1, \ldots, a_m\}$. After removing $B_1, \ldots, B_s$ and $C_1, \ldots, C_t$ from $A_1, \ldots, A_n$, each of the remaining $n - s - t$ sets has a non empty intersection with A_1 (that is, it contains a certain a_i), in which suppose there are x_i sets $(1 \leq i \leq m)$ containing a_i. Then

$$x_1 + \cdots + x_m \geq n - s - t.$$

Let $x_1 = \max x_i$. Then from the above the equality, $x_1 \geq \frac{n-s-t}{m}$. That is, among the remaining $n - s - t$ sets, at least $\frac{n-s-t}{m}$ sets contain a_1. Also $A_1 \subseteq C_i (i = 1, \ldots, t)$, so all $C_1, \ldots, C_t$ contain a_1. Therefore, the number of sets containing a_1 is at least

$$\frac{n-s-t}{m} + t = \frac{n - s + (m-1)t}{m}$$

$$\geq \frac{n-s+t}{m} \quad (\text{using } m \geq 2)$$

$$\geq \frac{n}{m} \quad (\text{using the proven } t \geq s).$$

Solution 18

Graph Theory Problems

1. It can't be. Refer to Example 1.
2. Use vertices to represent people. If two people are friends, form an edge between the corresponding vertices. Then a graph G is obtained. Prove that at least two vertices in G have the same degree. Because the degrees are not negative and less than or equal to $n - 1$, the degrees can only be one of $0, 1, 2, \ldots, n - 1$. But obviously and $n - 1$ can't appear at the same time, so the degrees can only be $0, 1, 2, \ldots, n - 2$ or $1, 2, \ldots, n - 1$. Therefore, at least two of the n vertices have the same degree.
3. If a graph G is constructed as in the previous problem, then this problem will be converted to proving: If G has $2n$ vertices and the degree of each vertex is greater than or equal to n, then there is a quadrilateral in G.

   If $G = K_{2n}$, then the conclusion is obviously true. If $G \neq K_{2n}$, then there are vertices v_1 and v_2 unconnected. Because $d(v_1) + d(v_2) \geq 2n$, there must be two of the rest vertices in G connected with both v_1 and v_2, denoted as v_3 and v_4. Then v_1, v_2, v_3, and v_4 form a quadrilateral.

4. Color the longest side of each triangle red (if it is an isosceles triangle or an equilateral triangle, then there may be two or three longest sides), and then color the remaining sides blue. From Example 5, there must exist a homochromatic triangle. In this triangle, there must exist a longest side, which is red. Therefore, all three sides of this triangle are red. Also its shortest side is red, so it is the longest side of a triangle.

   Please note that triangles here have dual meanings: On one hand, triangles are in the sense of graph theory, while on the other hand, triangles are as in general geometry (we can talk about the lengths of their sides, etc.).

5. (i) May not. An example can be constructed as follows: Take five complete graphs K_{100}, and use each vertex to represent one person. If two people don't know each other, form an edge between the corresponding vertices (note that this definition is different from the previous one). Since each vertex of K_{100} is connected with other vertices in this complete graph, but not with vertices of the other four complete graphs, each person knows 400 people. If we take any six vertices, then there must be two vertices in the same complete graph, and those people represented by the two vertices don't know each other.

   (ii) It can be deduced from Example 6 (take $r = 6$): We still use vertices to represent people, and if two people know each other, then connect these two corresponding vertices by an edge. The construction method here is just the opposite of that in (i), but it is only for the convenience of narration, and there is no essential difference.

6. Color the edges of the pentagon red, and color the edges of the pentagram (connected to the vertices) blue.

7. This problem is equivalent to proving that $R(3,3,3) \leq 17$ (refer to Remark 4). Color the edges of K_{17} with three colors (red, blue, and white). Take any vertex and 6 of the 16 edges connected by it must be in the same color. Suppose v and $v_1, v_2, \ldots, v_6$ are connected by the red edges. In the complete subgraph K_6 with $v_1, \ldots, v_6$ as the vertices, if there is a red edge $v_i v_j (1 \leq i \neq j \leq 6)$, then there is a red triangle $v v_i v_j$ in K_{17}. If there is no red edge in K_6, then its edge is colored with only two colors. From Example 5, there must be homochromatic triangles in it, so there are homochromatic triangles in K_{17}.

   Please note that it is possible to color edges of K_{16} with three colors, so that there is no homochromatic triangle in it (the readers can try it), thus $R(3,3,3) \geq 17$. By combining the results, we know that $(3,3,3) = 17$.

8. Let the vertices of G be $x_1, \ldots, x_n$, and $d(x_i) = d_i$ $(i = 1, \ldots, n)$. Let x_i and x_j be connected, A be the set of vertices connected with x_i except x_j and B be the set of vertices connected with x_j except x_i. Then the number of triangles with $x_i x_j$ as a side in the graph is

$$|A \bigcap B| = |A| + |B| - |A \bigcup B| \geq d_i + d_j - n$$

(since $|A| = d_i - 1$, $|B| = d_j - 1$, and $|A \bigcup B| \leq n - 2$). Because each triangle has three sides, the number of triangles in G is at least

$$k = \frac{1}{3} \sum_{x_i \text{ is connected to } x_j} (d_i + d_j - n).$$

Since each d_i appears d_i times in the above sum and G has m edges,

$$k = \frac{1}{3} \left(\sum_{i=1}^{n} d_i^2 - mn \right).$$

Because $\sum_{i=1}^{n} d_i = 2m$, from the Cauchy's inequality,

$$k \geq \frac{1}{3} \left(\frac{1}{n} \left(\sum_{i=1}^{n} d_i \right)^2 - mn \right) = \frac{1}{3} \left(\frac{4m^2}{n} - mn \right) = \frac{4m}{3n} \left(m - \frac{n^2}{4} \right).$$

The result of this problem implies Example 7 (refer to Proof 2 of Example 7).

Solution 19

Congruence (2)

1. Let q be any prime factor of $2^p + 1$. Consider the order k of 2 modulo q. From $2^p \equiv -1 \pmod q$, we get $2^{2p} \equiv 1 \pmod q$, so $k \mid 2p$. Because p is a prime number, $k = 1, 2, p$, or $2p$. Therefore, combining it with $2^p \equiv -1 \pmod q$, we easily know that $k = 2$ or $2p$. As in Example 2, it can be deduced that $q = 3$ or $q = 2px + 1$, where x is a positive integer.

2. Let q be any prime factor of $a^p - 1$. Then it is easy to prove that the order k of a modulo q is 1 or p. As in Example 2, it can be seen that $q \mid a - 1$ or $q = 2px + 1$ (x is a positive integer).

3. Let d_1 be a solution of $d_1 \equiv 1 \pmod 4$, $d_1 \equiv 0 \pmod 5$ and $d_1 \equiv 0 \pmod 7$, that is, d_1 satisfies $d_1 \equiv 1 \pmod 4$ and $35 \mid d_1$, so we can take $d_1 = -35$. Similarly, we can take $d_2 = 56$ satisfying $d_2 \equiv 1 \pmod 5$ and $28 \mid d_2$; take $d_2 = -20$ satisfying $d_2 \equiv 1 \pmod 7$ and $20 \mid d_3$. Hence the solution of the system of congruence equations is $x \equiv -35 \times 1 + 56 \times 2 - 20 \times 4 \equiv -3 \pmod{140}$.

4. Since there is an infinite number of prime numbers, we can take $2n$ prime numbers $p_1, \ldots, p_n$ and $q_1, \ldots, q_n$, which are different from each other. Because $p_i q_i$ $(1 \leq i \leq n)$ are pairwise coprime, it is known from the Chinese remainder theorem that the system of congruence equations

$$x \equiv -1 \pmod{p_1 q_1}, \ldots, x \equiv -n \pmod{p_n q_n}$$

has positive integer solutions. It is easy to know that $x + 1, \ldots, x + n$ meet the requirement.

5. First, prove by induction that $2^{\varphi(5^n)} - 1$ is divisible by 5^n, but not by 5^{n+1}, from which it is easy to prove the conclusion. In fact, suppose the order of 2 modulo 5^n is k. Then $k \mid \varphi(5^n)$. And the order of 2 modulo 5 is 4, so from $2^k \equiv 1 (\text{mod } 5)$ we know that $4 \mid k$, thus $k = 4 \times 5^{l-1} = \varphi(5^l)$. If $l < n$, then from $2^{\varphi(5^l)} \equiv 1 (\text{mod } 5^n)$, we obtain that $2^{\varphi(5^l)} \equiv 1 (\text{mod } 5^{l+1})$, which is contratictory to the above conclusion.

6. If all prime numbers not exceeding k are factors of m, then any reduced system modulo m satisfies the requirement.

   Consider the opposite case, and let N be the product of all prime numbers not exceeding k that are coprime with m. Then $N > 1$. Let r satisfy $(r, m) = 1$. Because $(m, N) = 1$, the system of congruence equations

$$x \equiv r(\text{mod } m), \ x \equiv 1(\text{mod } N)$$

   has solutions, and the prime factors of any solution are all greater than k (since they are coprime with N and m). When r takes $\varphi(m)$ numbers in any reduced system modulo m, the corresponding solutions x_r will form a reduced system that meets the requirement.

7. There is an infinite number of prime numbers, so we can take $(2n + 1)^2$ different prime numbers $p_{i,j}$ with $-n \leq i, j \leq n$. Then the systems of congruence equations

$$x \equiv i(\text{mod } p_{i,j}), \ -n \leq i, \ j \leq n,$$

   and

$$y \equiv j(\text{mod } p_{i,j}), \ -n \leq i, \ j \leq n$$

   have solutions $x = a$ and $y = b$. If the distance between an irreducible integer point (x', y') and the point (a, b) is less than or equal to n, then suppose $a - x' = i$ and $b - y' = j$, where $-n \leq i, j \leq n$, i.e., $x' = a - i$ and $y' = b - j$. But $p_{i,j} \mid a - i$ and $p_{i,j} \mid b - j$, so (x', y') is not an irreducible integer point.

8. Let p be any prime factor of m, and let $p^{\alpha} \parallel m$ with $\alpha \geq 1$. It is known from the condition of the problem that there is an infinite subset A_1 of A, in which none of the elements are the divisible by p. Hence there is an infinite number of elements in A_1, which have same remainder on division by mn. Let A_2 be the set of these numbers and suppose all the numbers satisfy $\equiv a(\text{mod } mn)$, where $p \nmid a$.

Because $(m, n) = 1$, we have $\left(p^\alpha, \frac{mn}{p^\alpha}\right) = 1$. From the Chinese remainder theorem, the system of congruence equations

$$\begin{cases} x \equiv a^{-1} (\mathrm{mod}\, p^\alpha), \\ x \equiv 0 \left(\mathrm{mod}\, \dfrac{mn}{p^\alpha}\right) \end{cases} \tag{1}$$

has an infinite number of positive integer solutions (where a^{-1} is the inverse of a modulo p^α). Take any positive integer solution x, and let B_p be a set of any x distinct elements in A_2. Then the sum S_p of the elements in B_p satisfies $S_p \equiv ax(\mathrm{mod}\, mn)$. Combining it with (1), we know that:

$$S_p \equiv ax \equiv 1(\mathrm{mod}\, p^\alpha), \ \ S_p \equiv \left(\mathrm{mod}\, \frac{mn}{p^\alpha}\right).$$

Now let $m = p_1^{\alpha_1} \cdots p_k^{\alpha_k}$. It can be seen from the above argument that for each p_i $(1 \le i \le k)$, a finite subset B_{p_i} of A can be selected, where $B_{p_i} \subset A \backslash B_{p_1} \bigcup \cdots \bigcup B_{p_{i-1}} (2 \le i \le k)$, such that the sum S_{p_i} of the elements in B_{p_i} satisfies

$$S_{p_i} \equiv 1(\mathrm{mod}\, p_i^{\alpha_i}), \ \ S_{p_i} \equiv \left(\mathrm{mod}\, \frac{mn}{p_i^{\alpha_i}}\right). \tag{2}$$

Consider the set $B = \bigcup_{i=1}^k B_{p_i}$. Then the sums of elements in B is $S = \sum_{i=1}^k S_{p_i}$ (because B_{p_i} are mutually disjoint). It can be seen from (2) that for $i = 1, 2, \ldots, k$,

$$S \equiv 1(\mathrm{mod}\, m), \quad \text{and} \quad S \equiv 0(\mathrm{mod}\, n).$$

Therefore the set B meets the requirement. (Please refer to the proof of the Chinese remainder theorem given in (5) of this chapter.)

9. Suppose there is at most a finite number of prime numbers that are congruent to 1 modulo p, denoted as $p_1, \ldots, p_k$. Let $M = p_1 \cdots p_k$. (If such prime numbers do not exist, then $M = 1$.) Let $a = pM$. Consider any prime factor q of

$$\frac{a^p - 1}{a - 1} = a^{p-1} + \cdots + a + 1. \tag{3}$$

Obviously, $q \mid a^p - 1$. Therefore, from the result of Exercise 2, we know that $a \equiv 1(\mathrm{mod}\, q)$, or $q \equiv 1(\mathrm{mod}\, p)$.

From the selection of q, we know that q exactly divides the right side of (1), that is $a^{p-1} + \cdots + a + 1 \equiv 0 \pmod{q}$. If $a \equiv 1 \pmod{q}$, then $a^{p-1} + \cdots + a + 1 \equiv 1 + \cdots + 1 + 1 = p \pmod{q}$, which implies that $p = q$. However, $p \nmid a^{p-1} + \cdots + a + 1$ since $p \mid a$, which leads to a contradiction. Hence $q \nmid a - 1$.

Therefore, $q \equiv 1 \pmod{p}$. Because $p_i \mid a (1 \leq i \leq k)$, this q is different from $p_1, \ldots, p_k$ (otherwise, q does not exactly divide the right side of (1)). This generates a prime number, which is different from $p_1, \ldots, p_k$, but $\equiv 1 \pmod{p}$, a contradiction.

Solution 2   A proof can be derived from the result of Example 1.

Suppose there exist prime numbers that are congruent to 1 modulo p, but there is only a finite number of such integers. Take any integer a that is greater than the largest one. Then particularly $a > p$. If such prime numbers do not exist, then take $a > p$. From Example 1 in this chapter,

$$p \mid \varphi((a!)^p - 1).$$

Let $(a!)^p - 1 = \prod p_i^{d_i}$. Then

$$\varphi((a!)^p - 1) = \prod p_i^{d_i - 1}(p_i - 1). \tag{4}$$

Because $a > p$, we have $p \nmid (a!)^p - 1$, that is, p is not equal to any p_i, so it is known from (2) that there exists i such that $p \mid p_i - 1$, namely $p_i \equiv 1 \pmod{p}$. But $p_i \mid (a!)^p - 1$, so $p_i > a$. Thus p_i is different from the finite number of the prime integers that are congruent to 1 modulo p mentioned earlier, which leads to a contradiction.

Solution 20

Indeterminate Equations (2)

1. Let the integers $c > b > a$ be the lengths of the three sides of a right triangle. If $\frac{1}{2}ab = 2u^2$ is twice a square number, then $ab = (2u)^2$. Let $(a, b) = d$, $a = a_1 d$, and $b = b_1 d$. Then $(a_1, b_1) = 1$. Thus $a_1 b_1$ is a square number, so $a_1 = r^2$ and $b_1 = s^2$ (r and s are integers). Also since $d \mid c$, let $c = dc_1$, and then from $a^2 + b^2 = c^2$, we deduce that $r^4 + s^4 = c_1^2$, which is contradictory to the conclusion of Example 2.

   If the area of the isosceles triangle whose side lengths are integers a, a, and b is a square number u^2, then Heron's formula gives that $b^4 + (2u)^4 = (4ab)^2$, which is contradictory to the result of Example 2.

2. The first conclusion does not need to use Example 4. If $4k + 3 = x^2 + y^2$, then we can use modulo 4 to obtain a contradiction.

   To prove the second assertion, we note that from Exercise 5 in Chapter 7, there is an infinite number of prime numbers $\equiv -1 \pmod 4$. Take any two such prime numbers p and $q(p \neq q)$. Then $pq \equiv 1 \pmod 4$. It is also known from Example 4 that pq can't be expressed as the sum of the squares of two integers.

   More simply, take any two prime numbers p and $q(p \neq q)$ that are congruent to 1 modulo 4. For example, take $p = 3$ and $q = 7$. Then $(pq)^n \equiv 1 \pmod 4$ for any positive odd number n. It is known from Example 4 that $(pq)^n$ cannot be expressed as the sum of the squares of two integers.

3. Note that $\binom{x}{2k} a^{2k-2} = \frac{x(x-1)}{2} \binom{x-2}{2k-2} \frac{2a^{2k-2}}{2k(2k-1)}$. Let the highest power of 2 contained in $\frac{x(x-1)}{2}$ be 2^r. Then the left side of the equality is divisible by 2^{r+1}, a contradiction.

4. The number x must be odd. Since the equation can be rearranged into $y^2 + 1 = (x+2)((x-1)^2 + 3)$, it is easy to obtain a contradiction from the lemma in this chapter. Refer to Example 5.

5. As shown in Example 5, $x \equiv 1 \pmod 4$. The equation can be rearranged into

$$y^2 + 4^2 = x^3 + 3^3 = (x+3)(x^2 - 3x + 3^2).$$

6. Let $\frac{a^2+b^2}{ab+1} = q$ (q is a positive integer). Then

$$a^2 + b^2 = q(ab + 1). \tag{1}$$

Regard (1) as a binary equation with regard to a and b. Obviously, $a = b$ will imply that $q = 1$. Let $a > b$, and in these solutions, take a solution (a_0, b_0) that minimizes a. Consider the quadratic equation with regard to x

$$x^2 - qb_0 x + b_0^2 - q = 0. \tag{2}$$

It already has a solution $x = a_0$, so the other solution is $a_1 = qb_0 - a_0$, which is an integer. If q is not a square number, then $a_0 a_1 = b_0^2 - q \neq 0$, so $a_1 \neq 0$. If a_1 is a negative number, then

$$a_1^2 - qb_0 a_1 + b_0^2 - q > -qb_0 a_1 - q \geq q - q = 0,$$

which is contradictory to that $x = a_1$ is a root of (2). Therefore, if q is not a square number, then a_1 is a positive integer. That is, (b_0, a_1) is a positive integer solution of (1) such that

$$a_1 = \frac{b_0^2 - q}{a_0} < \frac{b_0^2 - 1}{b_0} < b_0$$

and $b_0 < a_0$, so (b_0, a_1) is a "smaller" solution of (1), a contradiction. Therefore q must be a square number.

7. Prove by contradiction. Assume that the conclusion is wrong. We choose the smallest of all Pythagorean triangles with a square area. Suppose its side lengths are $x < y < z$, and $\frac{1}{2}xy$ is a square number. Similar to Example 2, now $(x, y) = 1$. Because $x^2 + y^2 = z^2$, there exist

integers $a > b > 0$ with $(a, b) = 1$, where one of a and b is odd and the other is even, such that (we may assume that y is even)

$$x = a^2 - b^2, \; y = 2ab, \; z = a^2 + b^2.$$

Since $\frac{1}{2}xy = (a-b)(a+b)ab$ is a square number, and the positive integers $a - b$, $a + b$, a and b are pairwise coprime, it follows that they are all square numbers, that is

$$a = p^2, \; b = q^2, \; a + b = u^2, \; a - b = v^2. \tag{3}$$

Hence $u^2 - v^2 = 2q^2$, i.e., $(u - v)(u + v) = 2q^2$. Because both u and v are odd numbers, $(u - v, u + v) = 2$. Hence one of $u - v$ and $u + v$ is $2r^2$, the other is $(2s)^2$, and $q^2 = 4r^2s^2$. On the other hand, from (3) we obtain that

$$p^2 = a = \frac{1}{2}(u^2 + v^2)$$

$$= \frac{1}{4}[(u + v)^2 + (u - v)^2]$$

$$= \frac{1}{4}[(2r^2)^2 + (2s)^4] = r^4 + 4s^4.$$

Thus the triangle with r^2, $2s^2$, and p as its side lengths is a right triangle, and its area is $\frac{1}{2}r^2 \cdot 2s^2 = (rs)^2$, which is a square number. But

$$(rs)^2 = \frac{q^2}{4} = \frac{b}{4} < (a^2 - b^2)ab = \frac{1}{2}xy.$$

Thus, we create a Pythagorean triangle with a smaller area. This is a contradiction.

8. Suppose the equation has a group of positive integer solutions x, y and z. Then

$$x^4 > x^4 - y^4 = z^2 > 0.$$

Take $a = x^4 - y^4 (= (x^2)^2 - (y^2)^2)$, $b = 2x^2y^2$, and $c = x^4 + y^4$. Then a, b, and c are all positive integers, and they can be treated as the three sides of a right triangle. But the area of this triangle is $\frac{1}{2}ab = (xyz)^2$ that is a perfect square, which is contradictory to the conclusion of the previous problem.

We can also use the infinite descent method directly: Assume that the equation has a positive integer solution. We choose a solution to minimize x, and then easily know that $(x, y) = 1$.

With modulo 4, we know that x is odd. If y is odd, then we know from Pythagorean theorem that there exist positive integers a and b with $(a, b) = 1$ such that

$$x^2 = a^2 + b^2, \ y^2 = a^2 - b^2, \ z = 2ab,$$

from which $a^4 - b^4 = (xy)^2$. But $a < x$, a contradiction to the minimality of x.

If y is even, then there exist positive integers a and b with $(a, b) = 1$, where one of a and b is odd and the other is even, such that

$$x^2 = a^2 + b^2, \ y^2 = 2ab, \ z = a^2 - b^2.$$

Because a and b have equal status, we can suppose a is even and b is odd. Thus we know from $y^2 = 2ab$ and $(a, b) = 1$ that there exist positive integers p and q with $(p, q) = 1$ and q being odd, satisfying $a = 2p^2$ and $b = q^2$, so

$$x^2 = 4p^4 + q^4, \ y = 2pq.$$

From the above formula and Pythagorean theorem, there are coprime positive integers r and s such that

$$2p^2 = 2rs, \ q^2 = r^2 - s^2, \ x = r^2 + s^2.$$

From $rs = p^2$ and $(r, s) = 1$, we know that $r = u^2$ and $s = v^2$, where $(u, v) = 1$. Therefore, $q^2 = r^2 - s^2$ becomes $u^4 - v^4 = q^2$. But $u = \sqrt{r} < x$, which contradicts the minimality of x.

9. (i) If there exist positive integers x, y, and z satisfying $x^4 + y^4 = 2z^2$, square both sides, and we get

$$z^4 - (xy)^4 = \left(\frac{x^4 - y^4}{2} \right)^2.$$

It can be seen from the original equation that x and y have the same parity, so the right side of the above equality is the square of an integer. From the result of Exercise 8, there must be that $\frac{x^4 - y^4}{2} = 0$, that is, $x = y$, a contradiction.

(ii) Similar to the method in (i). If the said equation has a positive integer solution, square both sides of the equation, and then it can be converted into

$$z^4 + (xy)^4 = \left(\frac{x^4 + y^4}{2} \right)^2.$$

The right side of the above equation is the square of a positive integer, which is contradictory to the conclusion of Example 2 in this chapter.

Comprehensive Exercises

1. There are $\binom{5}{1}$ ways to choose a girl first. There are $\dfrac{\binom{5}{1}\binom{4}{2}}{3} = 10$ ways if the other two are both girls; there are $\dfrac{\binom{5}{1}\binom{8}{1}\binom{4}{1}}{2} = 80$ ways if the other two are one boy and one girl; there are $\binom{5}{1}\binom{8}{2} = 140$ ways if the other two are both boys. Overall, there are $10 + 80 + 140 = 230$ ways in total.

2. According to the inclusion and exclusion principle, there are $\mathrm{P}_6^6 - 2\mathrm{P}_5^5 + \mathrm{P}_4^4 = 504$ kinds of permutations.

3. Consider the set M of n-digit numbers consisting of 1 and 2. For $a \in M$, let $M(a)$ denote the set of numbers b in M such that b and a have at most one different digit. Obviously, there are exactly n numbers different from a in $M(a)$, and $a \in M(a)$, so $|M(a)| = n + 1$.

   Suppose $a_1, \ldots, a_k \in M$ satisfy that for $i \neq j$, at least three digits are different between a_i and a_j. Then $M(a_i) \cap M(a_j) = \varnothing (i \neq j)$. Hence there are exactly $k(n+1)$ numbers in $M(a_1) \bigcup \cdots \bigcup M(a_k)$, and $|M| = 2^n$. Therefore $k(n+1) \leq 2^n$.

4. The heights of men and women arranged from high to low are represented by $b_1 \geq \cdots \geq b_{17}$ and $g_1 \geq \cdots \geq g_{17}$ respectively. If there is an i such that $|b_i - g_i| > 10$, then without loss of generality we may assume that $b_i - g_i > 10$. When $j \geq i$, we have $b_i - g_j > 10$; when $k \leq i$, there is $b_k - g_j > 10$. Originally, $b_1, \ldots, b_i$ need to match i women, but now there are at most $i - 1$ women $g_1, \ldots, g_{i-1}$, a contradiction.

5. There are n chess players in the game. Suppose the player who has played the least number of games has played m games. Now we use two ways to estimate the sum S of the points (in the game). On one hand, $S \le kn$; on the other hand, for each game, the total score is increased by 1 point, and one game is played by two people, so the number of games is at least $\frac{mn}{2}$ (the minimality of m was used), thus $S \ge \frac{mn}{2}$. Therefore $m \le 2k$.

6. Consider the player A who won the most. It is known that there must be a player C who beat A, and among those who lost to A, there must be someone who beat C (otherwise, C won more times than A). Suppose this person is B. Then A, B, and C meet the requirement.

7. $\sum_{i=1}^{n} a_i^3 - \sum_{i=1}^{n} a_i = \sum_{i=1}^{n} (a_i^3 - a_i)$. (Refer to Remark 2 of Chapter 6.)

8. Let the sequence be $\{a_n\}(n \ge 1)$. Then a_1, a_2, and a_3 are congruent to 1, 2 and 1 (modulo 3) respectively, and $S(n)$ denotes the sum of the digits of n. From $a_1 \equiv a_3 \,(\mathrm{mod}\,3)$, we obtain $S(a_1) \equiv S(a_3)(\mathrm{mod}\,3)$. Hence $a_2 = a_1 + S(a_1) \equiv a_3 + S(a_3) = a_4(\mathrm{mod}\,3)$. It can be seen that $\{a_n\}(n \ge 1)$ modulo 3 is periodic, and since $123456 \equiv 0(\mathrm{mod}\,3)$, it is not in the sequence.

9. There are many solutions to this problem. For example, from $\binom{2n+1}{n+1} = \frac{2n+1}{n+1}\binom{2n}{n}$, we get $(n+1)\binom{2n+1}{n+1} = (2n+1)\binom{2n}{n}$, and $(n+1, 2n+1) = 1$, so $n+1 \,\Big|\, \binom{2n}{n}$.

We can also derive more directly from the above result:

$$\frac{2n+1}{n+1}\binom{2n}{n} = \binom{2n+1}{n+1} = \binom{2n}{n+1} + \binom{2n}{n}.$$

Therefore, $\frac{1}{n+1}\binom{2n}{n} = \binom{2n}{n} - \binom{2n}{n+1}$ is the difference between two integers, so it is an integer.

10. The difficulty of this problem is to find out that it is actually an algebraic problem: By the binomial theorem, we know that

$$(x+y)^n + (x-y)^n = 2 \sum_{0 \le i \le \frac{n}{2}} \binom{n}{2i} y^{2i} x^{n-2i}.$$

From this, the sum is

$$\frac{1}{2}\left(\left(1+\sqrt{3}\right)^{2n} + \left(1-\sqrt{3}\right)^{2n}\right) = \frac{1}{2}\left(\left(4+2\sqrt{3}\right)^{n} + \left(4-2\sqrt{3}\right)^{n}\right)$$

$$= 2^{n-1}\left(\left(2+\sqrt{3}\right)^{n} + \left(2-\sqrt{3}\right)^{n}\right)$$

$$= 2^{n}\sum_{0\le l\le \frac{n}{2}}\binom{n}{2i}3^{i}\cdot 2^{(n-2i)},$$

which is obviously a multiple of 2^{n}.

11. Since $m\binom{m+n}{m} = (m+n)\binom{m+n-1}{m-1}$ and $(m, m+n) = 1$, we obtain $m\left|\binom{m+n-1}{m-1}\right.$.

12. By the division with remainder, $xm = \left[\frac{xm}{n}\right]n + r_x$ (refer to Remark 5 of Chapter 6). For $1 \le x, y \le n-1$ and $x \ne y$, since $(m, n) = 1$, it's easy to know the remainders $r_x \ne r_y$, so $r_1, \ldots, r_{n-1}$ is an arrangement of $1, 2, \ldots, n-1$. Thus, the result is easy to prove.

13. Let $f(x) = x^5 + x^4 + 1$, ω be a cubic root of unity, and $\omega \ne 1$. Then $f(\omega) = f(\omega^2) = 0$, so $x^2 + x + 1|f(x)$. In particular, $n^2 + n + 1|f(n)$. Because $f(n) > n^2 + n + 1$ $(n > 1)$, the number $f(n)$ is not prime when $n > 1$.

14. We can prove (a stronger conclusion): $a_n \not\equiv 0 \pmod 4$ (refer to Remark 2 of Chapter 9.) In fact, it is easy to know by mathematical induction that $\{a_n\}$ modulo 4 is periodic, and its recurring period can be determined. Modulo 3 can also solve this problem.

15. For $h \in H$, let $x = (1-h)^{-1}$. Then $h = \frac{x-1}{x}$, and $h^n = -1$ is equivalent to $(x-1)^n + x^n = 0$, i.e., $2x^n + \sum_{k=1}^{n}(-1)^k\binom{n}{k}x^{n-k} = 0$.

Let $S = \{x = (1-h)^{-1}|h \in H\}$. Then S corresponds to H one by one, so the sum is equal to $\sum_{x\in S}x^2 = \binom{n}{2}^2 - 2(-1)^2\frac{\binom{n}{2}}{2} = \frac{n(2-n)}{4}$ (using Vieta's theorem).

16. For any k, we have $4 = k^2 - (k+1)^2 - (k+2)^2 + (k+3)^2$. Therefore, if n has such a representation, then

$$n + 4 = \varepsilon_1 1^2 + \varepsilon_2 2^2 + \cdots + \varepsilon_m m^2 + (m+1)^2$$

$$- (m+2)^2 - (m+3)^2 + (m+4)^2,$$

that is, $n + 4$ also has such a representation. Because 1, 2, 3 and 4 all can be expressed in the form, it follows that every integer n can be so expressed.

17. Let $a_1 < \cdots < a_n$ be already determined. We prove that we can find a_{n+1} such that $A_{n+1} = a_1^2 + \cdots + a_n^2 + a_{n+1}^2$ is divisible by $B_{n+1} = a_1 + \cdots + a_n + a_{n+1}$ (and $a_{n+1} > a_n$). It can be seen from $A_{n+1} = A_n + (a_{n+1} - B_n)(a_{n+1} + B_n) + B_n^2$ that if $A_n + B_n^2$ is divisible by B_{n+1}, then A_{n+1} is divisible by B_{n+1}. Hence just take $a_{n+1} = A_n + B_n^2 - B_n$ (now $A_n + B_n^2 = B_{n+1}$, and obviously $a_{n+1} > a_n$).

18. Let $p_1, \ldots, p_n$ be prime numbers of the form $4k + 3$ that are different from each other (there is an infinite number of them, see Exercise 5 in Chapter 7). From the Chinese remainder theorem, the system of congruence equations

$$x \equiv p_i - i \pmod{p_i^2} \; (i = 1, 2, \ldots, n)$$

has solutions. If we take a positive solution x, then p_i just appears exactly once in $x + i$, so $x + i$ can't be expressed as the sum of the squares of two integers $(1 \le i \le n)$ (refer to Remark 8 of Chapter 19 and Example 4 of Chapter 20).

19. Let the number of 1 in M be greater than $\frac{1}{2}(4n - 3)^2$. If M does not contain a $2 \times n$ sub-table in which all elements are 1, then we use two methods to calculate the number of 2×1 sub-tables T in which all elements are 1.

Calculate by rows. Because there is no $2 \times n$ sub-table of M containing 1 only, the number of T contained in the $2 \times (4n - 3)$ table composed of any two rows is less than or equal to $n - 1$.

Therefore, the number of T contained in M is less than or equal to $(n - 1) \binom{4n - 3}{2}$.

Calculate by columns. Suppose there are d_j ones $(1 \le j \le 4n - 3)$ in column j in M, then the number of T contained in M is

$$\sum_{j=1}^{4n-3} \binom{d_j}{2} = \frac{1}{2} \sum_{j=1}^{4n-3} d_j^2 - \frac{1}{2} \sum_{j=1}^{4n-3} d_j.$$

We have assumed that $\sum_{j=1}^{4n-3} d_j > \frac{1}{2}(4n-3)^2$, so we can suppose

$$\sum_{j=1}^{4n-3} d_j = \frac{1}{2}(4n-3)^2 + \frac{1}{2}.$$

It is easy to know that when the difference between these d_j is the smallest, $\sum_{j=1}^{4n-3} d_j^2$ reaches its minimum, and since

$$\frac{1}{2}(4n-3)^2 + \frac{1}{2} = 8n^2 - 12n + 5 = (2n-2)(4n-3) + 2n - 1,$$

we see that when there are $2n-2$ numbers of $2n-2$ and $2n-1$ numbers of $2n-1$ among these d_j the numbers $\sum d_j^2$ is minimal, that is

$$\text{the number of } T \geq \binom{2n-2}{2}(2n-2) + \binom{2n-1}{2}(2n-1).$$

Hence

$$(n-1)\binom{4n-3}{2} \geq (2n-2)\binom{2n-2}{2} + (2n-1)\binom{2n-1}{2},$$

which is impossible.

20. First prove (i) $B(2n+1) = B(2n)$ and (ii) $B(2n) = B(2n-1) + B(n)$.
    (i) is obvious, because $2n+1 = 2^{i_1} + 2^{i_2} + \cdots + 2^{i_r} + 1$ ($i_1 \geq i_2 \geq \cdots \geq i_r \geq 0$) uniquely corresponds to a partition of $2n$: $2n = 2^{i_1} + 2^{i_2} + \cdots + 2^{i_r}$.

    To prove (ii), write the set of partitions of $2n$ (as the sum of powers of 2) as $A_n \bigcup B_n$, where the elements in A_n are the following partitions:

$$2n = 2^{i_1} + 2^{i_2} + \cdots + 2^{i_r} + 1, \ i_1 \geq i_2 \geq \cdots \geq i_r \geq 0,$$

while the elements in B_n are the following partitions:

$$2n = 2^{i_1} + 2^{i_2} + \cdots + 2^{i_r}, \ i_1 \geq i_2 \geq \cdots \geq i_r \geq 1.$$

Then $A_n \bigcap B_n = \emptyset$, and an element in A_n uniquely corresponds to a partition of $2n-1$, while an element in B_n uniquely corresponds to a partition of n. Hence (ii) holds. From (i) and (ii), it is easy to prove that $B(n)$ is even by mathematical induction.

21. From $r^{m-1} + \cdots + r + 1 = 0$, we obtain that

$$-1 = r\left(1 + r + \cdots + r^{2^k - 1}\right) = r\left(1 + r\right)\left(1 + r^2\right) \cdots \left(1 + r^{2^{k-1}}\right)$$

$$= \left(r + r^2\right)\left(1 + r^2\right) \cdots \left(1 + r^{2^{k-1}}\right). \tag{1}$$

(See Example 4 in Chapter 11.) Note that $r + r^2 = r^2 + r^{m+1} = r^2 + r^{2(2^{k-1}+1)}$. Therefore, each term on the right side of formula (1) is the sum of the squares (with integer coefficients), and so is their product. This generates the integer coefficient polynomials $f(x)$ and $g(x)$ such that $f^2(r) + g^2(r) = -1$ (refer to Exercise 8 in Chapter 12).

22. For any $m > 1$, we have $x_m \equiv 1 = x_1 \pmod{x_m - 1}$, and for any integers u and v, there is $(u - v) \mid p(u) - p(v)$. Hence $p(x_m) - p(x_1) \equiv 0 \pmod{x_m - 1}$, i.e., $x_{m+1} \equiv x_2 \pmod{x_m - 1}$. In this way, it can be seen that the sequence $\{x_n\}$ modulo $x_m - 1$ is periodic, which is $x_1, x_2, \ldots, x_{m-1}$.

From the known condition, for $N = x_m - 1$, there exists x_k such that $x_m - 1 \mid x_k$. Combining it with the previous result, we can suppose $1 \le k \le m - 1$. But $\{x_n\}$ is strictly increasing and $x_m - 1 \ge x_{m-1}$, so $k = m - 1$, i.e., $x_m - 1 \mid x_{m-1}$, hence $x_{m-1} \ge x_m - 1$. Thus $x_m - 1 = x_{m-1}$, i.e., $p(x_{m-1}) - 1 = x_{m-1}$. Therefore $p(x) = x + 1$ has an infinite number of different roots, and consequently $p(x) = x + 1$.

23. When m is odd, $a^m - 1 = (a - 1)(a^{m-1} + \cdots + a + 1)$, where the latter factor is odd, so $2^m \mid a^m - 1$ implies that $2^m \mid a - 1$. There are only finitely many such m. When m is even, we only need to prove (replace m with $2m$) that there is at most a finite number of m such that $2^m \mid \left(a^2\right)^m - 1$. This is equivalent to replacing a with a^2 in the original problem, so we can assume that the given a satisfies $a \equiv 1 \pmod 8$.

Let $a - 1 = 2^l a_1$, $2 \nmid a_1$, and $l \ge 3$. The key is to prove that when $m \ge l$, the order of a modulo 2^m is 2^{m-l}. For this purpose, it can be proved by mathematical induction first that when $m \ge l$, the integer $a^{2^{m-l}} - 1$ is divisible by 2^m, but not by 2^{m+1} (leaving the details to the readers).

Let r be the order of a modulo 2^m ($m \ge l$). Then $r \mid 2^{m-l}$, so $r = 2^t$. If $t < m - l$, then from $a^{2^t} \equiv 1 \pmod{2^m}$, we deduce that $a^{2^t} - 1$ is divisible by 2^{t+l+1}, which is contrardictory to the previous conclusion, so $t = m - l$, i.e., $r = 2^{m-l}$.

Now it is not difficult to prove the conclusion of this problem. If m satisfies $2^m \mid a^m - 1$ and $m \geq l$, then $2^{m-l} \mid m$. Hence either $m < l$ or $2^{m-l} \leq m$, and there are obviously finitely many such m.

24. It can be seen from the formula (6) in Chapter 14 that

$$\sum_{\varepsilon_j = 0,1} (-1)^{\varepsilon_1 + \cdots + \varepsilon_n} (\varepsilon_1 x_1 + \cdots + \varepsilon_n x_n)^i$$

$$= \begin{cases} 0 & \text{if } i < n; \\ (-1)^n n! x_1 \cdots x_n, & \text{if } i = n. \end{cases}$$

Take $x_1 = 1, x_2 = 2, \ldots, x_n = 2^{n-1}$. Then when $\varepsilon_j = 0$ and 1, the expression $\varepsilon_1 + 2\varepsilon_2 + \cdots + 2^{n-1}\varepsilon_n$ gives exactly the values $0, 1, 2, \ldots, 2^n - 1$. Thus when $i < n$, the sum is 0; when $i = n$, the sum is $(-1)^n n! 2^{\frac{n(n-1)}{2}}$ (when $i < n$, it is not difficult to prove the result by induction on i).

25. Let $f(k) = (k + a_1) \cdots (k + a_n)$, which can be regarded as the degree-n polynomial in k, (with the leading coefficient 1). Then

$$\sum_{k=0}^{n} (-1)^{n-k} \binom{n}{k} f(k) = n!$$

(see formula (5) in Chapter 14). Because $f(0)$ exactly divides $f(k)(k = 0, 1, \ldots, n)$, we have $f(0) \mid n!$, so $f(0) \leq n!$ (note that $f(0)$ is a positive number). But $f(0)$ is the product of n different positive integers, so $a_1, \ldots, a_n$ must be a permutation of $1, 2, \ldots, n$.

26. The answer is $n = 2^k$, where k is a non-negative integer.

First, if n has an odd divisor r greater than 1, then since $2^r - 1$ exactly divides $2^n - 1$, the integer $2^r - 1$ exactly divides $m^2 + 9$. Because $3 \nmid 2^r - 1$ and $2^r - 1 \equiv -1 \pmod 4$, the number $2^r - 1$ must have a prime divisor $p > 3$ that is $\equiv -1 \pmod 4$, so $m^2 + 3^2 \equiv 0 \pmod p$. Thus p exactly divides m and 3, so $p = 3$, a contradiction (refer to Remark 8 in Chapter 20). Therefore, n must be a power of 2.

Now we prove that $n = 2^k$ meets the requirement. Let $F_i = 2^{2^i} + 1$ $(i \geq 0)$. Then (see the solution of Example 1 in Chapter 6)

$$2^{2^k} - 1 = F_{k-1} F_{k-2} \cdots F_1 F_0.$$

Note that the congruence equation

$$x^2 \equiv -1 \pmod{F_i}$$

has a solution $x \equiv 2^{2^{i-1}} \pmod{F_i}$, where $i \geq 1$. And $(F_i, F_j) = 1$ for $i \neq j$, (Example 4 in Chapter 6). Hence by the Chinese remainder theorem, the system of congruence equations

$$x \equiv 2^{2^{i-1}} \pmod{F_i}, \quad i = 1, 2, \ldots, k-1$$

has a solution x_0. Obviously, x_0 satisfies $x_0^2 \equiv -1 \pmod{F_i}(i = 1, \ldots, k-1)$, so $x_0^2 + 1$ is divisible by $F_{k-1}F_{k-2}\cdots F_1$ (we used the fact that F_i are pairwise coprime once again).

We take $m = 3x_0$. Then $m^2 + 9 = 9(x_0^2 + 1)$ is divisible by $F_{k-1}\cdots F_1 F_0 = 2^{2^k} - 1$.

27. For any integer $x > 1$, we use p_x to represent the minimum prime factor of x. The proof needs the following lemma:

Lemma If p is prime, which makes $p \mid 2^x + 1$, and $p < p_x$, then $p = 3$.

Proof. Obviously, p is an odd prime number. Assume that the order of 2 modulo p is d. Then it is known from $2^{p-1} \equiv 1 \pmod{p}$ that $d \mid p - 1$. Also $p \mid 2^x + 1$, so $2^{2x} \equiv 1 \pmod{p}$. Thus $d \mid 2x$, so $d \mid (2x, p - 1)$. But $p < p_x$, hence $p - 1$ cannot have prime divisors that exactly divide x, which implies that $(2x, p - 1) = 2$, i.e., $d \mid 2$. Therefore, $d = 2$, so $p = 3$.

Now let's solve the problem. If there are integers a, b, and c that meet the requirement of the problem, then a, b, and c are all odd numbers, and p_a, p_b, and p_c are different from each other. We may assume that p_a is the smallest of the three. Because $a \mid 2^b + 1$, we have $p_a \mid 2^b + 1$. However, since $p_a < p_b$, the lemma above implies that $p_a = 3$. Denote $a = 3a'$.

Let p be the minimum prime factor of a', b, and c. We prove that $p \mid b$. Suppose p exactly divides a'. It can be seen from $(a, c) = 1$ and $p \leq p_c$ that $p < p_c$. But $a \mid 2^c + 1$, so $p \mid 2^c + 1$. Then from the above lemma, $p = 3$, so $9 \mid a$. Thus, $9 \mid 2^c + 1$, hence $9 \mid 2^{2c} - 1$. But it is easy to verify that the order of 2 modulo 9 is 6, thus $6 \mid 2c$, i.e., $3 \mid c$, which is contradictory to that a and c are coprime. Therefore, $p \nmid a'$. Similarly, we can prove that $p \nmid c$. Consquently, $p \mid b$.

Suppose the order of 2 modulo p is d. Similarly as the lemma above, it can be proved that $d \mid (2a, p-1)$, i.e., $d \mid (6a', p-1)$. (Use $p \mid b$ and $b \mid 2^a + 1$.) From the selection of p, we see that $p - 1$ and a' are coprime, from which $(6a', p-1)$ exactly divides 6, i.e., $d \mid 6$. Thus $p \mid 2^6 - 1$, hence $p = 7$. However, since

$$2^a + 1 = \left(2^3\right)^{a'} + 1 \equiv 2 (\mathrm{mod}\,7),$$

we see that p does not exactly divide $2^a + 1$ a contradiction. □

28. Suppose $a_1, \ldots, a_{k+1}$ are not congruent to each other modulo $n + k$. We consider the following three groups of numbers (here $S = a_1 + \cdots + a_{k+1}$):

(i) $a_1, a_2, \ldots, a_{k+1}$;
(ii) $S - a_1, S - a_2, \ldots, S - a_{k+1}$;
(iii) $S, S + a_{k+2}, S + a_{k+2} + a_{k+3}, \ldots, S + a_{k+2} + \cdots + a_n$.

There are $(k+1) + (k+1) + n - (k+2) + 2 = n + k + 2$ numbers, so there must be two numbers congruent modulo $n + k$. Obviously, the numbers in (i) are not congruent to each other, and neither are the numbers in (ii). Note that we can assume that no two numbers are congruent to each other in (iii), no number in (i) is congruent to a number in (iii), and also no number in (ii) is congruent to a number in (iii), since otherwise, the conclusion is true. Therefore, there must be two numbers in (i) and (ii) that are congruent modulo $n + k$. If there exist $i \neq j$ such that a_i and $S - a_j$ are congruent modulo $n + k$, then the conclusion is obviously true. Thus, there must exists an i such that a_i and $S - a_i$ are congruent modulo $n + k$ ($1 \le i \le k + 1$).

If there are at most two a_i satisfying $a_i \equiv S - a_i (\mathrm{mod}\, n + \mathrm{k})$, delete such a_i from (i). Then there are at least $n + k$ numbers left. Since now there is no number in (i) that is congruent to a number in (ii) modulo $n + k$, no two in the remaining numbers that are congruent modulo $n + k$, so there are exact $n + k$ numbers left. In particular, one of them is congruent to modulo $n + k$. Therefore, the conclusion is true.

If at least three a_i satisfy the above congruence, supposed as a_1, a_2, and a_3, then $2a_1 \equiv 2a_2 \equiv 2a_3 (\mathrm{mod}\, n + k)$. If $n + k$ is odd, then this gives $a_1 \equiv a_2 \equiv a_3 (\mathrm{mod}\, n + k)$, which is contradictory to the previous assumption about $a_1, \ldots, a_{k+1}$; if $n + k$ is even, then $a_1 \equiv a_2 \equiv a_3 \left(\mathrm{mod}\, \frac{n+k}{2}\right)$, i.e., $a_1 = a_2 + \frac{n+k}{2} M_1$ and $a_1 = a_3 + \frac{n+k}{2} M_2$, where M_1 and M_2 are both odd (otherwise $a_1 \equiv a_2$ or $a_1 \equiv a_3 (\mathrm{mod}\, n + k)$). Hence

$$a_2 = a_3 + \frac{n+k}{2}(M_2 - M_1) \equiv a_3 (\mathrm{mod}\, n + k),$$

which leads to a contradiction.

29. Let $t = \frac{r^2+s^2+k}{rs}$. From the recurrence relation in the problem (for $n \geq 2$), we obtain that

$$a_{n+2}a_n = a_{n+1}^2 + k, \quad a_{n+1}a_{n-1} = a_n^2 + k.$$

Subtracting one equality from the other in the above gives that

$$\frac{a_{n+2} + a_n}{a_{n+1}} = \frac{a_{n+1} + a_{n-1}}{a_n}.$$

From this, we can recursively obtain (for $n \geq 1$) that

$$\frac{a_{n+2} + a_n}{a_{n+1}} = \cdots = \frac{a_3 + a_1}{a_2} = \frac{r^2 + s^2 + k}{rs} = t,$$

so $a_{n+2} = ta_{n+1} - a_n$. Therefore, if t is an integer, then all a_n are integers by mathematical induction.

Conversely, if all a_n are integers, then t is a rational number. Let $t = \frac{u}{v}$, where $(u, v) = 1$ and $v > 0$. From $a_3 = \frac{u}{v}a_2 - a_1$, and note that a_1, a_2 and a_3 are integers, we see that $v | a_2$. Similarly, $v \mid a_i$ (for $i = 2, 3, \ldots$). Then from $a_4 = \frac{u}{v}a_3 - a_2$, and that v exactly divides a_2, a_3, and a_4, we know that $v^2 \mid a_3$. Similarly, $v^2 \mid a_i$ for $i = 3, 4, \ldots$. In this way, it is easy to prove that $v^n \mid a_{n+i}$ $(i = 1, 2, \ldots)$ for any positive integer n.

Now, if $v \neq 1$, then we can take a natural number N large enough to make $v^N > |k|$. In the equality $k = a_{N+1}a_{N+3} - a_{N+2}^2$, the right side is divisible by v^N, but obviously the left side is not divisible by v^N, which is a contradiction. Therefore $v = 1$, i.e., t is an integer.

30. We first prove that when $m = 5^n$, there exists an n-digit number that is a multiple of 5^n and its digits are all odd.

Prove by mathematical induction. When $n = 1$, the one-digit number 5 meets the requirement. Assume that $\overline{a_k a_{k-1} \cdots a_1}$ is a k-digit number that meets the requirement for $n = k$. We need to prove that there exists $a_{k+1} \in \{1, 3, 5, 7, 9\}$ such that the $(k+1)$-digit number

$$\overline{a_{k+1}a_k \cdots a_1} = a_{k+1} \times 10^k + \overline{a_k \cdots a_1}$$

is a multiple of 5^{k+1}. Since $\overline{a_k \cdots a_1}$ is a multiple of 5^k, the right side of the above formula is $5^k(a_{k+1} \times 2^k + l)$, where l is odd. It is easy to know that 2^k modulo 5 is 2, 4, 3, and 1 periodically when $k = 1, 2, \ldots$. Therefore, it is easy to verify that for any odd number l and positive integer k, there always exists a_{k+1} such that $a_{k+1} \times 2^k + l$ is divisible by 5. Thus, $\overline{a_{k+1}a_k \cdots a_1}$ is divisible by 5^{k+1}.

Now let m be any odd number. Decompose m into $5^k \cdot m_1$, where $k \geq 0$ and m_1 is a positive integer coprime to 10. From the above conclusion, we can first construct a multiple of 5^k, namely $\overline{a_k a_{k-1} \cdots a_1}$, each digit of which is odd. (When $k = 0$, take this number as 1.) Consider the sequence $\{A_n\}$ ($n \geq 1$), where A_n is the (nk-digit) number generated by writing the number $\overline{a_k \cdots a_1}$ (from left to right) n times repeatedly, that is, A_n has the form

$$\overline{a_k \cdots a_1 a_k \cdots a_1 \cdots a_k \cdots a_1}.$$

There must be two numbers in $\{A_n\}$, the difference between which is divisible by m_1; this difference can be expressed in the form $A_i \times 10^j$ ($i \geq 1$ and $j \geq 1$). Because m_1 and 10 are coprime, m_1 exactly divides A_i. And A_i is a multiple of 5^k, so A_i is a multiple of $5^k \times m_1$ (we used the fact that m_1 and 5 are coprime once again). Of course, the digits of A_i are all odd.

31. Let $\sum S(a^*)$ denote the sum of the corresponding $S(a^*)$ of the $n!$ permutations a^*. If the conclusion is wrong, then we can generate a contradiction by calculating $\sum S(a^*)$.

   On one hand, $n!$ does not exactly divide $S(a^*) - S(b^*)$ for $a^* \neq b^*$ and there are $n!$ different permutations, so the remainders of $S(a^*)$ modulo $n!$ are exactly $0, 1, \ldots, n! - 1$. Therefore

$$\sum S(a^*) \equiv \frac{(n! - 1)n!}{2} \pmod{n!}. \tag{2}$$

   On the other hand, in the $n!$ permutations $a^* = (a_1, \ldots, a_n)$, there are $(n-1)!$ permutations of $a_1 = k$, where $k = 1, 2, \ldots, n$. Therefore, the coefficient of c_1 in $\sum S(a^*)$, is

$$(n-1)!(1 + 2 + \cdots + n) = \frac{(n+1)!}{2}.$$

The same conclusion is true for the rest of c_i, from which

$$\sum S(a^*) = \frac{(n+1)!}{2} \sum_{i=1}^{n} c_i. \tag{3}$$

Combining (1) with (2), we obtain that

$$\frac{(n+1)!}{2} \sum_{i=1}^{n} c_i \equiv \frac{(n! - 1)n!}{2} \pmod{n!}.$$

Because n is odd, the left side of the above formula is divisible by $n!$, while the right side is not a multiple of $n!$, for $n > 1$, a contradiction.

32. We only need to consider the problem modulo 2. For any ternary integer array (A, B, C), there are eight possibilities (A_j, B_j, C_j) in the sense of modulo 2, where A_j, B_j, and C_j are 0 or 1($j = 0, 1, \ldots, 7$), and $A_0 = B_0 = C_0 = 0$. It is easy to verify that for any ternary integer array (a_i, b_i, c_i) (here at least one of a_i, b_i, and c_i is an odd number), there are exactly four even numbers and four odd numbers among the eight numbers

$$a_i A_j + b_i B_j + c_i C_j \ (j = 0, 1, \ldots, 7).$$

Hence there are exactly three even numbers and four odd numbers in the seven numbers

$$a_i A_j + b_i B_j + c_i C_j \ (j = 1, \ldots, 7).$$

From this, we know that there are exactly $4n$ odd numbers in the following numbers

$$a_i A_j + b_i B_j + c_i C_j \ (j = 1, \ldots, 7, \ i = 1, \ldots, n).$$

However, the above numbers can be divided into seven groups. The numbers in group j are $a_i A_j + b_i B_j + c_i C_j \ (i = 1, \ldots, n)$ with $1 \leq j \leq 7$. Therefore, there must be a group, in which the number of odd integers greater than or equal to $\frac{4}{7}n$.

33. Let $S = \{a_1, a_2, \ldots, a_{2000}\}$. We only need to prove that S has two non-empty subsets C and D satisfying the conditions (i), (ii), and (iii). This is because if there exist such C and D, then by taking $A = C/D$ and $B = D/C$, we see that A and B meet the requirement in the problem.

To prove it, let each non-empty subset E of S correspond to an (ordered) ternary array:

$$E \to (f_0(E), f_1(E), f_2(E)),$$

where $f_i(E)$ denotes the sum of the ith powers of the elements in E ($i = 0, 1$, and 2). We prove that this mapping is not an injection, i.e., there are subsets C and D such that $f_i(C) = f_i(D) \ (i = 0, 1, \text{ and } 2)$. Such C and D meet the above requirement (refer to Remark 5 in Chapter 3).

Because $|S| = 2000$ and the maximum number in S is less than 10^{100}, for any E,

$$f_0(E) \leq 2000, \ f_1(E) \leq 2000 \times 10^{100}, \ f_2(E) \leq 2000 \times 10^{200}.$$

Therefore, the number of different images does not exceed

$$(2000)^3 \times 10^{300} < \left(2^{11}\right)^3 \times \left(2^{10}\right)^{100} = 2^{1033}.$$

If the above mapping is an injection, then the number of non-empty subsets of S should not exceed the number of image elements, that is, $2^{2000} - 1 < 2^{1033}$ a contradiction.

34. Let $S_{n-1} = \{a_1, a_2, \ldots, a_k\}$, where $a_1 < a_2 < \cdots < a_k$. It is obvious that the minimum number in S_n is a_1, and the maximum number is $a_k + 1$. We note that $a_i \notin S_n$ for $i \geq 2$ if and only if $a_i - 1 \in S_{n-1}$, i.e., $a_{i-1} = a_i - 1$. Therefore,

$$\sum_{a \in S_n} x^a \equiv (1 + x) \sum_{a \in S_{n-1}} x^a \pmod{2}.$$

From this, we recursively obtain that

$$\sum_{a \in S_n} x^a \equiv (1 + x)^n \sum_{a \in S_0} x^a \pmod{2}.$$

On the other hand, from the congruence $(1 + x)^2 \equiv 1 + x^2 \pmod{2}$ and by induction on n, it is easy to get that $(1 + x)^{2^n} \equiv x^{2^n} + 1 \pmod{2}$. Therefore, if N is a power of 2, then

$$\sum_{a \in S_n} x^a \equiv (1 + x^N) \sum_{a \in S_0} x^a \pmod{2}.$$

Finally, take N as a number greater than the maximum number in S_0. Then from the above formula, we obtain that

$$S_N = S_0 \bigcup \{N + a : a \in S_0\}.$$

35. The given integers are treated as the vertices of a graph. If two integers are coprime to each other, then an edge is connected between the corresponding vertices of the two integers. Our conclusion is that if the number of edges in the graph is greater than 94, then there must be a cyclic subgraph of length 4, that is, there are vertices a, b, c, and d such that a is connected to b, b is connected to c, c is connected to d, and d is connected to a.

We prove the following more general result of graph theory:
If a graph G has n vertices and does not contain a cyclic subgraph of length 4, then the number m of edges in G satisfies

$$m \leq \frac{n}{4}(1 + \sqrt{4n - 3}).$$

To prove it, let V be the set of vertices of G, and for any $u \in V$, let $d(u)$ denote the degree of u (i.e., the number of edges incident to u). The main point of the proof is to count the number of the figures in G as shown below (here x is connected with y and z with $y \neq z$, regardless of whether y and z are connected).

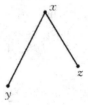

For each $u \in V$, there are exactly $\binom{d(u)}{2}$ above figures, so the number of such figures is $\sum_{u \in V} \binom{d(u)}{2}$. On the other hand, every pair of vertices $\{y, z\}$ has at most one x adjacent to both y and z (otherwise, there is a cyclic subgraph of length 4 in G). Therefore (refer to Remark 4 in Chapter 3),

$$\sum_{u \in V} \binom{d(u)}{2} \leq \binom{n}{2},$$

that is

$$\sum_{u \in V} d^2(u) \leq n(n-1) + \sum_{u \in V} d(u).$$

From Cauchy's inequality,

$$\left(\sum_{u \in V} d(u) \right)^2 \leq n \sum_{u \in V} d^2(u).$$

Also we know well that $\sum_{u \in V} d(u) = 2m$, so

$$4m^2 \leq n^2(n-1) + 2nm.$$

Therefore, $m \leq \frac{n}{4}(1 + \sqrt{4n-3})$.

36. We prove that there is a sequence $\{a_n\}$ of positive integers such that $a_1^2 + \cdots + a_n^2$ are all the squares of odd numbers. Take $a_1 = 1$. If $a_1, \ldots, a_n$ have been determined, let $a_1^2 + \cdots + a_n^2 = (2k+1)^2$ and take $a_{n+1} = 2k^2 + 2k$. Then from $(2k+1)^2 + (2k^2 + 2k)^2 = (2k^2 + 2k + 1)^2$, we know that $a_1^2 + \cdots + a_n^2 + a_{n+1}^2$ is also the square of an odd number.

37. For $i = 1, \ldots, k-1$, the product $(i+1) \cdots \cdots (i+k)$ of k consecutive integers is divisible by $k!$, so $i^k \equiv b_1 i^{k-1} + \cdots + b_k \pmod{k!}$, where $b_1, \ldots, b_k$ are all integers. Thus, from the condition, we know that

$$\sum_{i=1}^{n} a_i i^k \equiv \sum_{i=1}^{n} (b_1 a_i i^{k-1} + \cdots + b_k a_i)$$

$$= b_1 \left(\sum_{i=1}^{n} a_i i^{k-1} \right) + \cdots + b_k \left(\sum_{i=1}^{n} a_i \right)$$

$$\equiv 0 \pmod{k!}.$$

38. Suppose for $x = 0, 1, \ldots$, the values of the non-constant integer coefficient polynomial $f(x) = a_n x^n + \cdots + a_1 x + a_0$ have only a finite number of different prime factors $p_1, \ldots, p_k$. Then $a_0 \neq 0$. Take $x = p_1 \cdots p_k a_0 t$, and let the integer t be sufficiently large. Then $f(x)$ can be expressed as follows:

$$f(x) = a_0(p_1 \cdots p_k A_t + 1),$$

where A_t is an integer that depends on t, and when t is large enough, $|A_t| > 1$. Hence $p_1 \cdots p_k A_t + 1$ has a prime factor p, and obviously p is different from $p_1, \ldots, p_k$. This contradicts the previous assumption.

39. We can take $f(x) = x^n + 210(x^{n-1} + \cdots + x^2) + 105x + 12$. According to Eisenstein's criterion (take $p = 3$), $f(x)$ can't be decomposed into the product of two non-constant integer coefficient polynomials. ($210 = 2 \times 3 \times 5 \times 7$, $105 = 3 \times 5 \times 7$ and $12 = 2^2 \times 3$, so 3 exactly divides the coefficients of $x, \ldots, x^{n-1}$, 3^2 can't exactly divide the constant term, and 3 can't exactly divide the leading coefficient 1.)

For any integer x, because $f(x) = x(x^{n-1}+105)+210(x^{n-1}+\cdots+x)+12$, the value of $f(x)$ is always even. Therefore, $|f(x)|$ can't take odd prime values.

If $f(x) = 2$ has an integer solution x, then

$$f_1(x) = x^n + 210(x^{n-1} + \cdots + x^2) + 105x + 10$$

has an integer root. However, according to Eisenstein's criterion (take $p = 5$), $f_1(x)$ cannot be decomposed into the product of two non-constant integer coefficient polynomials, let alone an integer coefficient linear factor. This contradicts that $f_1(x) = 0$ has an integer root.

If $f(x) = -2$ has an integer root x, then

$$f_2(x) = x^n + 210(x^{n-1} + \cdots + x^2) + 105x + 14$$

has an integer root. However, according to Eisenstein's criterion (take $p = 7$), $f_2(x)$ cannot be decomposed into the product of two non-constant integer coefficient polynomials, a contradiction. Therefore, $|f(x)|$ can't take even prime 2.

40. Since there is an infinite number of prime numbers, we can (inductively) select an infinite prime number sequence $\{p_n\}(n = 0, 1, \ldots)$, which satisfies $p_n > p_1 + \cdots + p_{n-1}$ for all $n \geq 1$. We can see from Exercise 7 in Chapter 13 that the polynomial

$$f_n(x) = p_0 x^n + \cdots + p_{n-1}x + p_n \text{ (for all } n \geq 1)$$

can't be decomposed into the product of two non-constant integer coefficient polynomials, and the same is true with $x^n f_n\left(\frac{1}{x}\right)$. That is, $p_n x^n + p_{n-1}x^{n-1} + \cdots + p_1 x + p_0$ can't be decomposed into the product of two non-constant integer coefficient polynomials for any $n \geq 1$.

This problem can also be solved by using Example 6 in Chapter 13 to construct a sequence that meets the requirement. Let's take $\{p_n\}$ as an infinite prime sequence, which satisfies $p_0 < p_1 < \cdots < p_n < \cdots$. Then we can see from Example 6 in Chapter 13 that the polynomial

$$f_n(x) = p_0 x^n + p_1 x^{n-1} + \cdots + p_n \ (n \geq 1)$$

can't be decomposed into the product of two non-constant integer coefficient polynomials, and the same is true with $x^n f_n\left(\frac{1}{x}\right)$. That is, $p_n x^n + \cdots + p_1 x + p_0$ can't be decomposed into the product of two non-constant integer coefficient polynomials for all $n \geq 1$ (refer to Remark 4 in Chapter 13).

41. If the degree of $f(x)$ is less than or equal to $p - 2$, then by taking the $(p-1)$st order difference of $f(x)$, we see that (see (4) in Chapter 14)

$$\sum_{k=0}^{p-1} (-1)^k \binom{p-1}{k} f(k) = 0. \tag{4}$$

Since p is prime, $\binom{p-1}{k} \equiv (-1)^k \pmod{p}$ (Example 4 in Chapter 8), thus the left side of (1) $\equiv \sum_{k=0}^{p-1} f(k) \pmod{p}$. Therefore,

$$\sum_{k=0}^{p-1} f(k) \equiv 0 \pmod{p}. \tag{5}$$

But $f(k)(k = 0, 1, \ldots, p-1)$ modulo p only takes values 0 or 1, and not all values are 0 and not all values are 1 (because $f(0) = 0$ and $f(1) = 1$). There are p terms on the left side of (2), so (1) can't be true.

42. Let $n + 1 = 2^l q$, where q is odd. Let $x - 1 = y$. Then

$$f(x) = f(1 + y)$$
$$= y^m g(1 + y)$$
$$= a_m y^m + a_{m+1} y^{m+1} + \cdots. \tag{6}$$

Here $a_m, a_{m+1}, \ldots$ are all integers.

On the other hand, since the coefficients of $f(x)$ are all ± 1, so with the mod 2 operation

$$f(x) \equiv x^n + x^{n-1} + \cdots + x + 1$$
$$= \frac{x^{n+1} - 1}{x - 1}$$
$$= \frac{1}{y}((1 + y)^{2^l q} - 1)$$
$$\equiv \frac{1}{y}((1 + y^{2^l})q - 1). \text{ (Use Exercise 3 of Chapter 11.)}$$
$$= y^{2^l - 1} + \cdots \pmod 2. \text{ (Use the binomial theorem.)} \tag{7}$$

Comparing (1) and (2), we know that $2^l - 1 \geq m$. Because $m \geq 2^k$, we have $l \geq k + 1$. Therefore $2^{k+1} \mid n + 1$.